# Lecture Notes in Mathematics

A collection of informal reports and seminars
Edited by A. Dold, Heidelberg and B. Eckmann, Z/

T0213195

99

# Category Theory, Homology Theory and their Applications III

Proceedings of the Conference held at the Seattle Research Center of the Battelle Memorial Institute, June 24 – July 19, 1968
Volume Three

1969

Springer-Verlag Berlin · Heidelberg · New York

Library of Congress Catalog Card Number 75-75931 Printed in Germany. Title No. 3705

# Preface

This is the third and last part of the Proceedings of the Conference on Category Theory, Homology Theory and their Applications, held at the Seattle Research Center of the Battell Memorial Institute during the summer of 1968. The first part, comprising 12 papers, was published as Volume 86 in the Lecture Notes series; the second part, also comprising 12 papers, as Volume 92.

It is again a pleasure to express to the administrative and clerical staff of the Seattle Research Center the appreciation of the contributors to this volume, and of the organizing committee of the conference, for their invaluable assistance in the preparation of the manuscripts.

Cornell University, Ithaca, March, 1969 Peter Hilton

Table of Contents

# LECTURES ON GENERALISED COHOMOLOGY*

by

J. F. Adams

## LECTURE 1. THE UNIVERSAL COEFFICIENT THEOREM AND THE KUNNETH THEOREM

It is an established practice to take old theorems about ordinary homology, and generalise them so as to obtain theorems about generalised homology theories. For example, this works very well for duality theorems about manifolds. We may ask the following question. Take all those theorems about ordinary homology which are standard results in everyday use. Which are the ones which still lack a fully satisfactory generalisation to generalised homology theories? I want to devote this lecture to such problems.

As my candidates for theorems which need generalising, I offer you the universal coefficient theorem and the Künneth theorem. I will first try to formulate the conclusions which these theorems should have in the generalised case. I will then make some comments on these formulations, and discuss a certain number of cases in which they are known to be true. I will then comment on the connection between one form of the universal coefficient theorem and the "Adams

* Note. These lectures are not arranged in the order in which they were originally given.

spectral sequence". After that I will give some proofs under suitable assumptions. Finally I will show that certain results of Conner and Floyd [14] can be related to the universal coefficient theorem.

In discussing the universal coefficient theorem and the Künneth theorem, we will write $E_*$ and $F_*$ for generalised homology theories and $E^*$, $F^*$ for generalised cohomology theories. In order to avoid tedious notation for relative groups, we will suppose that they are "reduced" theories, defined on some category of spaces with base-point. Thus we can replace the pair $X, X'$ by the space with base-point $X/X'$. In particular, the coefficient groups for $E_*$ are the groups $E_*(S^0)$, and similarly for the other theories.

The universal coefficient theorem should address itself to the following problems.

(1) Given $E_*(X)$, calculate $F_*(X)$.

(2) Given $E_*(X)$, calculate $F^*(X)$.

(3) Given $E^*(X)$, calculate $F^*(X)$.

(4) Given $E^*(X)$, calculate $F_*(X)$.

The last two problems correspond to the "upside-down universal coefficient theorems" in ordinary homology.

It will surely be necessary to assume some relation between $E_*$ (or $E^*$) and $F_*$ (or $F^*$). To begin with, we must suppose given enough products. For example, we need products in order to give sense to the Tor and Ext functors which occur in our statements. We postpone all further discussion of data; the first step is to formulate the conclusions which our generalised theorems ought to assert. We suggest the following.

(UCT1)

Suppose given product maps

$$\mu\colon E_*(X) \otimes E_*(S^0) \longrightarrow E_*(X)$$

$$\nu\colon E_*(X) \otimes F_*(S^0) \longrightarrow F_*(X)$$

satisfying suitable axioms. Then there is a spectral sequence

$$\mathrm{Tor}^{E_*(S^0)}_{p,*}(E_*(X),\ F_*(S^0)) \underset{p}{\Longrightarrow} F_*(X)\ .$$

The edge-homomorphism

$$E_*(X) \otimes_{E_*(S^0)} F_*(S^0) \longrightarrow F_*(X)$$

is induced by $\nu$ .

(UCT2)

Suppose given product maps

$$\mu\colon E_*(S^0) \otimes E_*(X) \longrightarrow E_*(X)$$

$$\nu\colon E_*(X) \otimes F^*(X) \longrightarrow F^*(S^0)$$

satisfying suitable axioms. Then there is a spectral sequence

$$\mathrm{Ext}^{p,*}_{E_*(S^0)}(E_*(X),\ F^*(S^0)) \underset{p}{\Longrightarrow} F^*(X)\ .$$

The edge-homomorphism

$$F^*(X) \longrightarrow \mathrm{Hom}^*_{E_*(S^0)}(E_*(X), F^*(S^0))$$

is induced by $\nu$ .

(UCT3)

Suppose given product maps

$$\mu: E^*(X) \otimes E^*(S^0) \longrightarrow E^*(X)$$

$$\nu: E^*(X) \otimes F^*(S^0) \longrightarrow F^*(X)$$

satisfying suitable axioms. Then there is a spectral sequence

$$\mathrm{Tor}^{E^*(S^0)}_{p,*}(E^*(X), F^*(S^0)) \underset{p}{\Longrightarrow} F^*(X) .$$

The edge-homomorphism

$$E^*(X) \otimes_{E^*(S^0)} F^*(S^0) \longrightarrow F^*(X)$$

is induced by $\nu$ .

(UCT4)

Suppose given product maps

$$\mu: E^*(S^0) \otimes E^*(X) \longrightarrow E^*(X)$$

$$\nu: E^*(X) \otimes F_*(X) \longrightarrow F_*(S^0)$$

satisfying suitable axioms. Then there is a spectral sequence

$$\mathrm{Ext}^{p,*}_{E^*(S^0)}(E^*(X), F_*(S^0)) \underset{p}{\Longrightarrow} F_*(X) .$$

The edge-homomorphism

$$F_*(X) \longrightarrow \mathrm{Hom}^*_{E^*(S^0)}(E^*(X), F_*(S^0))$$

is induced by $\nu$ .

Note 1. We should spell out some of the axioms on the product maps. We will obviously assume that the product maps have the correct behavior with respect to induced homomorphisms and with respect to suspension. We will assume that the map $\mu$, for $X = S^0$, makes $E_*(S^0)$ (in cases 1 and 2) or $E^*(S^0)$ (in cases 3 and 4) into a graded ring with unit. We will assume that the map $\mu$ makes $E_*(X)$ (in cases 1 and 2) or $E^*(X)$ (in cases 3 and 4) into a graded module over $E_*(S^0)$ or $E^*(S^0)$. This module is a left module in cases 2 and 4, a right module in cases 1 and 3. We will assume that the map $\nu$, for $X = S^0$, makes $F_*(S^0)$ (in cases 1 and 4) or $F^*(S^0)$ (in cases 2 and 3) into a graded module over $E_*(S^0)$ or $E^*(S^0)$. This module is a left module in all four cases. This is sufficient to give sense to the Tor and Ext functors in the statements. Again, in cases 1 and 3 we will assume that the product maps

$$\nu: E_*(X) \otimes F_*(S^0) \longrightarrow F_*(X)$$

$$\nu: E^*(X) \otimes F^*(S^0) \longrightarrow F^*(X)$$

factor through $E_*(X) \otimes_{E_*(S^0)} F_*(S^0)$ and $E^*(X) \otimes_{E^*(S^0)} F^*(S^0)$ respectively. In cases 2 and 4 we convert the maps $\nu$ into maps

$$F^*(X) \longrightarrow \mathrm{Hom}^*(E_*(X), F^*(S^0))$$

$$F_*(X) \longrightarrow \mathrm{Hom}^*(E^*(X), F_*(S^0))$$

and assume that these actually map into $\mathrm{Hom}^*_{E_*(S^0)}(E_*(X), F^*(S^0))$ and $\mathrm{Hom}^*_{E^*(S^0)}(E^*(X), F_*(S^0))$ respectively. All these four conditions may be viewed as associativity conditions on our products. They give sense to the statements about the edge-homomorphisms.

Note 2. The case of representable functors is particularly important. In this case we suppose given a ring-spectrum E and a spectrum F which is a left module-spectrum over the ring-spectrum E. We take $E_*$ and $E^*$ to be the functors determined by E, as in [31]; we take $F_*$ and $F^*$ to be the functors determined by F. In this case we obtain all the products required for the statements UCT 1-4. For example, in cases 2 and 4 the products $\nu$ are Kronecker products. All these products satisfy all the assumptions mentioned in Note 1.

As examples of ring-spectra E, we have MU, and the BU spectrum, and the sphere spectrum S. We also have examples of module-spectra. Any spectrum is a module-spectrum over S; and BU is a module-spectrum over MU, this being the case explored by Conner and Floyd [14].

Note 3. As remarked above, we have yet to discuss the data which might suffice to prove these statements, or the lines of proof which might establish them. The assumptions in Note 1 are intended simply to give meaning to the statements.

Note 4. By assuming extra data, we might expect

to make all these spectral sequences into spectral sequences of modules over $E_*(S^0)$ or $E^*(S^0)$. The extra data would be modelled on the case in which we start from a ring-spectrum $E$ which is commutative, and a module-spectrum $F$ over $E$. For example, we would take the ring $E_*(S^0)$ or $E^*(S^0)$ to be anticommutative. We spare ourselves the details. If the basic results are proved in any reasonable way, it should not be hard to add such trimmings.

The Künneth theorem (for reduced functors) should address itself to the problem of computing $E_*$ and $E^*$ for the smash-product $X \wedge Y$ in terms of corresponding groups of $X$ and $Y$. (This corresponds to computing an unreduced theory on $X \times Y$.) We may obtain four statements by substituting in UCT 1 and 4 the functor $F_*(X) = E_*(X \wedge Y)$, and in UCT 2 and 3 the functor $F^*(X) = E^*(X \wedge Y)$. We obtain the following statements.

<u>(KT1)</u>

Suppose given an external product

$$\nu \colon E_*(X) \otimes E_*(Y) \longrightarrow E_*(X \wedge Y)$$

satisfying suitable axioms. Then there is a spectral sequence

$$\mathrm{Tor}^{E_*(S^0)}_{p,*}(E_*(X), E_*(Y)) \underset{p}{\Longrightarrow} E_*(X \wedge Y) .$$

The edge-homomorphism

$$E_*(X) \otimes_{E_*(S^0)} E_*(Y) \longrightarrow E_*(X \wedge Y)$$

is induced by $\nu$ .

(KT2)

Suppose given a product

$$\mu: E_*(S^0) \otimes E_*(X) \longrightarrow E_*(X)$$

and a slant product

$$\nu: E_*(X) \otimes E^*(X \wedge Y) \longrightarrow E^*(Y)$$

satisfying suitable axioms. Then there is a spectral sequence

$$\mathrm{Ext}^{p,*}_{E_*(S^0)}(E_*(X), E^*(Y)) \underset{p}{\Longrightarrow} E^*(X \wedge Y) .$$

The edge-homomorphism

$$E^*(X \wedge Y) \longrightarrow \mathrm{Hom}^*_{E_*(S^0)}(E_*(X), E^*(Y))$$

is induced by $\nu$ .

(KT3)

Suppose given an external product

$$\nu: E^*(X) \otimes E^*(Y) \longrightarrow E^*(X \wedge Y)$$

satisfying suitable axioms. Then there is a spectral sequence

$$\mathrm{Tor}^{E^*(S^0)}_{p,*}(E^*(X), E^*(Y)) \underset{p}{\Longrightarrow} E^*(X \wedge Y) .$$

The edge-homomorphism

$$E^*(X) \otimes_{E^*(S^0)} E^*(Y) \longrightarrow E^*(X \wedge Y)$$

is induced by $\nu$ .

(KT4)

Suppose given a product map

$$\mu: E^*(S^0) \otimes E^*(X) \longrightarrow E^*(X)$$

and a slant product

$$\nu: E^*(X) \otimes E_*(X \wedge Y) \longrightarrow E_*(Y)$$

satisfying suitable axioms. Then there is a spectral sequence

$$\mathrm{Ext}^{p,*}_{E^*(S^0)}(E^*(X), E_*(Y)) \underset{p}{\Longrightarrow} E_*(X \wedge Y) .$$

The edge-homomorphism

$$E_*(X \wedge Y) \longrightarrow \mathrm{Hom}^*_{E^*(S^0)}(E^*(X), E_*(Y))$$

is induced by $\nu$ .

Note 5. In KT 1 and 3 it is unnecessary to suppose given the product $\mu$, as it can be obtained by specialising the product $\nu$ to the case $Y = S^0$ .

Note 6. As each part of the "Künneth theorem" is obtained by transcribing the corresponding part of the "universal coefficient theorem", Notes 1, 3 and 4 above can also be transcribed. Note 1 yields the formal properties of our products $\mu$ and $\nu$ which we should assume in order to give sense to the statements.

Note 7. The case of representable functors is particularly important. In this case we suppose given a ring-spectrum E. We take $E_*$ and $E^*$ to be the functors determined by E, as in [31]. We then have four classical products - two external products and two slant products [31]. These products satisfy all the formal properties needed to give sense to our statements - see Note 6.

This provides some justification for stating the Künneth theorem in four parts. In fact, we have four products; from each product product we can construct an associated "edge-homomorphism"; the corresponding spectral sequence (if it applies) shows whether or not this homomorphism is an isomorphism.

Note 8. Since each part of the Künneth theorem is obtained by specialising the corresponding part of the universal coefficient theorem, the latter will presumably imply the former, once we get the data settled. (Of course, if we wished to stay inside ordinary homology we could not use this argument.) It should therefore be enough to discuss the universal coefficient theorem.

Note 9. It is almost certain that UCT 3 and UCT 4 will require some finiteness condition, because such a condition is needed for the "upside-down universal coefficient theorems" in ordinary homology. If X is a finite complex, then we can deduce UCT 3 from UCT 1 by S-duality. Let DX be the Spanier-Whitehead dual of X. Suppose given $E^*$, $F^*$ as in UCT 3. Then we can define theories $E_*$, $F_*$ on finite complexes by setting

$$E_*(X) = E^*(DX), \; F_*(X) = F^*(DX) \; ;$$

we extend to infinite complexes and spectra by direct limits. We obtain product maps

$$E_*(X) \otimes E_*(S^0) \longrightarrow E_*(X)$$

$$E_*(X) \otimes F_*(S^0) \longrightarrow F_*(X)$$

as required for UCT 1. Applying UCT 1 to DX, we obtain UCT 3 for X.

Similar remarks would apply to deduce UCT 4 from UCT 2, except that the definition

$$F^*(X) = F_*(DX)$$

will only define $F^*$ on finite complexes. At this point we do not know whether it will suffice for UCT 2 to have $F^*$ defined on so small a category. It therefore seems best to begin from a ring-spectrum E and a module-spectrum F. In this case $F^*$ will be defined on a sufficiently large category. We have isomorphisms

$$E_*(DX) \cong E^*(X)$$

$$F^*(DX) \cong F_*(X)$$

and these can be taken to throw the usual products

$$E_*(S^0) \otimes E_*(DX) \longrightarrow E_*(DX)$$

$$E_*(DX) \otimes F^*(DX) \longrightarrow F^*(S^0)$$

onto the usual products

$$E^*(S^0) \otimes E^*(X) \longrightarrow E^*(X)$$

$$E^*(X) \otimes F_*(X) \longrightarrow F_*(S^0) \; .$$

Applying UCT 2 to DX, we obtain UCT 4 for any finite complex X.

Of course, this method of deducing UCT 4 from UCT 2 only gives UCT 4 for representable functors. It is therefore necessary to note that UCT 4 for representable functors implies KT 4 for representable functors. Suppose we start from a ring-spectrum E. Then the functor

$$F_*(X) = E_*(X \wedge Y)$$

is representable; the representing spectrum is given by $F = E \wedge Y$. This spectrum can be made into a (left) module-spectrum over E in the obvious way; this results in a product

$$E^*(X) \otimes F_*(X) \longrightarrow F_*(S^0)$$

which coincides with the usual slant-product

$$E^*(X) \otimes E_*(X \wedge Y) \longrightarrow E_*(Y) \ .$$

If X is a finite complex, and we apply UCT 4 to X (with this E and F), we obtain KT 4 for X.

The result of this discussion is that to obtain all eight results, under suitable conditions, it should be enough to discuss UCT 1 and UCT 2.

<u>Note 10</u>. Our treatment leads to KT 3 with a finiteness assumption on X but none on Y. Since KT 3 is symmetrical between X and Y, it would be equally reasonable to make a finiteness assumption on Y but none on X. Some finiteness assumption is almost certainly necessary, because

it is so far the corresponding Künneth theorem in ordinary cohomology.

Our treatment leads to KT 4 with a finiteness assumption on $X$ but none on $Y$. Some finiteness assumption on $X$ is almost certainly necessary, for the usual reason. A finiteness assumption on $Y$ is very likely to be irrelevant. For example, suppose that $E^*(X)$ has a resolution by finitely-generated projectives over $E^*(S^0)$; e.g. this is so if $E = MU$ and $X$ is a finite complex (see Lecture 5). Then

$\mathrm{Ext}^{p,*}_{E^*(S^0)}(E^*(X), E_*(Y))$ passes to direct limits as we vary

$Y$; and KT 4 for this $X$ and general $Y$ follows from the case in which $Y$ is a finite complex.

It is now time to discuss some cases in which the statements we have formulated are known to be true.

<u>Note 11</u>. Certain special cases of the statements are classical theorems about ordinary homology.

<u>Note 12</u>. Suppose that $F_*(S^0)$ is flat over $E_*(S^0)$. Then UCT 1 asserts that the edge-homomorphism

$$\varepsilon\colon E_*(X) \otimes_{E_*(S^0)} F_*(S^0) \longrightarrow F_*(X)$$

is an isomorphism. This is certainly true when $X$ is a finite complex, because as we vary $X$, $\varepsilon$ is a natural transformation between homology functors which is iso for $X = S^0$. If we assume that $E_*$ and $F_*$ pass to direct limits as we vary X, then the same result holds when X is a CW-complex or a spectrum.

Since KT 1 is symmetrical between X and Y, it follows that KT 1 is true if either $E_*(X)$ or $E_*(Y)$ is flat.

Similar remarks apply to UCT 2 if $F^*(S^0)$ is injective, although this case hardly ever arises. One has to approach the case of infinite complexes X by discussing the case of infinite wedge-sums, as in [21].

The same approach does not immediatly prove UCT 1 under the assumption that $E_*(X)$ is flat, because we cannot vary F arbitrarily without losing the products we need. (See Note 14 below.) However, UCT 1 and UCT 2 are trivially true if X is a wedge-sum of spheres; we will use this later.

Note 13. If E is the sphere-spectrum S then any spectrum is a module over S. In this case all the results are true and easy to prove. This will appear as a special case in Note 15 below.

Note 14. Next I have to recall that in the definition of a ring-spectrum, one is allowed various homotopies; for example, the product is supposed to be homotopy-associative. If we do not wish to allow any homotopies, we speak of a strict ring-spectrum. The sphere S is a strict ring-spectrum; otherwise it is usually laborious to show that a given spectrum is a strict ring-spectrum. It has been proved by E. Dyer and D. Kahn (to appear) that if $E$ is a strict ring-spectrum, then KT 1 holds. Their argument also shows that if $E$ is a strict ring-spectrum and $F$ is a strict module-spectrum over $E$, then UCT 1 holds. The method amounts to constructing an E-free resolution of $F$; compare the last paragraph of Note 12 above.

This is at least a general theorem. It is likely that one could weaken the conditions on the spectra slightly, by analogy with the case of "$A_\infty$ H-spaces" [28]. Unfortunately, the method does not seem to prove any of the theorems involving Ext; this would require a different sort of resolution.

<u>Note 15</u>. If $E$ is the BU-spectrum and $X$, $Y$ are finite complexes then KT 3 is a result of Atiyah [6]. (Of course in this case $Tor_p = 0$ for $p > 1$.) By combining the idea of Atiyah's proof with S-duality, one can obtain a proof of UCT 1 and UCT 2 (and hence of all the rest) for various spectra for which the method happens to work. The spectra $E$ to which the method applies include BO, BU, MO, MU, MSp, S and the Eilenberg-MacLane spectrum $K(Z_p)$.

This method is already known to E. Dyer, and perhaps to many other workers in the field. Since giving the original lecture I have heard that L. Smith has applied the method (a) to consider UCT 1 for the case $E = MU$, $F = K(Z)$ and (b) to consider KT 1 for the case $E = MU$; I am grateful to him for sending me a preprint.

This method is very practical when it works. It definitely doesn't work for $E = K(Z)$. Thus it fails to generalise the classical theorems for ordinary homology.

Note 16. Atiyah [6, footnote on p. 245] has indicated an example in which the edge-homomorphism is not monomorphic; and presumably further such examples can be found. They do not contradict our thesis, because they presumably give examples in which the differentials of the relevant spectral sequence are non-zero.

Next I want to comment on the connection between UCT 2 and the "Adams spectral sequence" [1,2,15]. For this I need some standard ideas from homological algebra, and I give them now in order to avoid interrupting the discussion later.

Let $A$ be an algebra over a ground ring $R$, and let $M$ be an $R$-module. Then $A \otimes_R M$ may be made into an $A$-module by giving it the obvious structure maps; and we have

$$\mathrm{Hom}_A(A \otimes_R M, N) \cong \mathrm{Hom}_R(M, N) .$$

(Hence the same thing is true for Ext .) $A \otimes_R M$ is called an "extended" module. Similarly, let $C$ be a coalgebra over a ring $R$, and let $M$ be an $R$-module. Then $C \otimes_R M$ may be made into a C-comodule by giving it the obvious structure maps; and we have

$$\mathrm{Hom}_C(L, C \otimes_R M) \cong \mathrm{Hom}_R(L,M) .$$

(Hence the same thing is true for Ext .) $C \otimes_R M$ is called an extended comodule.

In the applications everything will be graded. Also C will be a bimodule over R and the two actions of R on C will be quite distinct; but this does not affect the truth of the clichés presented above.

Let $[X,Y]_*$ be the set of stable homotopy classes of maps from X to Y. I shall argue in Lecture 2 that the most plausible generalisation of the "Adams spectral sequence" would give the following statement.

<u>(ASS)</u>

Under suitable assumptions, there is a spectral sequence

$$\mathrm{Ext}^{p,*}_{E_*(E)}(E_*(X), E_*(Y)) \underset{p}{\Longrightarrow} [X,Y]_* .$$

The edge-homomorphism

$$[X,Y]_* \longrightarrow \mathrm{Hom}^*_{E_*(E)}(E_*(X), E_*(Y))$$

assigns to each map f its induced homomorphism $f_*: E_*(X) \longrightarrow E_*(Y)$.

Here E is (as usual) a ring-spectrum. The functors Hom and Ext are defined by considering $E_*(X)$ and

$E_*(Y)$ as comodules with respect to the coalgebra $E_*(E)$. We use $E_*(S^0)$ as the ground ring for our comodules etc. The necessary details are given in Lecture 3.

This result refers to $[X,Y]_*$ for a general $Y$. If we assume that $Y$ is $F$, a left module-spectrum over $E$, then $[X,Y]_*$ becomes $F^*(X)$, and we may hope that this extra data will simplify the computation of the $E_2$ term. We will now make this more precise. In Lecture 3 we will define a product map

$$m: E_*(E) \otimes_{E_*(S^0)} F_*(S^0) \longrightarrow E_*(F) .$$

This map is not one of those we have so far considered, but it is related to the map $\nu$ of UCT 1 by the following commutative diagram.

$$\begin{array}{ccc}
E_*(E) \otimes_{E_*(S^0)} F_*(S^0) & \xrightarrow{m} & E_*(F) \\
\Big\downarrow {\scriptstyle c\otimes 1} & & \Big\downarrow {\scriptstyle \tau_*} \\
E_*(E) \otimes_{E_*(S^0)} F_*(S^0) & \xrightarrow{\nu} & F_*(E)
\end{array}$$

Here $\tau_*$ is the isomorphism induced by the switch map $\tau: E \wedge F \longrightarrow F \wedge E$, and similarly for $c$. In Lecture 3 we shall assume that the relevant action of $E_*(S^0)$ on $E_*(E)$ makes $E_*(E)$ into a flat module. So if UCT 1 applies to $\nu$, it will show that $\nu$ is an isomorphism, and hence $m$ is an isomorphism. In any case, for each $E$ and $F$ we can check once for all whether this is so. If it is, then the results of Lecture 3 show that $E_*(F)$ is an extended comodule; that is, the isomorphism $m$ throws the diagonal $\psi \otimes 1$ for $E_*(E) \otimes_{E_*(S^0)} F_*(S^0)$ onto the diagonal $\psi$ for $E_*(F)$. In this case we have

$$\mathrm{Ext}^{p,*}_{E_*(E)}(E_*(X), E_*(F)) \cong \mathrm{Ext}^{p,*}_{E_*(S^0)}(E_*(X), F_*(S^0)) .$$

Since $F_*(S^0) \cong F^*(S^0)$ (as modules over $E_*(S^0)$), the statement ASS specialises to UCT 2. (Checking reveals that the edge-homomorphism behaves correctly.)

Since $F_*(X)$ admits an interpretation in terms of stable homotopy, one may ask whether UCT 1 can be related to ASS. Further thought reveals that this is unlikely, as the spectral sequence of UCT 1 involves a filtration starting from 0 and increasing indefinitely, while ASS involves a filtration starting from the whole group $[X,Y]_*$ and decreasing indefinitely. In particular, the edge-homomorphisms run in opposite directions.

I can now explain one motivation for interest in UCT 2. I would like to see further results of the general form of ASS; compare Novikov [23, 24]. It seems that UCT 2 is a special case which sufficiently exhibits many of the difficulties. I would therefore like to see new proofs of UCT 2, as general as possible, in the hope that they may generalise to proofs of ASS.

I will now turn to give further details of the

method mentioned in Note 15. For this purpose I will assume once for all that in what follows the functors $E_*$ and $F_*$ or $F^*$ satisfy Milnor's additivity axiom on wedge-sums [21]. The first step is to deal with a special case which is very restrictive, but important for the applications.

Let $X$ be a CW-complex or a connected spectrum. We assume that the spectral sequence

$$H_*(X;E_*(S^0)) \Longrightarrow E_*(X)$$

is trivial, that is, its differentials are zero. We observe that this spectral sequence is a spectral sequence of modules over $E_*(S^0)$; in the case of UCT 1 it is a spectral sequence of right modules, and in the case of UCT 2 it is a spectral sequence of left modules. The module structure of the $E^2$ term $H_*(X;E_*(S^0))$ is the obvious one. We assume that for each $p$, $H_p(X;E_*(S^0))$ is projective as a module over $E_*(S^0)$ (on the left or right as the case may be). Note that for this purpose it is not necessary to assume that $H_p(X)$ is free; for example, if $E_0(S^0)$ is a (commutative) principal ideal ring it will be sufficient if $H_p(X;E_0(S^0))$ is free. Then we conclude:

Proposition 17

With these assumptions, $E_*(X)$ is projective and $X$ satisfies UCT 1 or UCT 2 (as the case may be). That is, the map

$$E_*(X) \otimes_{E_*(S^0)} F_*(S^0) \longrightarrow F_*(X)$$

or

$$F^*(X) \longrightarrow \mathrm{Hom}^*_{E_*(S^0)}(E_*(X), F^*(S^0))$$

is iso.

Proof. Let $E^r_{p,q}(0)$, $E^r_{p,q}(1)$ and $E_r^{p,q}(2)$ be the spectral sequences

$$H_*(X;E_*(S^0)) \Longrightarrow E_*(X)$$

$$H_*(X;F_*(S^0)) \Longrightarrow F_*(X)$$

$$H^*(X;F^*(S^0)) \Longrightarrow F^*(X)$$

It follows immediately from the assumptions on the spectral sequence $E^*_{**}(0)$ that $E_*(X)$ is projective.

The products $\nu$ yield homomorphisms

$$E^r_{p,*}(0) \otimes_{E_*(S^0)} F_*(S^0) \longrightarrow E^r_{p,*}(1)$$

$$E_r^{p,*}(2) \longrightarrow \mathrm{Hom}_{E_*(S^0)}(E^r_{p,*}(0), F^*(S^0))$$

as the case may be. These homomorphisms send $d^r \otimes 1$ into $d^r$, or $d_r$ into $(d^r)^*$, as the case may be. (These assertions need detailed proof from the definitions of the spectral sequences, but it can be done using only formal properties of the products $\nu$ and the fact that $\otimes$ is right exact while Hom is left exact.) Because of the assumption that the spectral sequence $E^*_{**}(0)$ is trivial (which is essential here), the groups

$E^r_{p,*}(0) \otimes_{E_*(S^0)} F_*(S^0)$ (for $r \geq 2$), equipped with the boundaries $d^r \otimes 1$, form a (trivial) spectral sequence $E^r_{p,q}(3)$. Similarly, the groups $\mathrm{Hom}^*_{E_*(S^0)}(E^r_{p,*}(0), F^*(S^0))$, equipped with the boundaries $(d^r)^*$, form a (trivial) spectral sequence $E_r^{p,q}(4)$. We now have a map of spectral sequences

$$E^r_{p,q}(3) \longrightarrow E^r_{p,q}(1)$$

or

$$E_r^{p,q}(2) \longrightarrow E_r^{p,q}(4)$$

as the case may be. For $r = 2$ it becomes the obvious map

$$H_p(X;E_*(S^0)) \otimes_{E_*(S^0)} F_*(S^0) \longrightarrow H_p(X;F_*(S^0))$$

or

$$H^p(X;F^*(S^0)) \longrightarrow \mathrm{Hom}^*_{E_*(S^0)}(H_p(X;E_*(S^0)), F^*(S^0))$$

as the case may be. But since we are assuming that $H_p(X;E_*(S^0))$ is projective over $E_*(S^0)$ for each $p$, a theorem on ordinary homology shows that for $r = 2$ the map is iso. Therefore it is iso for all finite $r$, and the spectral sequence $E^r_{p,q}(1)$ or $E_r^{p,q}(2)$ is trivial.

We next deduce that the map

$$E^\infty_{p,*}(0) \otimes_{E_*(S^0)} F_*(S^0) \longrightarrow E^\infty_{p,*}(1)$$

or

$$E_\infty^{p,*}(2) \longrightarrow \mathrm{Hom}^*_{E_*(S^0)}(E^\infty_{p,*}(0), F^*(S^0))$$

is iso. (If $X$ is not finite-dimensional, this needs

properties of $E_*$ and $F_*$ or $F^*$ with respect to limits, but these follow from the axiom on wedge-sums.)

Let us now introduce notation for the filtration subgroups or quotient groups, as the case may be; say

$$G_{p,*}(0) = \mathrm{Im}(E_*(X^p) \longrightarrow E_*(X))$$

$$G_{p,*}(1) = \mathrm{Im}(F_*(X^p) \longrightarrow F_*(X))$$

$$G^{p,*}(2) = \mathrm{Coim}(F^*(X) \longrightarrow F^*(X^p)) .$$

The product $\nu$ yields us homomorphisms

$$G_{p,*}(0) \otimes_{E_*(S^0)} F_*(S^0) \longrightarrow G_{p,*}(1)$$

$$G^{p,*}(2) \longrightarrow \mathrm{Hom}^*_{E_*(S^0)}(G_{p,*}(0), F^*(S^0))$$

as the case may be. (Again, the verification uses only formal properties of the products $\nu$ and the fact that $\otimes$ is right exact while Hom is left exact.) Consider the following commutative diagrams.

$$\begin{array}{ccccccccc}
0 \longrightarrow & G_{p-1,*}(0) \otimes F_*(S^0) & \longrightarrow & G_{p,*}(0) \otimes F_*(S^0) & \longrightarrow & E^\infty_{p,*}(0) \otimes F_*(S^0) & \longrightarrow 0 \\
 & \downarrow & & \downarrow & & \downarrow & \\
0 \longrightarrow & G_{p-1,*}(1) & \longrightarrow & G_{p,*}(1) & \longrightarrow & E^\infty_{p,*}(1) & \longrightarrow 0
\end{array}$$

$$\begin{array}{ccccc}
0 \longrightarrow & E_{\infty}^{p,*}(2) & \longrightarrow & G^{p,*}(2) & \longrightarrow \\
& \downarrow & & \downarrow & \\
0 \longrightarrow & \mathrm{Hom}^*(E^{\infty}_{p*}(0),F_*(S^0)) & \longrightarrow & \mathrm{Hom}^*(G_{p*}(0),F^*(S^0)) & \longrightarrow
\end{array}$$

$$\begin{array}{ccc}
\longrightarrow & G^{p-1,*}(2) & \longrightarrow 0 \\
& \downarrow & \\
\longrightarrow & \mathrm{Hom}^*(G_{p-1*}(0),F^*(S^0)) & \longrightarrow 0
\end{array}$$

Here all the $\otimes$'s and Hom's are taken over $E_*(S^0)$. The first and last rows are exact because $E^{\infty}_{p,*}(0)$ is projective. An easy induction over $p$, using the short five lemma, now shows that

$$G_{p,*}(0) \otimes_{E_*(S^0)} F_*(S^0) \longrightarrow G_{p,*}(1)$$

or

$$G^{p,*}(2) \longrightarrow \mathrm{Hom}^*_{E_*(S^0)}(G_{p,*}(0),F^*(S^0))$$

is iso.

In the case of UCT 1, we now pass to direct limits and see that

$$E_*(X) \otimes_{E_*(S^0)} F_*(S^0) \longrightarrow F_*(X)$$

is iso. In the case of UCT 2, we first observe that the spectral sequence $E_r^{p,q}(2)$ satisfies the Mittag-Leffler condition for spectral sequences, and therefore

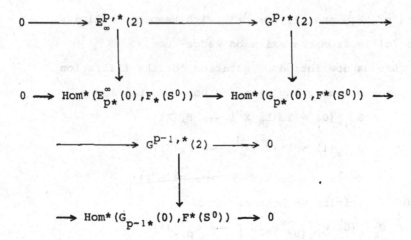

$$F^*(X) = \varprojlim_p G^{p,*}(2) .$$

Because

$$G_{p,*}(0) = G_{p-1,*}(0) \oplus E^{\infty}_{p,*}(0)$$

and

$$E_*(X) = \varinjlim_p G_{p,*}(0)$$

we have

$$\mathrm{Hom}^*_{E_*(S^0)}(E_*(X), F^*(S^0)) = \varprojlim_p \mathrm{Hom}^*_{E_*(S^0)}(G_{p,*}(0), F^*(S^0)) .$$

We can thus pass to inverse limits and see that

$$F^*(X) \longrightarrow \mathrm{Hom}^*_{E_*(S^0)}(E_*(X), F^*(S^0))$$

is iso. This proves Proposition 17.

We next need two further lemmas. For this purpose we assume that we can work in a suitable category in which we can do stable homotopy theory [7, 8, 25]. We assume that the theories $E_*$ and $F_*$ or $F^*$ are defined on this category, and that $E_*$ is represented by an object $E$ in this category. The next two lemmas are stated for $E$, but they also apply to any other object (such as $F$, if we have an $F$.) We assume that $E$ is the direct limit of a given system of finite CW-complexes $E_\alpha$.

Lemma 18

For any object $X$ and any class $e \in E_p(X)$ there is an $E_\alpha$ and a class $f \in E_p(S^p \wedge DE_\alpha)$ and a map

$g\colon S^p \wedge DE_\alpha \longrightarrow X$ such that $e = g_* f$.

Proof. Take a class $e \in E_p(X)$. Then there is a finite subcomplex $X' \subset X$ and a class $e' \in E_p(X')$ such that $i_* e' = e$. We may interpret $e'$ as a class in $E^{-p}(DX')$; so $e'$ may be represented by a map $h\colon DX' \longrightarrow S^{-p}E$. Since $DX'$ is a finite complex and $E$ is the direct limit of the $E_\alpha$, we can factor $h$ in the form

$$DX' \xrightarrow{k} S^{-p} \wedge E_\alpha \longrightarrow S^{-p} \wedge E .$$

That is, there is a class $f$ in $E^{-p}(S^{-p} \wedge E_\alpha)$ such that $k^* f = e'$. Dualising back, $f$ may be interpreted as a class in $E_p(S^p \wedge DE_\alpha)$, and we obtain a map

$$Dk\colon S^p \wedge DE_\alpha \longrightarrow X'$$

such that $(Dk)_* f = e'$. We have only to take

$$g = i(Dk)\colon S^p \wedge DE_\alpha \longrightarrow X .$$

This proves Lemma 18.

Lemma 19

For any object $X$ there exists an object of the form

$$W = \bigvee_\beta S^{p(\beta)} \wedge DE_{\alpha(\beta)}$$

and a map $g\colon W \longrightarrow X$ such that

$$g_*\colon E_*(W) \longrightarrow E_*(X)$$

is epi.

The construction is immediate from Lemma 18, by allowing the class  e  in Lemma 18 to run over a set of generators for  $E_*(X)$.

We now introduce the sort of resolution we need. By a "resolution of  X  with respect to  $E_*$"  we shall mean a diagram of the following form, with the properties listed below.

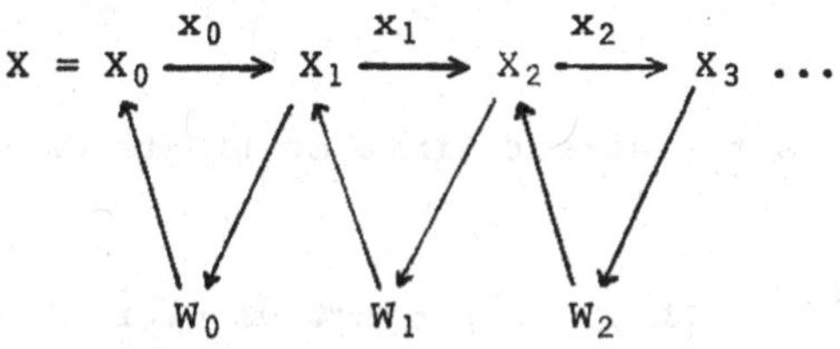

$$X = X_0 \xrightarrow{\ x_0\ } X_1 \xrightarrow{\ x_1\ } X_2 \xrightarrow{\ x_2\ } X_3 \ \ldots$$

$$W_0 \qquad W_1 \qquad W_2$$

(i)   The triangles

$$X_r \xrightarrow{\ x_r\ } X_{r+1}$$

$$W_r$$

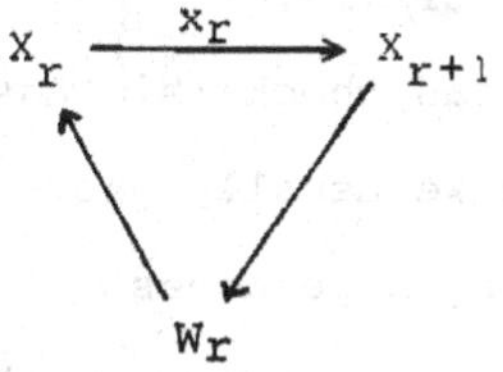

are exact (cofibre) triangles.

(ii)   For each  r,

$$(x_r)_* : E_*(X_r) \longrightarrow E_*(X_{r+1})$$

is zero.

(iii)   For each  r,  $E_*(W_r)$  is projective over  $E_*(S^0)$.

(iv)   For each  r,  $W_r$  satisfies UCT 1 or UCT 2, i.e. the map

$$E_*(X) \otimes_{E_*(S^0)} F_*(S^0) \longrightarrow F_*(X)$$

or

$$F^*(X) \longrightarrow \mathrm{Hom}^*_{E_*(S^0)}(E_*(X), F^*(S^0))$$

is iso.

In order to prove the existence of such resolutions, we introduce the following hypothesis.

Assumption 20

$E$ is the direct limit of finite CW-complexes $E_\alpha$ for which

(i) $E_*(DE_\alpha)$ is projective over $E_*(S^0)$, and

(ii) $DE_\alpha$ satisfies UCT 1 or UCT 2, as the case may be, for the theory $F_*$ or $F^*$.

In theory we can check this assumption for given $E$ and $F$. In practice we usually prove it using Proposition 17, which requires strong hypotheses on $DE_\alpha$ but none on $F$. In practice $E$ is a ring-spectrum, so the use of Proposition 17 involves checking the following two conditions.

(i) The spectral sequence

$$H^*(E_\alpha; E^*(S^0)) \Longrightarrow E^*(E_\alpha)$$

is trivial, and

(ii) For each $p$, $H^p(E_\alpha; E^*(S^0))$ is projective as a module over $E^*(S^0)$.

Examples.

(i) $E = S$, the sphere spectrum. Take $E_\alpha = S^n$; the conditions are trivially satisfied, and of course Assumption 20 is very easily verified directly.

(ii) $E = K(Z_p)$. The conditions of Proposition 17 are satisfied by any $X$. It is sufficient to let $E_\alpha$ run over any system of finite complexes whose limit is $K(Z_p)$.

(iii) $E = MO$. It is well known that

$$MO \simeq \bigvee_i S^{n(i)} K(Z_2) \simeq \prod_i S^{n(i)} K(Z_2) .$$

The conditions of Proposition 17 are satisfied by any $X$. It is sufficient to let $E_\alpha$ run over any system of finite complexes whose limit is $MO$.

(iv) $E = MU$. We have $H^p(MU; MU^q(S^0)) = 0$ unless $p$ and $q$ are even. Therefore the spectral sequence

$$H^*(MU; MU^*(S^0)) \Longrightarrow MU^*(MU)$$

is trivial. Again, $H^p(MU; MU^*(S^0))$ is free over $MU^*(S^0)$. It is sufficient to let $E_\alpha$ run over a system of finite complexes which approximate $MU$ in the sense that

$$i_*: H_p(E_\alpha) \longrightarrow H_p(MU)$$

is iso for $p \leq n$, while $H_p(E_\alpha) = 0$ for $p > n$.

(v) $E = MSp$. A simple adaptation of the method of S. P. Novikov [23, 24] from the unitary to the symplectic case shows that the spectral sequence

$$H^*(MSp; MSp^*(S^0)) \Longrightarrow MSp^*(MSp)$$

is trivial. Again, $H^p(MSp; MSp^*(S^0))$ is free over $MSp^*(S^0)$. The rest of the argument is as in (iv).

(vi) $E = \underline{BU}$. Let us recall that in the spectrum $\underline{BU}$ every even term is the space BU. We have $H^p(BU; \underline{BU}^q(S^0)) = 0$ unless p and q are even. Therefore the spectral sequence

$$H^*(BU; \underline{BU}^*(S^0)) \Longrightarrow \underline{BU}^*(BU)$$

is trivial. Again, $H^p(BU; \underline{BU}^*(S^0))$ is free over $\underline{BU}^*(S^0)$. It is sufficient to let $E_\alpha$ run over a system of finite complexes which approximate as in (iv) to the different spaces BU of the spectrum $\underline{BU}$.

(vii) $E = \underline{BO}$. Let us recall that in the spectrum $\underline{BO}$ every eighth term is the space BSp. I claim that the spectral sequence

$$H^*(BSp; \underline{BO}^*(S^0)) \Longrightarrow \underline{BO}^*(BSp)$$

is trivial. In fact, for each class $h \in H^{8p}(BSp(m))$ we can construct a real representation of $Sp(m)$ whose Chern character begins with h; for each class $h \in H^{8p+4}(BSp(m))$ we can construct a symplectic representation of $Sp(m)$ whose Chern character begins with h. The rest of the argument is as for (vi).

(viii) Cobordism and K-theory with coefficients. The reader will find further examples in Lecture 4.

Assumption 20 allows us to use the method of

Atiyah [6].

The next lemma will construct the resolutions we require; but we state it in a more general form, so that it will also allow us to compare resolutions. We suppose given a diagram of the following form.

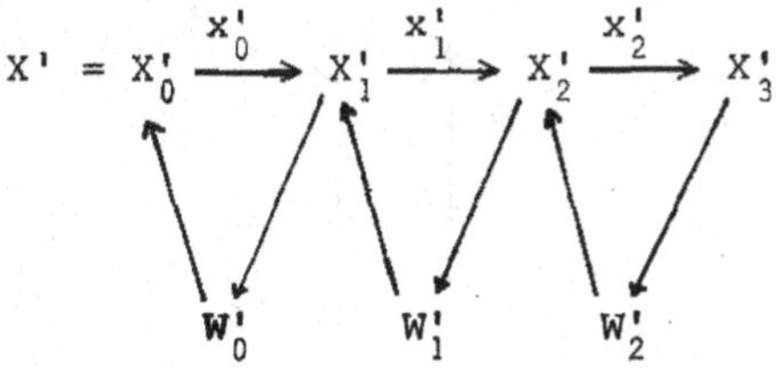


Here the triangles are supposed to be exact (cofibre) triangles, and

$$(x_r')_*: E_*(X_r') \longrightarrow E_*(X_{r+1}')$$

is zero for each $r$. We also suppose given a map $f: X \longrightarrow X'$.

<u>Lemma 21</u>

Under these conditions we can construct a resolution of $X$ with respect to $E_*$ which admits a map over $f$, in the sense that we can construct the following diagram so that the prisms are maps of exact (cofibre) triangles.

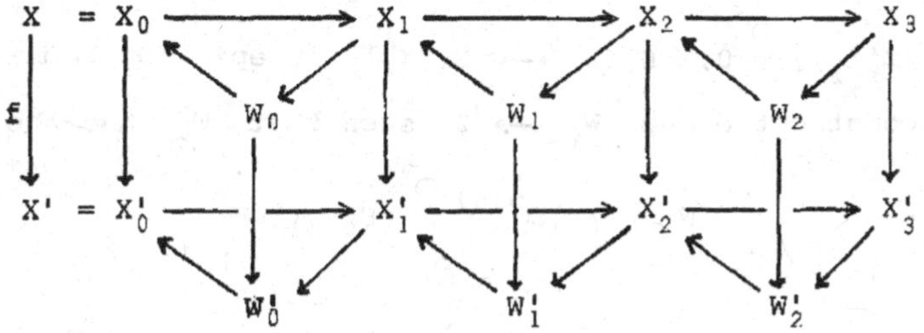

In order to construct a resolution of $X$ with respect to $E_*$, we need only apply Lemma 21 to the case in which all the objects $X'_r$ and $W'_r$ are trivial.

Proof of Lemma 21. As an inductive hypothesis, suppose the diagram constructed up to the following map.

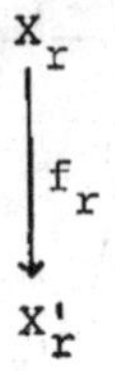

Form the following cofibre triangle.

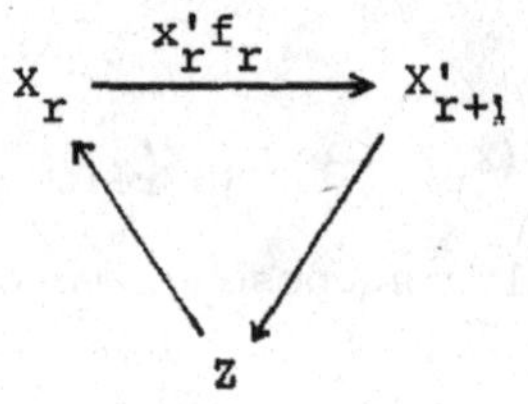

Then we have the following commutative square.

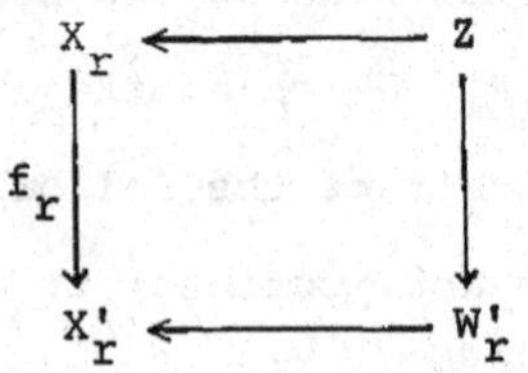

Since $(x'_r f_r)_* = 0$, $E_*(Z) \longrightarrow E_*(X_r)$ is epi. By Lemma 19 we can construct a map $W_r \longrightarrow Z$ such that $W_r$ has the form

$$W_r = \bigvee_\beta S^{p(\beta)} \wedge DE_{\alpha(\beta)}$$

and $E_*(W_r) \longrightarrow E_*(Z)$ is epi. We now have the following commutative square.

$$\begin{array}{ccc} X_r & \longleftarrow & W_r \\ {\scriptstyle f_r}\downarrow & & \downarrow \\ X'_r & \longleftarrow & W'_r \end{array}$$

Here $E_*(W_r) \longrightarrow E_*(X_r)$ is epi. Form the following cofibre triangle.

$$\begin{array}{ccc} X_r & \xrightarrow{\;x_r\;} & X_{r+1} \\ & \nwarrow \quad \swarrow & \\ & W_r & \end{array}$$

This triangle can be mapped in the required way, and we have $(x_r)_* = 0$. This completes the induction.

We have constructed a resolution, because $W_r$ inherits the property that $E_*(W_r)$ is projective from its summands $S^p \wedge DE_\alpha$, and similarly for UCT 1, UCT 2 (see Assumption 20). This proves Lemma 21.

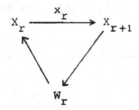

We will now construct the spectral sequences of UCT 1 and UCT 2, using Lemma 21 and the assumption that $E_*$ and $F_*$ or $F^*$ are defined on a sufficiently large category in which we can do stable homotopy theory. Take a resolution of $X$ with respect to $E_*$, as provided by Lemma 21. By applying the functor $F_*$ or $F^*$, we obtain a spectral sequence. Now the sequence

$$0 \longleftarrow E_*(X) \longleftarrow E_*(W_0) \longleftarrow E_*(W_1) \longleftarrow E_*(W_2) \longleftarrow \cdots$$

is a resolution of $E_*(X)$ by projective modules over $E_*(S^0)$. Since the $W_r$ satisfy UCT 1 or UCT 2, the $E^1$-term of the spectral sequence is obtained by taking this projective resolution and applying $\otimes_{E_*(S^0)} F_*(S^0)$ or $\mathrm{Hom}_{E_*(S^0)}(\ \ , F^*(S^0))$. Therefore the $E^2$-term is the required Tor or Ext.

We have to show that the spectral sequence is independent of the choice of resolution. Suppose given two resolutions, as follows.

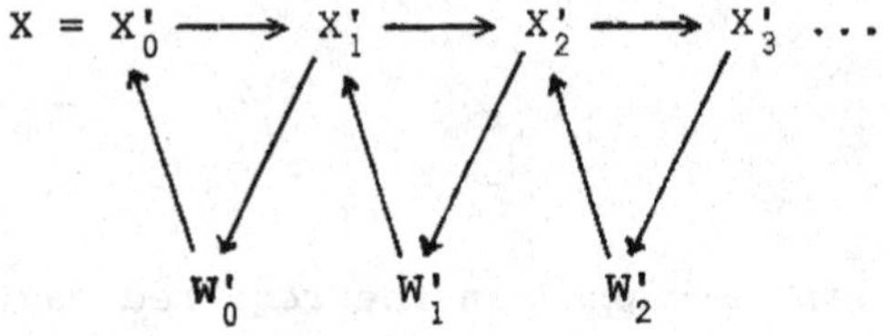

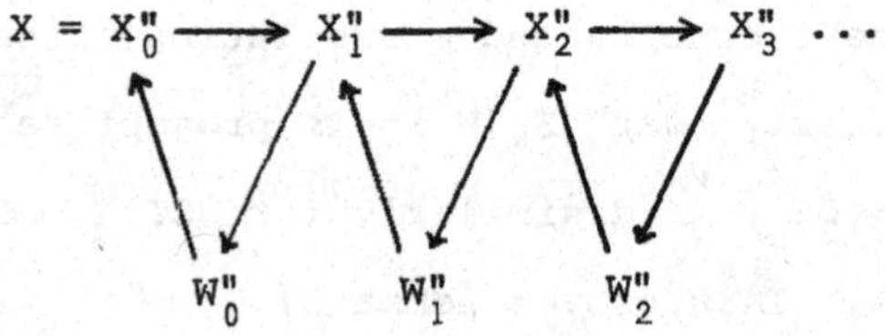

Then we can form the following diagram.

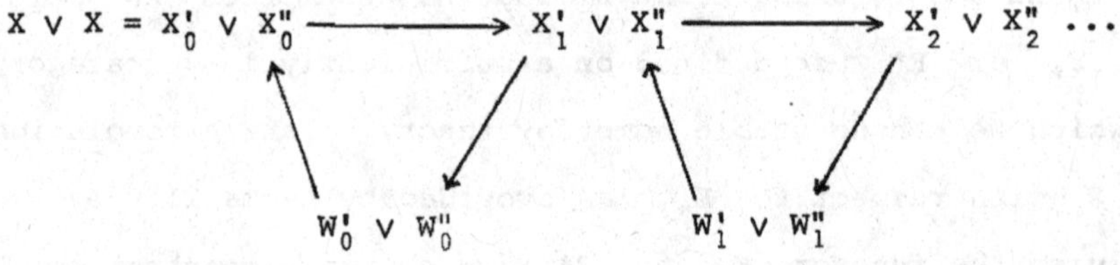

We can now apply Lemma 21 to the map $X \longrightarrow X \vee X$ of type (1,1). We obtain a third resolution and a third spectral sequence which admits comparison maps to or from the first two spectral sequences. ("To" for $F_*$, "from" for $F^*$.) But both these comparison maps are iso for $r = 2$ by the comparison theorem of homological algebra; therefore they are iso for all finite $r$.

It remains to discuss the convergence of these spectral sequences. Given a resolution of $X$, we can construct a direct limit $X_\infty$ of the objects $X_r$ (by forming a "telescope" or iterated mapping-cylinder). The object $X_\infty$ has the property that

$$E_*(X_\infty) = \varinjlim_r E_*(X_r) = 0 .$$

In the case of UCT 1, for example, the spectral sequence converges in a perfectly satisfactory manner to $F_*(X_\infty, X_0)$. We therefore face the following question.

Problem 22

When can we assert that $E_*(X) = 0$ implies $F_*(X) = 0$ or $F^*(X) = 0$ ?

This is of course a special case of UCT 1 or UCT 2. When the answer is affirmative, we have (for example) $F_*(X_\infty) = 0$, $F_*(X_\infty, X_0) \cong F_*(X)$ and the spectral sequence of UCT 1 converges in a satisfactory way to $F_*(X)$.

Unfortunately the present state of our knowledge

on Problem 22 appears to be far from satisfactory*. Of course we know special cases; for example, if $E = S$, then $S_*(X) = 0$ implies that $X$ is contractible, and so $F_*(X) = 0$, $F^*(X) = 0$. Again, if $E_*(X) = 0$, then as we vary $Y$, $E_*(X \wedge Y)$ is a homology functor of $Y$ with zero coefficient groups, therefore zero. Thus the spectral sequence of KT 1 always converges.

At this point we pause to show that our spectral sequences can behave well even in cases which are known to be somewhat pathological.

Example 23. We consider UCT 2 for the case in which $X$ is $K(Z)$, while $E$ and $F$ are the spectrum $\underline{BU}$. We can compute the ordinary homology of the spectrum $\underline{BU}$ by considering that of the space BU and passing to a direct limit; we find

$$H_n(\underline{BU}) = \begin{cases} Q & \text{if } n \text{ is even} \\ 0 & \text{if } n \text{ is odd} . \end{cases}$$

By George Whitehead's remark [31], this is equivalent to

$$\underline{BU}_n(K(Z)) = \begin{cases} Q & \text{if } n \text{ is even} \\ 0 & \text{if } n \text{ is odd} . \end{cases}$$

Now owing to the favourable structure of the ring $\underline{BU}_*(S^0)$, the computation of Ext over this ring reduces to computing Ext over $Z$. We find

---

* Note added in proof. A satisfactory answer to Problem 22 is now available.

$$\mathrm{Ext}^{p,q}_{\underline{BU}_*(S^0)}(\underline{BU}_*(K(Z)),\underline{BU}^*(S^0)) = \begin{cases} \mathrm{Ext}_Z(Q,Z) & \text{if } p = 1 \text{ and } q \text{ is even} \\ 0 & \text{otherwise.} \end{cases}$$

This agrees with the result of Hodgkin and Anderson [5, 17].

We will now make some comments on the situation whose exploration was pioneered by Conner and Floyd [14]. We assume that we have representing objects $E$ and $F$, that $E$ satisfies Assumption 20 and that $F$ satisfies the following hypothesis.

Assumption 24

$F$ is the direct limit of finite CW-complexes $F_\alpha$ for which

(i) $E_*(DF_\alpha)$ is projective over $E_*(S^0)$, and

(ii) $DF_\alpha$ satisfies UCT 1 for the theory $F_*$.

(Compare Assumption 20.) In practice we generally verify this assumption by using Proposition 17, as for Assumption 20.

Examples.

(i) $E = MU$, $F = \underline{BU}$. In the spectrum $\underline{BU}$ every even term is the space $BU$. For the space $BU$ we have $H^p(BU;MU^q(S^0)) = 0$ unless $p$ and $q$ are even. Therefore the spectral sequence

$$H^*(BU;MU^*(S^0)) \Longrightarrow MU^*(BU)$$

is trivial. Again, $H^p(BU;MU^*(S^0))$ is free over $MU^*(S^0)$. As in Example (vi) on Assumption 20, it is sufficient to let $F_\alpha$ run over a system of finite complexes which approximate to the different spaces BU of the spectrum $\underline{BU}$ in the sense that

$$i_*: H_p(F_\alpha) \longrightarrow H_p(BU)$$

is iso for $p \leq n$, while $H_p(F_\alpha) = 0$ for $p > n$.

(ii) $E = MSp$, $F = \underline{BO}$. In the spectrum $\underline{BO}$ every eighth term is the space BSp. It follows from the work of Conner and Floyd [14] that the spectral sequence

$$H^*(BSp;MSp^*(S^0)) \Longrightarrow MSp^*(BSp)$$

is trivial. Again, $H^p(BSp;MSp^*(S^0))$ is free over $MSp^*(S^0)$. The rest of the argument is as in (i).

With these assumptions (especially 20 and 24) we have the following results for any X.

<u>Proposition 25</u>

We have

$$\mathrm{Tor}^{p*}_{E_*(S^0)}(E_*(X),F_*(S^0)) = 0 \quad \text{for} \quad p > 0 .$$

The spectral sequence of UCT 1 collapses, and its edge-homomorphism

$$E_*(X) \otimes_{E_*(S^0)} F_*(S^0) \longrightarrow F_*(X)$$

is iso.

Compare Conner and Floyd [14, pp. 60, 63]; but these authors state their theorem with the variance of UCT 3, and use finiteness assumptions.

Proof. It follows from Lemma 19 that given any object X, there exists an object W of the form

$$W = \bigvee_{\beta} S^{p(\beta)} \wedge DE_{\alpha(\beta)} \vee \bigvee_{\gamma} S^{p(\gamma)} \wedge DF_{\alpha(\gamma)}$$

and a map $g: W \longrightarrow X$ such that both

$$g_*: E_*(W) \longrightarrow E_*(X)$$

and

$$g_*: F_*(W) \longrightarrow F_*(X)$$

are epi. Arguing as in Lemma 21, we can now construct a resolution of X with respect to $E_*$ which has the following extra properties.

(i) The objects $W_r$ have the form

$$W_r = \bigvee_{\beta} S^{p(\beta)} \wedge DE_{\alpha(\beta)} \vee \bigvee_{\gamma} S^{p(\gamma)} \wedge DF_{\alpha(\gamma)} .$$

(ii) Not only the homomorphisms

$$(x_r)_*: E_*(X_r) \longrightarrow E_*(X_{r+1})$$

but also the homomorphisms

$$(x_r)_*: F_*(X_r) \longrightarrow F_*(X_{r+1})$$

are zero for all r.

Then the sequence

$$0 \longleftarrow E_*(X) \longleftarrow E_*(W_0) \longleftarrow E_*(W_1) \longleftarrow E_*(W_2) \ \ldots$$

is a resolution of $E_*(X)$ by projectives over $E_*(S^0)$. Consider the following diagram.

$$\begin{array}{ccccc}
E_*(W_0) \otimes_{E_*(S^0)} F_*(S^0) & \longleftarrow & E_*(W_1) \otimes_{E_*(S^0)} F_*(S^0) & \longleftarrow & \\
\downarrow \nu_0 & & \downarrow \nu_1 & & \\
F_*(W_0) & \longleftarrow & F_*(W_1) & \longleftarrow &
\end{array}$$

$$\begin{array}{cc}
\longleftarrow & E_*(W_2) \otimes_{E_*(S^0)} F_*(S^0) \ldots \\
& \downarrow \nu_2 \\
\longleftarrow & F_*(W_2) \ \ldots
\end{array}$$

The homomorphisms $\nu_r$ are iso. The lower row is exact by construction. Therefore the upper row is exact, and

$$\mathrm{Tor}^{p,*}_{E_*(S^0)}(E_*(X), F^*(S^0)) = 0 \quad \text{for } p > 0 .$$

We can now consider the following diagram.

$$\begin{array}{ccccccc}
0 & \longleftarrow & E_*(X) \otimes_{E_*(S^0)} F_*(S^0) & \longleftarrow & E_*(W_0) \otimes_{E_*(S^0)} F_*(S^0) & \longleftarrow & \\
& & \downarrow \nu & & \downarrow \nu_0 & & \\
0 & \longleftarrow & F_*(X) & \longleftarrow & F_*(W_0) & \longleftarrow &
\end{array}$$

$$\begin{array}{ll}
\longleftarrow & E_*(W_1) \otimes_{E_*(S^0)} F_*(S^0) \\
 & \quad\quad\quad \Big\downarrow \nu_1 \\
\longleftarrow\!\!\!-\!\!\!-\!\!\!-\!\!\!-\!\!\!- & F_*(W_1)
\end{array}$$

The upper row is exact because $\otimes$ is right exact, and the lower row is exact by construction. The maps $\nu_0$ and $\nu_1$ are iso. Therefore $\nu$ is iso. This completes the proof of Proposition 25.

Since we now know what happens to UCT 1 in this situation, it is natural to ask what happens to UCT 2. For this we need slightly more data. We suppose given two ring-spectra $E$, $F$ and a map $i: E \longrightarrow F$ of ring-spectra. (For example, $E = MU$ and $F = \underline{BU}$, or $E = MSp$ and $F = \underline{BO}$.) We suppose given also a spectrum $G$ which is a module-spectrum over $F$, and therefore a module-spectrum over $E$ via $i$. (For example, $G = F$.) (It would presumably be sufficient to suppose given enough products in homology and cohomology, but let us spare ourselves the details.) We suppose that the pair of theories $(E,G)$ satisfies Lemma 21, so that we can construct a spectral sequence for computing $G_*$ or $G^*$ from $E_*$ as in UCT 1 or UCT 2; we also suppose that the pair of theories $(F,G)$ satisfies Lemma 21, so that we can construct a spectral sequence for computing $G_*$ or $G^*$ from $F_*$ as in UCT 1 or UCT 2.

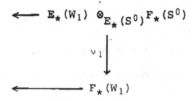

<u>Proposition 26</u>

(i) The spectral sequence for computing $G_*$ from $E_*$ coincides with the spectral sequence for computing $G_*$ from $F_*$.

(ii) The spectral sequence for computing $G^*$ from $E_*$ coincides with the spectral sequence for computing $G^*$ from $F_*$.

<u>Note</u>. By specialising Proposition 26(i) to the case $G = F$, we obtain a result agreeing with Proposition 25; for of course the spectral sequence for computing $F_*$ from $F_*$ collapses.

Proposition 26 will follow almost immediately from the following lemma.

<u>Lemma 27</u>

(i) If $E_*(W)$ is projective over $E_*(S^0)$, then $F_*(W)$ is projective over $F_*(S^0)$.

(ii) If

$$E_*(W) \otimes_{E_*(S^0)} G_*(S^0) \longrightarrow G_*(W)$$

is iso, then

$$F_*(W) \otimes_{F_*(S^0)} G_*(S^0) \longrightarrow G_*(W)$$

is iso.

(iii) If

$$G^*(W) \longrightarrow \mathrm{Hom}^*_{E_*(S^0)}(E_*(W), G^*(S^0))$$

is iso, then

$$G^*(W) \longrightarrow \mathrm{Hom}^*_{F_*(S^0)}(F_*(W), G^*(S^0)) \quad \text{is iso.}$$

Proof.

(i) $F_*(W) \cong E_*(W) \otimes_{E_*(S^0)} F_*(S^0)$, by Proposition 25. So if $E_*(W)$ is projective over $E_*(S^0)$, $F_*(W)$ is projective over $F_*(S^0)$.

(ii) Consider the following commutative diagram.

$$\begin{array}{ccc}
E_*(W) \otimes_{E_*(S^0)} F_*(S^0) \otimes_{F_*(S^0)} G_*(S^0) & \xrightarrow{1 \otimes \nu} & E_*(W) \otimes_{E_*(S^0)} G_*(S^0) \\
\Big\downarrow{\scriptstyle \nu \otimes 1} & & \Big\downarrow{\scriptstyle \nu} \\
F_*(W) \otimes_{F_*(S^0)} G_*(S^0) & \xrightarrow{\nu} & G_*(W)
\end{array}$$

The left-hand column is iso by Proposition 25, the right-hand column is iso by assumption, and the top row is trivially iso. Therefore the bottom row is iso.

(iii) Consider the following commutative diagram.

$$\begin{array}{ccc}
G^*(W) & \xrightarrow{\nu} & \mathrm{Hom}_{F_*(S^0)}(F_*(W), G^*(S^0)) \\
\Big\downarrow{\scriptstyle \nu} & & \Big\downarrow{\scriptstyle \nu^*} \\
 & & \mathrm{Hom}_{F_*(S^0)}\Big(F_*(S^0) \otimes_{E_*(S^0)} E_*(W),\ G^*(S^0)\Big) \\
 & & \Big\| \\
\mathrm{Hom}_{E_*(S^0)}(E_*(W), G^*(S^0)) & \xrightarrow{\nu_*} & \mathrm{Hom}_{E_*(S^0)}\Big(E_*(W),\ \mathrm{Hom}_{F_*(S^0)}(F_*(S^0),\ G^*(S^0))\Big)
\end{array}$$

The result follows as in part (ii).

Proof of Proposition 26. Take any resolution of $X$ over $E_*$, say the following.

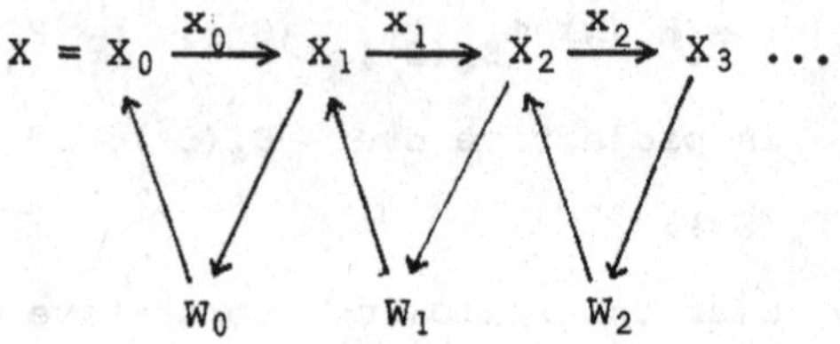

Here the objects $W_r$ are supposed to satisfy UCT 1 or UCT 2 with respect to the functors $E_*$ and $G_*$ or $G^*$. We will show that it qualifies as a resolution of $X$ over $F_*$. In fact, since

$$(x_r)_*: E_*(X_r) \longrightarrow E_*(X_{r+1})$$

is zero, the homomorphism

$$(x_r)_*: F_*(X_r) \longrightarrow F_*(X_{r+1})$$

is zero by Proposition 25. The remaining statements which need to be checked are provided by Lemma 27. Proposition 26 follows immediately.

Example. For any $X$ we have

$$\mathrm{Ext}^{p,*}_{MU_*(S^0)}(MU_*(X), \underline{BU}^*(S^0)) = 0 \quad \text{for} \quad p > 1 .$$

This follows immediately from Proposition 26, since the result is trivial for

$$\mathrm{Ext}^{p,*}_{\underline{BU}_*(S^0)}(\underline{BU}_*(X), \underline{BU}^*(S^0))^{\dagger}.$$

The following result is required for use in Lecture 3.

Lemma 28

If $E = \underline{BO}$, $\underline{BU}$, $MO$, $MU$, $MSp$, $S$ or $K(Z_p)$ then $E_*(E)$ is flat as a module over $E_*(S^0)$.

Proof. The cases $E = MO$, $S$ and $K(Z_p)$ are trivial. In the cases $E = MU$, $MSp$ we can apply the spectral sequence

$$H_*(E; E_*(S^0)) \Longrightarrow E_*(E)$$

to show that $E_*(E)$ is projective over $E_*(S^0)$; in the case $E = MSp$ this involves remarking that the spectral sequence is trivial, by duality with the spectral sequence

$$H^*(MSp; MSp^*(S^0)) \Longrightarrow MSp^*(MSp)$$

which is known to be trivial (see Assumption 20, Example (v)). In the cases $E = \underline{BU}$, $\underline{BO}$ we apply this argument to the spaces $BU$, $BSp$ to show that the modules $\underline{BU}_*(BU)$, $\underline{BO}_*(BSp)$ are projective (compare Assumption 20, examples (vi), (vii)). We then remark that a direct limit of projective modules is flat. This proves Lemma 28.

---

† Note added in proof. I have been asked to say explicitly at this point that UCT2 gives the following exact sequence.

$$0 \to \mathrm{Ext}^{1*}_{MU_*(S^0)}\left(MU_*(X), \underline{BU}^*(S^0)\right) \to \underline{BU}^*(X) \to \mathrm{Hom}^*_{MU_*(S^0)}\left(MU_*(X), \underline{BU}^*(S^0)\right) \to 0$$

## LECTURE 2. THE ADAMS SPECTRAL SEQUENCE

In this lecture I want to discuss the prospects of setting up an "Adams spectral sequence" [1, 2, 15] using a generalised homology or cohomology theory. Everything is to be taken as provisional, or as work in progress, and no proofs will be given.

I shall assume that we can work in some stable category, like those supplied by Boardman [7, 8] and Puppe [25]. I shall also suppose that we are given a homology or cohomology functor to use in our constructions. I will suppose that this functor takes values in an abelian category. As long as we are talking generalities, we can then suppose that the functor is covariant; because if it is contravariant, we can replace the abelian category by its opposite. We will write $E_*$ for this homology functor.

I suggest that we now adopt a construction reminiscent of those constructions for Ext which avoid using projectives and injectives. More precisely, I suggest we proceed as follows. Suppose given two objects $X, Y$ in our stable category. Consider diagrams of the following form.

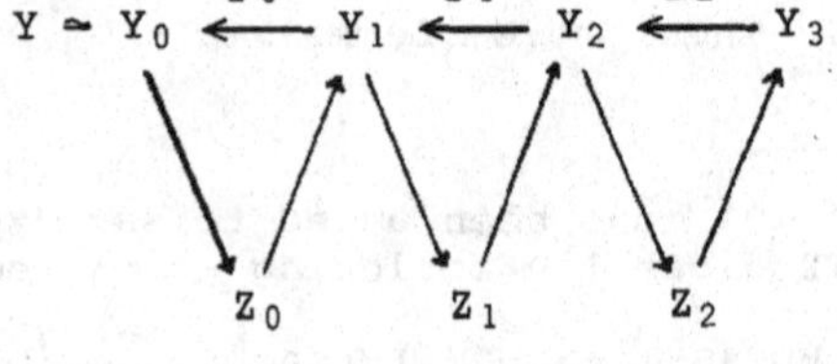

Here the notation $Y \simeq Y_0$ means a homotopy equivalence; and the triangles are supposed to be exact (cofibre) triangles in our stable category. We restrict attention to the diagrams such that

$$E_*(y_r) = 0: E_*(Y_{r+1}) \longrightarrow E_*(Y_r)$$

for each $r \geq 0$; this is the crucial condition. In this case the sequence

$$0 \longrightarrow E_*(Y) \longrightarrow E_*(Z_0) \longrightarrow E_*(Z_1) \longrightarrow E_*(Z_2) \longrightarrow \cdots$$

is exact. We call such diagrams "filtrations" of $Y$. If we wish, we can suppose without loss of generality that each $y_r$ is an inclusion map (replace $Y_0$ by a "telescope").

By mapping $X$ into such a filtration of $Y$ we get a spectral sequence; but this is not yet the spectral sequence we seek. However, we can take all possible filtrations of $Y$ and consider them as the objects of a directed category (in the sense of Grothendieck). (Since I am omitting proofs, I will omit certain details as to how this is done, although they were given in the original lecture.) From each filtration we get a spectral sequence, and we can now take the direct limit of all these spectral sequences; this is the spectral sequence I suggest. Let us call it $SS(X,Y;E_*)$.

I will also omit some arguments in favour of this definition, although they were given in the original lecture.

At this level one should already be able to set up

some formal properties of the spectral sequence. For example, suppose that we have a functor $T$ from one abelian category to another, and that both $E_*$ and $TE_*$ are homology functors. (For examples, see Lecture 1, Proposition 25, or Lecture 4.) Then there clearly is a homomorphism

$$SS(X,Y;E_*) \longrightarrow SS(X,Y,TE_*) ,$$

because every diagram which qualifies as a filtration for $E_*$ also qualifies as a filtration for $TE_*$. (Compare Lecture 1, Proposition 26.) If $E_*$ and $F_*$ are homology functors which mutually determine each other in this way, then

$$SS(X,Y;E_*) \cong SS(X,Y;F_*) .$$

(For examples, see Lecture 4.)

We can now raise the following question. Suppose that $X$ and $Y$ are finite complexes, and that we consider only filtrations in which each $Y_r$ is equivalent to a finite complex. Do these yield in the limit the same spectral sequence as if we did not restrict the filtrations? This is probably true if the homology theory $E_*$ has sufficiently strong finiteness properties.

We can now consider the behaviour of our constructions under S-duality. Do we have

$$SS(X,Y,E_*) \cong SS(DY,DX,E_*D) \;?$$

(Note that $E_*D$ is a cohomology theory defined on finite complexes.) This problem leads one to consider also a "dual" approach to the construction.

We consider diagrams of the following form.

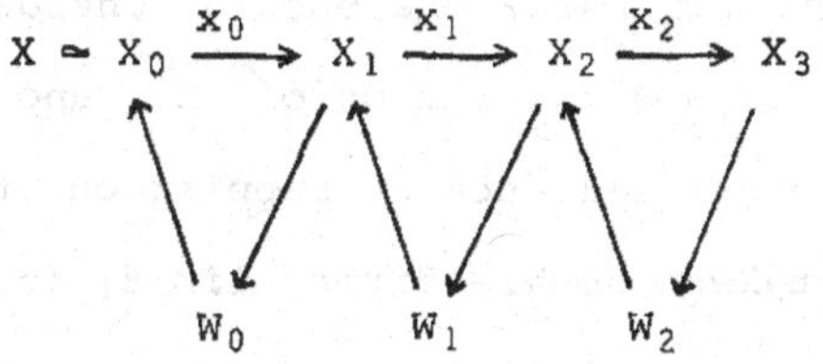


As above, the notation $X \simeq X_0$ means a homotopy equivalence, and the triangles are supposed to be exact (cofibre) triangles in our stable category. We restrict attention to the diagrams such that

$$E_*(x_r) = 0: E_*(X_r) \longrightarrow E_*(X_{r+1})$$

for each $r \geq 0$. In this case the sequence

$$0 \longleftarrow E_*(X) \longleftarrow E_*(W_0) \longleftarrow E_*(W_1) \longleftarrow E_*(W_2) \longleftarrow \ldots$$

is exact. We call such diagrams "filtrations" of $X$. If we wish, we can suppose without loss of generality that each $x_r$ is an inclusion map (replace $X_0$ by a "telescope").

By mapping such a filtration of $X$ into $Y$ we get a spectral sequence. The suggestion would be to vary the filtration (inversely) and take a direct limit of the resulting spectral sequences. Does this give the same spectral sequence as before?

Evidently the situation is like that in homological algebra; there we can define $\mathrm{Ext}^*(L,M)$ by resolving $L$, or by resolving $M$, and we want to show that the result is the same. The proof there, as we know, is to resolve both of them, and show that that gives the same result as resolving either one. Similarly here; one should consider a filtration of X, and also a filtration of Y, and one should try to get a spectral sequence by mapping one to the other. Then one should take a double direct limit, and show that this gives the same spectral sequence as one obtains by filtering either X or Y alone. I haven't tried to write down any details about this.

If one can attain this sort of manipulative ability, one ought to be able to set up various formal properties of the spectral sequences without further assumptions on $E_*$. For example, there should be a pairing

$$SS(Y,Z;E_*) \otimes SS(X,Y;E_*) \longrightarrow SS(X,Z;E_*)$$

which on the $E_\infty$ level is given by composition.

The next step would be to compute the $E_2$ term of our spectral sequence. We are supposing that $E_*$ takes values in an abelian category, so we can define Ext by classifying long exact sequences. It is reasonable to hope that we can define a homomorphism from the $E_2$ term to $\text{Ext}^{**}(E_*(X),E_*(Y))$. The question would be, when can we prove that this homomorphism is an isomorphism? For this purpose one obviously needs to choose the right category, so as to obtain the right Ext groups. More precisely, we need to arrange a very close correspondence between the algebra and the geometry, so that there is some algebraic situation which gives us a legitimate calculation of the Ext groups and which can be realised geometrically.

At this point all suggestions for proceeding assume that our functor is represented by a spectrum $E$.

(i) The original formulation asks us to work in cohomology, and consider $E^*(X)$, $E^*(Y)$ as modules over the ring $E^*(E)$ of cohomology operations [1, 2, 23, 24]. This approach has various disadvantages.

(a) In the generalised case $E^*(E)$ is a topologised ring, and $E^*(X)$, $E^*(Y)$ are topologised modules over the topologised ring $E^*(E)$. We have to take account of the topology [24]. Topologised modules usually fail to form an abelian category, owing to the existence of maps $f: L \longrightarrow M$ which are isomorphisms of the module structure, and continuous, but such that $f^{-1}$ is not continuous.

(b) We cannot assert that $E^q(E) = 0$ for $q < 0$; we may have non-zero cohomology operations which lower dimension by any prescribed amount, as well as ones which raise it. Similar remarks apply to our modules. Both (a) and (b) mean that our constructions and calculations lose a certain element of finiteness which is present in the classical case.

(c) By means of examples (which I will now omit, although they were given in the original lecture) we see that even in the classical case of ordinary cohomology with $Z_p$ coefficients, approach (i) only works under finiteness assumptions on $Y$. In the generalised case, we may see this as follows.

We wish to consider filtrations of $Y$ in which each object $Z_r$ is "free"; in particular, $E^*(Z_r)$ should be "free" in some sense applicable to topologised modules, and we should have

$$[X,Z_r]_* \cong \mathrm{Hom}_{E^*(E)}(E^*(Z_r),E^*(X)) .$$

Since we wish to know about maps from $X$ to $Z_r$ and from $Z_r$ to $E$, this means in practice that we must stick to the case in which $Z_r$ is both a sum and a product of suspensions $S^nE$ of $E$. And again, this means in practice that we must stick to the case in which $E$ is connected and $Z_r$ is a countable sum,

$$Z_r = \bigvee_{i=1}^{\infty} S^{n(i)}E ,$$

in which $n(i) \longrightarrow \infty$ as $i \longrightarrow \infty$. In other words, we are compelled to prove or assume that $E^*(Y)$ admits a resolution by "free" topologised graded modules which have only a finite number of "generators" in dimensions less than $n$ (for each $n$). Although Novikov [24] arranges his work somewhat differently, it is essentially for this purpose that he relies on finiteness properties of $E^*$ which are true in the case $E = MU$ (see Lecture 5). The corresponding properties are unknown for $E = MSp$, and definitely false for $E = S$, although the Adams spectral sequence works for these spectra in some cases at least.

It may be seen from the examples that trouble (c) arises from a double dualisation. The spectral sequence is covariant in Y, but by taking $E^*(Y)$ we are taking a contravariant functor of Y, and then by taking $\mathrm{Ext}^{**}_{E^*(E)}(E^*(Y), E^*(X))$ we are taking a contravariant functor of $E^*(Y)$. This leads to the next approach.

(ii) The next approach would ask us to follow Cartan and Douady [15], and work in homology, considering $E_*(X)$ and $E_*(Y)$ as modules over the ring $E^*(E)$. In the classical case $E = K(Z_p)$ this works quite well. This is partly owing to the fact that $E_*(E)$ is then an injective module over the ring $E^*(E)$; but this fails to generalise to cases in which $E_*(S^0)$ is not a field. In general the ring $E^*(E)$ retains its previous disadvantages, and this approach suffers from being a compromise or half-way house between (i) and (iii). The way ahead appears to lie in a more whole-hearted acceptance of the idea that homology is better than cohomology.

(iii) My final suggestion is that we should work wholly in homology, and consider $E_*(X)$, $E_*(Y)$ as comodules with respect to the coalgebra $E_*(E)$. We use $E_*(S^0)$ as the ground ring for our comodules etc. The necessary details are given in Lecture 3. Of course, we need some data for this; in fact, we need to assume that E is a ring-spectrum and $E_*(E)$ is flat over $E_*(S^0)$. This is true for the spectra mentioned in Lecture 1, Lemma 28. Everything now works much better. The comodules $E_*(X)$, $E_*(Y)$ and the coalgebra $E_*(E)$ are discrete; in typical cases we have $E_q(X) = 0$ for sufficiently large negative q, and $E_q(E) = 0$ for $q < 0$. The comodules form an abelian category. Our constructions and calculations regain that element of finiteness which we lost before.

In order to compute $\mathrm{Ext}^{**}_{E_*(E)}(E_*(X), E_*(Y))$, it is sufficient to take a resolution of $E_*(X)$ by comodules which are projective over $E_*(S^0)$, and a resolution of $E_*(Y)$ by extended comodules; the latter play the part of "relative injectives". Both sorts of resolution can be constructed geometrically. For the first, we require a filtration of X such that $E_*(W_r)$ is projective over $E_*(S^0)$ for each r. Such a filtration can be constructed by Lemma 21 of Lecture 1. Moreover, we see that such filtrations are cofinal in the set of all filtrations of X. For the second, we require a filtration of Y such that $E_*(Z_r)$ is an extended comodule for each r. Such a filtration can be constructed in the following way. Let the structure maps of the ring spectrum E be $\mu: E \wedge E \longrightarrow E$ and $i: S^0 \longrightarrow E$. Suppose we have constructed $Y_r$; the induction starts with $Y_0 = Y$. Take

$Z_r = E \wedge Y_r$, and form the map

$$Y_r \simeq S^0 \wedge Y_r \xrightarrow{i \wedge 1} E \wedge Y_r = Z_r .$$

Then $E_*(Y_r) \longrightarrow E_*(Z_r)$ is mono, since it is defined to be $\pi_*(E \wedge Y_r) \longrightarrow \pi_*(E \wedge E \wedge Y_r)$, and this has a one-sided inverse induced by $E \wedge E \wedge Y_r \xrightarrow{\mu \wedge 1} E \wedge Y_r$. The comodule $E_*(Z_r)$ is extended, by the results of Lecture 3. Form the following cofibre triangle.

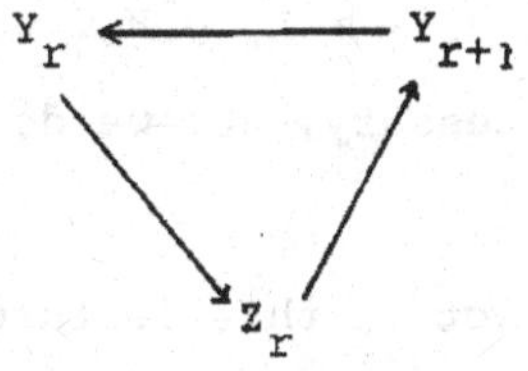


Then $E_*(Y_{r+1}) \longrightarrow E_*(Y_r)$ must be zero. This completes the induction. By adding a few details, we see that such filtrations are cofinal in the set of all filtrations of $Y$.

We may say that at the present time approach (iii) seems to be promising.

The final step, of course, would be to discuss the convergence of the spectral sequence. I would like to defer this question.

## LECTURE 3 HOPF ALGEBRA AND COMODULE STRUCTURE

In the classical case of ordinary cohomology with coefficients $Z_p$, the mod $p$ Steenrod algebra $A^*$ is a Hopf algebra, and it acts on the left on the cohomology of any space, so that we have an action map $A^* \otimes H^* \longrightarrow H^*$. If we dualise by applying $\mathrm{Hom}_{Z_p}(\ ,Z_p)$, we see that the dual $A_*$ of the Steenrod algebra is also a Hopf algebra; and if the homology $H_*$ of a space is locally finitely generated, we have a coaction map $H_* \longrightarrow A_* \otimes H_*$. (The finiteness condition is actually unnecessary, but we do not need to spend time on that here.)

It is the object of this lecture to see how the material mentioned above generalises to the case of a generalised homology theory. We will begin by stating our assumptions; then we will list the structure maps we propose to introduce, and list their principal formal properties. Next we will give the definitions of the structure maps, and comment on the proofs of the formal properties. Then we give two propositions which relate $A_*$ to $A^*$ in the generalised case. Finally, we use these two propositions to show that if we specialise to the classical case of ordinary cohomology with $Z_p$ coefficients, all our structure maps specialise to those classically considered.

It will be convenient to write as if we are working in a stable category in which we have smash-products with the usual properties; but if the reader objects to this, our statements can be "demythologised" by known methods. We shall suppose given a ring-spectrum $E$, so that we are given a product map $\mu: E \wedge E \longrightarrow E$ and a unit map $i: S^0 \longrightarrow E$. These are supposed to have the usual properties; that is, the following diagrams are homotopy-commutative.

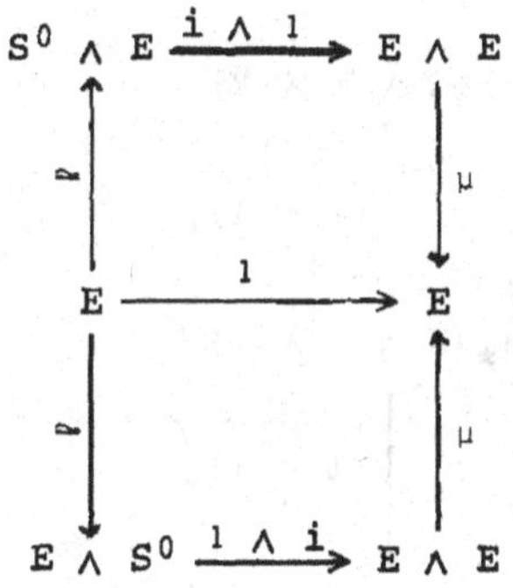


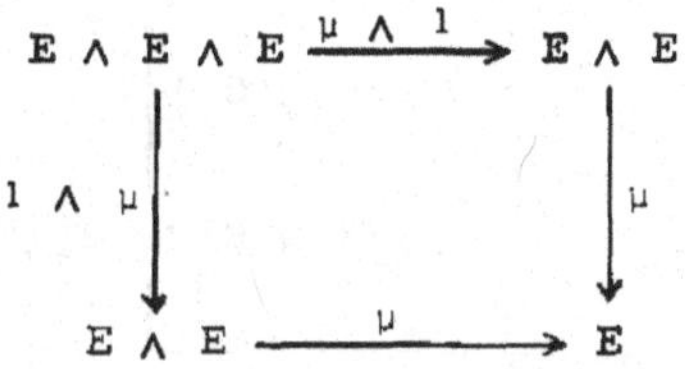


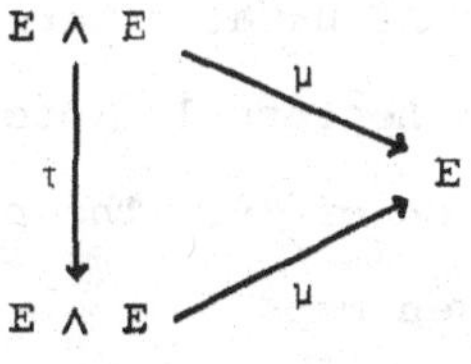


Here $\tau$ is the usual switch map.

We recall that the homology groups of a spectrum $X$ with coefficients in $E$ are given by

$$E_n(X) = [S^n, E \wedge X] = \pi_n(E \wedge X) .$$

The classical case is given by taking $E$ to be the Eilenberg-MacLane spectrum $K(Z_p)$. The analogue of $A_*$ in the generalised case is therefore $E_*(E) = \pi_*(E \wedge E)$, the homology of $E$ with coefficients in $E$. The analogue of $Z_p$ is $E_*(S^0) = \pi_*(E)$. Since $E$ is a ring-spectrum, we have various products. More precisely, suppose given a pairing $\mu: E \wedge F \longrightarrow G$ of spectra. Then we shall have to consider three products, which appear in the following commutative diagram.

$$\begin{array}{ccc}
\pi_*(E \wedge X) \otimes \pi_*(F \wedge Y) & \xrightarrow{\nu} & \pi_*(G \wedge X \wedge Y) \\
\uparrow \scriptstyle{\tau_* \otimes 1} & & \uparrow \scriptstyle{(\tau \wedge 1)_*} \\
\pi_*(X \wedge E) \otimes \pi_*(F \wedge Y) & \xrightarrow{m} & \pi_*(X \wedge G \wedge Y) \\
\downarrow \scriptstyle{1 \otimes \tau_*} & & \downarrow \scriptstyle{(1 \wedge \tau)_*} \\
\pi_*(X \wedge E) \otimes \pi_*(Y \wedge F) & \xrightarrow{\nu'} & \pi_*(X \wedge Y \wedge G)
\end{array}$$

Here the product $\nu$ is the usual external homology product, as used (for example) in Lecture 1, Note 7. The product $\nu'$ is a back-to-front version of $\nu$. The product $m$ is defined as follows. Suppose given maps

$$f: S^p \longrightarrow X \wedge E, \qquad g: S^q \longrightarrow F \wedge Y .$$

Then $m(f \otimes g)$ is the following composite.

$$S^p \wedge S^q \xrightarrow{f \wedge g} X \wedge E \wedge F \wedge Y \xrightarrow{1 \wedge \mu \wedge 1} X \wedge G \wedge Y .$$

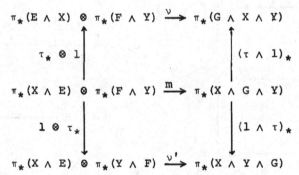

Since it is important for us in this lecture to keep factors in their correct order, we will use m as our basic product. By taking $X = S^0$ or $Y = S^0$, we obtain the following special cases.

$$m: \pi_p(E) \otimes \pi_q(F \wedge Y) \longrightarrow \pi_{p+q}(G \wedge Y)$$

$$m: \pi_p(X \wedge E) \otimes \pi_q(F) \longrightarrow \pi_{p+q}(X \wedge G)$$

$$m: \pi_p(E) \otimes \pi_q(F) \longrightarrow \pi_{p+q}(G) .$$

In particular, $\pi_*(E)$ is an anticommutative ring with unit. For any $Y$, $\pi_*(E \wedge Y)$ is a left module over $\pi_*(E)$; the product map

$$m: \pi_*(E) \otimes \pi_*(E \wedge Y) \longrightarrow \pi_*(E \wedge Y)$$

is the usual one, and coincides with the map $\mu$ considered in UCT 2 (see Lecture 1, Note 2). For any $X$, $\pi_*(X \wedge E)$ is a right module over $\pi_*(E)$. The product

$$m: \pi_*(X \wedge E) \otimes \pi_*(E \wedge Y) \longrightarrow \pi_*(X \wedge E \wedge Y)$$

factors to give a map

$$\pi_*(X \wedge E) \otimes_{\pi_*(E)} \pi_*(E \wedge Y) \longrightarrow \pi_*(X \wedge E \wedge Y),$$

which we also call m.

We have product maps

$$m: \pi_*(E) \otimes \pi_*(E \wedge E) \longrightarrow \pi_*(E \wedge E)$$

$$m: \pi_*(E \wedge E) \otimes \pi_*(E) \longrightarrow \pi_*(E \wedge E),$$

and thus $\pi_*(E \wedge E)$ becomes a bimodule over $\pi_*(E)$. It should be noted that the two actions of $\pi_*(E)$ on $\pi_*(E \wedge E)$

are in general quite distinct; this is the main difference between the generalised case and the classical case, in which we have only one action of $Z_p$ on $A_*$. The presence of these two actions means that the generalised case demands a little more care than the classical case.

We now assume that $\pi_*(E \wedge E)$ is flat as a right module over $\pi_*(E)$ (using the right action). By using the switch map

$$\tau: E \wedge E \longrightarrow E \wedge E$$

to interchange the two factors, we check that it is equivalent to assume that $\pi_*(E \wedge E)$ is flat as a left module over $\pi_*(E)$ (using the left action). This hypothesis is somewhat restrictive, but it is satisfied in many important cases, notably the cases

$$E = \underline{BO},\ \underline{BU},\ MO,\ MU,\ MSp,\ S \text{ and } K(Z_p)$$

(see Lecture 1, Lemma 28).

With this hypothesis, we will see that $\pi_*(E \wedge E)$ is a Hopf algebra in a fully satisfactory sense, and that for any spectrum $X$, $\pi_*(E \wedge X)$ is a comodule over the coalgebra $\pi_*(E \wedge E)$. We will now make this more precise by listing the structure maps we shall introduce, and giving their principal properties.

The structure maps comprise a product map

$$\phi: \pi_*(E \wedge E) \otimes \pi_*(E \wedge E) \longrightarrow \pi_*(E \wedge E),$$

two "unit" maps

$$\eta_L: \pi_*(E) \longrightarrow \pi_*(E \wedge E)$$

$$\eta_R: \pi_*(E) \longrightarrow \pi_*(E \wedge E)$$

a counit map

$$\varepsilon: \pi_*(E \wedge E) \longrightarrow \pi_*(E)$$

a canonical anti-automorphism

$$c: \pi_*(E \wedge E) \longrightarrow \pi_*(E \wedge E)$$

a diagonal map

$$\psi = \psi_E: \pi_*(E \wedge E) \longrightarrow \pi_*(E \wedge E) \otimes_{\pi_*(E)} \pi_*(E \wedge E)$$

and for each spectrum $X$, a coaction map

$$\psi = \psi_X: \pi_*(E \wedge X) \longrightarrow \pi_*(E \wedge E) \otimes_{\pi_*(E)} \pi_*(E \wedge X) .$$

(The diagonal map $\psi_E$ is obtained by specialising the coaction map $\psi_X$ to the case $X = E$.)

It is important to note that in the tensor-product $\pi_*(E \wedge E) \otimes_{\pi_*(E)} \pi_*(E \wedge X)$, the action of $\pi_*(E)$ on the left-hand factor $\pi_*(E \wedge E)$ is the right action. (The action of $\pi_*(E)$ on the right-hand factor $\pi_*(E \wedge X)$ is the usual left action.) This is exactly what we need to use the tensor-product notation in a systematic way.

The tensor-product $\pi_*(E \wedge E) \otimes_{\pi_*(E)} \pi_*(E \wedge X)$ can be considered as a left module over $\pi_*(E)$, by using the left action of $\pi_*(E)$ on $\pi_*(E \wedge E)$; that is,

$$\lambda(e \otimes x) = (\lambda e) \otimes x$$

$(\lambda \in \pi_*(E),\ e \in \pi_*(E \wedge E),\ x \in \pi_*(E \wedge X))$ .

The coaction map $\psi_X$ is a map of left modules over $\pi_*(E)$.

In particular, the previous two paragraphs apply to the case $X = E$. Here the tensor-product $\pi_*(E \wedge E) \otimes_{\pi_*(E)} \pi_*(E \wedge E)$ can also be considered as a right module over $\pi_*(E)$, by using the right action of $\pi_*(E)$ on the right-hand factor. The diagonal map $\psi_E$ is a map of bimodules over $\pi_*(E)$.

The behaviour of the other structure maps with respect to the actions of $\pi_*(E)$ will emerge from the properties given below. The tensor-product on which the product map $\phi$ is defined can be taken over the integers.

The principal properties of these structure maps are as follows. The product map $\phi$ is associative, anticommutative and has a unit element 1. The maps $\eta_L$, $\eta_R$, $\varepsilon$ and $c$ are homomorphisms of graded rings with unit. The left action of $\pi_*(E)$ on $\pi_*(E \wedge E)$ is given by

$$\lambda e = \phi((\eta_L\lambda) \otimes e) \qquad (\lambda \in \pi_*(E),\ e \in \pi_*(E \wedge E)) .$$

Similarly, the right action of $\pi_*(E)$ on $\pi_*(E \wedge E)$ is given by

$$e\lambda = \phi(e \otimes (\eta_R\lambda)) \qquad (e \in \pi_*(E \wedge E),\ \lambda \in \pi_*(E)) .$$

We have

$$\varepsilon\eta_L = 1,\ \varepsilon\eta_R = 1,\ c\eta_L = \eta_R,\ c\eta_R = \eta_L,$$

$$\varepsilon c = \varepsilon,\ c^2 = 1 .$$

These properties determine the behaviour of $\phi$, $\eta_L$, $\eta_R$, $\varepsilon$ and $c$ with respect to the actions of $\pi_*(E)$. In particular, $\varepsilon$ is a map of bimodules.

The coaction map is natural for maps of $X$. The coaction map is associative, in the sense that the following diagram is commutative.

$$\begin{array}{ccc}
\pi_*(E\wedge X) & \xrightarrow{\psi_X} & \pi_*(E\wedge E) \otimes_{\pi_*(E)} \pi_*(E\wedge X) \\
\downarrow \psi_X & & \downarrow 1\otimes\psi_X \\
\pi_*(E\wedge E) \otimes_{\pi_*(E)} \pi_*(E\wedge X) & \xrightarrow{\psi_E\otimes 1} & \pi_*(E\wedge E) \otimes_{\pi_*(E)} \pi_*(E\wedge E) \otimes_{\pi_*(E)} \pi_*(E\wedge X)
\end{array}$$

(Note that $1 \otimes \psi_X$ is defined because $\psi_X$ is a map of left modules over $\pi_*(E)$, and $\psi_E \otimes 1$ is defined because $\psi_E$ is a map of right modules over $\pi_*(E)$.) In particular, we can specialise this diagram to the case $X = E$, and we see that the diagonal map is associative.

The behaviour of the diagonal with respect to the product is given by the following commutative diagram.

$$\begin{array}{ccc}
\pi_*(E\wedge E) \otimes \pi_*(E\wedge E) & \xrightarrow{\phi} & \pi_*(E\wedge E) \\
\downarrow \psi_E \otimes \psi_E & & \downarrow \psi_E \\
[\pi_*(E\wedge E) \otimes_{\pi_*(E)} \pi_*(E\wedge E)] \otimes [\pi_*(E\wedge E) \otimes_{\pi_*(E)} \pi_*(E\wedge E)] & \xrightarrow{\Phi} & \pi_*(E\wedge E) \otimes_{\pi_*(E)} \pi_*(E\wedge E)
\end{array}$$

Here the map $\Phi$ is defined by

$$\Phi(e \otimes f \otimes g \otimes h) = (-1)^{pq}\phi(e \otimes g) \otimes \phi(f \otimes h)\,,$$

where $f \in \pi_p(E \wedge E)$, $g \in \pi_q(E \wedge E)$. It has to be verified that this formula does give a well-defined map of the product of tensor products over $\pi_*(E)$, but this can be done using the facts stated above.

The behaviour of the diagonal map on the unit is given by $\psi_E(1) = 1 \otimes 1$. It follows that we have

$$\psi_E\eta_L\lambda = (\eta_L\lambda) \otimes 1 \;,\; \psi_E\eta_R\lambda = 1 \otimes (\eta_R\lambda) \quad (\lambda \in \pi_*(E))\;.$$

The behaviour of the diagonal map with respect to the counit is given by the following commutative diagram.

$$\begin{array}{ccc}
\pi_*(E \wedge X) & \xrightarrow{\psi_X} & \pi_*(E \wedge E) \otimes_{\pi_*(E)} \pi_*(E \wedge X) \\
\Big\downarrow 1 & & \Big\downarrow \varepsilon \otimes 1 \\
\pi_*(E \wedge X) & \xleftarrow{\cong} & \pi_*(E) \otimes_{\pi_*(E)} \pi_*(E \wedge X)
\end{array}$$

Here the bottom arrow is given by the usual left action of $\pi_*(E)$ on $\pi_*(E \wedge X)$. The map $\varepsilon \otimes 1$ is defined because $\varepsilon$ is a map of right modules over $\pi_*(E)$. Similarly, we have the following commutative diagram.

$$\begin{array}{ccc}
\pi_*(E \wedge E) & \xrightarrow{\psi_E} & \pi_*(E \wedge E) \otimes_{\pi_*(E)} \pi_*(E \wedge E) \\
\Big\downarrow 1 & & \Big\downarrow 1 \otimes \varepsilon \\
\pi_*(E \wedge E) & \xleftarrow{\cong} & \pi_*(E \wedge E) \otimes_{\pi_*(E)} \pi_*(E)
\end{array}$$

Here the bottom arrow is given by the right action of $\pi_*(E)$ on $\pi_*(E \wedge E)$. The map $1 \otimes \varepsilon$ is defined because $\varepsilon$ is a map of left modules over $\pi_*(E)$.

The behaviour of the diagonal with respect to the canonical anti-automorphism $c$ is given by the following commutative diagram.

$$\begin{array}{ccc}
\pi_*(E \wedge E) & \xrightarrow{\psi_E} & \pi_*(E \wedge E) \otimes_{\pi_*(E)} \pi_*(E \wedge E) \\
{\scriptstyle c}\downarrow & & \downarrow{\scriptstyle C} \\
\pi_*(E \wedge E) & \xrightarrow{\psi_E} & \pi_*(E \wedge E) \otimes_{\pi_*(E)} \pi_*(E \wedge E)
\end{array}$$

Here the map $C$ is defined by

$$C(e \otimes f) = (-1)^{pq} cf \otimes ce$$

$(e \in \pi_p(E \wedge E),\ f \in \pi_q(E \wedge E))$ .

It has to be verified that this formula does give a well-defined map of the tensor product over $\pi_*(E)$, but this can be done using the facts stated above.

The following commutative diagrams express that property of the canonical anti-automorphism which in the classical case is taken as its definition.

$$\begin{array}{ccc}
\pi_*(E \wedge E) & \xrightarrow{\varepsilon} & \pi_*(E) \\
{\scriptstyle \psi_E}\downarrow & & \downarrow{\scriptstyle \eta_L} \\
\pi_*(E \wedge E) \otimes_{\pi_*(E)} \pi_*(E \wedge E) & \xrightarrow{\phi(1 \otimes c)} & \pi_*(E \wedge E)
\end{array}$$

$$\begin{array}{ccc}
\pi_*(E \wedge E) & \xrightarrow{\quad \varepsilon \quad} & \pi_*(E) \\
\downarrow{\scriptstyle \psi_E} & & \downarrow{\scriptstyle \eta_R} \\
\pi_*(E \wedge E) \otimes_{\pi_*(E)} \pi_*(E \wedge E) & \xrightarrow{\phi(c\otimes 1)} & \pi_*(E \wedge E)
\end{array}$$

It has to be verified that $\phi(1\otimes c)$ and $\phi(c\otimes 1)$ do give well-defined maps of the tensor product over $\pi_*(E)$, but this can be done using the facts stated above.

This completes the list of properties of our structure maps. We also require one further formal property in order to show that certain comodules $E_*(X)$ are extended (see Lectures 1, 2). Let $F$ be a left module-spectrum over the ring-spectrum $E$; for example, we might have $F = E \wedge Y$. Then the following diagram is commutative.

$$\begin{array}{ccc}
\pi_*(E \wedge E) \otimes_{\pi_*(E)} \pi_*(F) & \xrightarrow{\quad m \quad} & \pi_*(E \wedge F) \\
\downarrow{\scriptstyle \psi_E \otimes 1} & & \downarrow{\scriptstyle \psi_F} \\
\pi_*(E\wedge E) \otimes_{\pi_*(E)} \pi_*(E\wedge E) \otimes_{\pi_*(E)} \pi_*(F) & \xrightarrow{1\otimes m} & \pi_*(E\wedge E) \otimes_{\pi_*(E)} \pi_*(E\wedge F)
\end{array}$$

The map $1 \otimes m$ is defined because $m$ is a map of left modules over $\pi_*(E)$.

We now give the definition of our structure maps. The product $\phi$ is given by either way of chasing round the following commutative square.

$$\begin{array}{ccc}
\pi_*(E \wedge E) \otimes \pi_*(E \wedge E) & \xrightarrow{\nu'} & \pi_*(E \wedge E \wedge E) \\
\downarrow \nu & & \downarrow (\mu \wedge 1)_* \\
\pi_*(E \wedge E \wedge E) & \xrightarrow{(1 \wedge \mu)_*} & \pi_*(E \wedge E)
\end{array}$$

(For $\nu$ and $\nu'$, see the discussion of products at the beginning of this lecture.) In other words, suppose given

$$f\colon S^p \longrightarrow E \wedge E, \ g\colon S^q \longrightarrow E \wedge E;$$

then $\phi(f \otimes g)$ is the following composite.

$$S^p \wedge S^q \xrightarrow{f \wedge g} E \wedge E \wedge E \wedge E \xrightarrow{1 \wedge \tau \wedge 1} E \wedge E \wedge E \wedge E \xrightarrow{\mu \wedge \mu} E \wedge E.$$

We have maps

$$E \simeq E \wedge S^0 \xrightarrow{1 \wedge i} E \wedge E$$

$$E \simeq S^0 \wedge E \xrightarrow{i \wedge 1} E \wedge E$$

which map $E$ into $E \wedge E$ as the left and right factors. We define $\eta_L$ and $\eta_R$ to be the corresponding induced homomorphisms. We define $\varepsilon$ and $c$ to be the homomorphisms induced by

$$\mu\colon E \wedge E \longrightarrow E$$

and

$$\tau\colon E \wedge E \longrightarrow E \wedge E .$$

It only remains to define $\psi_X$.

Lemma 1

If $\pi_*(X \wedge E)$ is flat as a right module over $\pi_*(E)$, then $m: \pi_*(X \wedge E) \otimes_{\pi_*(E)} \pi_*(E \wedge Y) \longrightarrow \pi_*(X \wedge E \wedge Y)$ is iso.

Proof. This is essentially the trivial case of KT 1 (see Lecture 1, Note 12). The map $m$ is a natural transformation between homology functors of $Y$ which is iso for $Y = S^0$; therefore it is iso for any finite complex $Y$. Pass to direct limits.

We now define

$$h: \pi_*(X \wedge Y) \longrightarrow \pi_*(X \wedge E \wedge Y)$$

to be the homomorphism induced by

$$X \wedge Y \simeq X \wedge S^0 \wedge Y \xrightarrow{1 \wedge i \wedge 1} X \wedge E \wedge Y .$$

The map $h$ is essentially the Hurewicz homomorphism in E-homology.

If $\pi_*(X \wedge E)$ is flat, we can consider the following composite.

$$\pi_*(X \wedge Y) \xrightarrow{h} \pi_*(X \wedge E \wedge Y) \xrightarrow{m^{-1}} \pi_*(X \wedge E) \otimes_{\pi_*(E)} \pi_*(E \wedge Y) .$$

We define $\psi = m^{-1}h$. In particular, since we are assuming that $\pi_*(E \wedge E)$ is flat, we can specialise to the case $X = E$; we take the resulting map $\psi$ for our coaction map $\psi_Y$. This completes the definition of the structure maps.

The proofs of all the formal properties are by

diagram-chasing. In proving any property of $\psi_X$, of course we have to make our diagram up out of two subdiagrams, one for h and one for m. For example, in proving that the coaction map is associative, we first prove two more elementary results; $\psi_X$ is natural for maps of X, and $\psi_F m = (1 \otimes m)(\psi_E \otimes 1)$ (which is the diagram required to prove that $E_*(F)$ is an extended comodule). We now set up the following diagram.

$$\begin{array}{ccc}
\pi_*(E \wedge X) & \xrightarrow{\psi_X} & \pi_*(E \wedge E) \otimes_{\pi_*(E)} \pi_*(E \wedge X) \\
\downarrow h & & \downarrow 1 \otimes h \\
\pi_*(E \wedge E \wedge X) & \xrightarrow{\psi_{E\wedge X}} & \pi_*(E \wedge E) \otimes_{\pi_*(E)} \pi_*(E \wedge E \wedge X) \\
\uparrow m & & \uparrow 1 \otimes m \\
\pi_*(E\wedge E) \otimes_{\pi_*(E)} \pi_*(E\wedge X) & \xrightarrow{\psi_E \otimes 1} & \pi_*(E\wedge E) \otimes_{\pi_*(E)} \pi_*(E\wedge E) \otimes_{\pi_*(E)} \pi_*(E\wedge X)
\end{array}$$

Here the top square is commutative because h is induced by a map

$$X = S^0 \wedge X \xrightarrow{i\wedge 1} E \wedge X ,$$

and $\psi_X$ is natural for maps of X. Similarly, the bottom square is commutative by the second result mentioned, taking $F = E \wedge X$. This gives the required result. The two subsidiary results are proved in the same way.

In proving the behaviour of the diagonal with respect to the product, it is convenient to prove a slightly more general result first. Suppose that $\pi_*(A \wedge E)$, $\pi_*(B \wedge E)$ and $\pi_*(A \wedge B \wedge E)$ are all flat; then the following diagram is commutative.

$$\begin{array}{ccc}
\pi_*(A\wedge X) \otimes \pi_*(B\wedge Y) & \longrightarrow & \pi_*(A\wedge B\wedge X\wedge Y) \\
\downarrow{\scriptstyle \psi \otimes \psi} & & \downarrow{\scriptstyle \psi} \\
[\pi_*(A\wedge E) \otimes_{\pi_*(E)} \pi_*(E\wedge X)] \otimes [\pi_*(B\wedge E) \otimes_{\pi_*(E)} \pi_*(E\wedge Y)] & \longrightarrow & \pi_*(A\wedge B\wedge E) \otimes_{\pi_*(E)} \pi_*(E\wedge X\wedge Y)
\end{array}$$

Here the upper horizontal map is the obvious product, and the lower horizontal map sends $e \otimes f \otimes g \otimes h$ into $(-1)^{pq}\nu'(e \otimes g) \otimes \nu(f \otimes h)$ (see the discussion of products at the beginning of this lecture). This diagram is proved commutative in the same way as before - separate $h$ and $m$. Next observe that since the functor $\pi_*(E \wedge E) \otimes_{\pi_*(E)}$ preserves exactness, applying it twice preserves exactness; that is, the right module

$$\pi_*(E \wedge E \wedge E) \cong \pi_*(E \wedge E) \otimes_{\pi_*(E)} \pi_*(E \wedge E)$$

is flat. So we may specialise to the case $A = B = E$. Now apply naturality to the map

$$A \wedge B = E \wedge E \xrightarrow{\mu} E \ ;$$

we see that the following diagram is commutative.

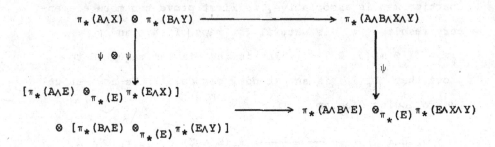

$$\begin{array}{ccc}
\pi_*(E\wedge X) \otimes \pi_*(E\wedge Y) & \xrightarrow{\nu} & \pi_*(E\wedge X\wedge Y) \\
\downarrow \psi_X \otimes \psi_Y & & \downarrow \psi_{X\wedge Y} \\
[\pi_*(E\wedge E) \otimes_{\pi_*(E)} \pi_*(E\wedge X)] \otimes [\pi_*(E\wedge E) \otimes_{\pi_*(E)} \pi_*(E\wedge Y)] & \longrightarrow & \pi_*(E\wedge E) \otimes_{\pi_*(E)} \pi_*(E\wedge X\wedge Y)
\end{array}$$

Here the lower horizontal map sends $e \otimes f \otimes g \otimes h$ into $(-1)^{pq}\phi(e \otimes g) \otimes \nu(f \otimes h)$. This diagram gives the behaviour of the coaction map with respect to the external homology product. Finally we specialise to the case $X = Y = E$ and apply naturality to the map

$$X \wedge Y = E \wedge E \xrightarrow{\mu} E .$$

We obtain the required commutative diagram.

The proof of the remaining formal properties does not call for any special comment.

We now turn to further formulae, involving cohomology, which will help to show that our definitions specialise correctly to the classical case. We recall that the cohomology groups of a spectrum $X$ with coefficients in $E$ are given by

$$E^{-n}(X) = [S^n \wedge X, E] .$$

We have a Kronecker product

$$E^{-p}(X) \otimes E_q(X) \longrightarrow \pi_{p+q}(E)$$

defined as follows. Suppose given maps

$$f\colon S^p\wedge X \longrightarrow E, \quad g\colon S^q \longrightarrow E\wedge X .$$

Then $\langle f,g\rangle$ is the following composite.

$$S^p\wedge S^q \xrightarrow{1\wedge g} S^p\wedge E\wedge X \xrightarrow{1\wedge \tau} S^p\wedge X\wedge E \xrightarrow{f\wedge 1} E\wedge E \xrightarrow{\mu} E .$$

In particular, we have the cohomology groups $E^*(E)$. Since these are defined in terms of maps from $E$ to $E$ (up to suspension), they act on the left on the homology and cohomology groups $E_*(X)$ and $E^*(X)$. The precise definitions are as follows. Suppose given maps

$$a\colon S^p\wedge E \longrightarrow E, \quad f\colon S^q \longrightarrow E\wedge X, \quad g\colon S^r\wedge X \longrightarrow E .$$

Then $af$ is

$$S^p\wedge S^q \xrightarrow{1\wedge f} S^p\wedge E\wedge X \xrightarrow{a\wedge 1} E\wedge X ,$$

and $ag$ is

$$S^p\wedge S^r\wedge X \xrightarrow{1\wedge g} S^p\wedge E \xrightarrow{a} E .$$

In this way $E^*(E)$ becomes a ring with unit, and $E_*(X)$, $E^*(X)$ become left modules over this ring.

We will show that the action of $E^*(E)$ on $E_*(X)$ is determined by the coaction map $\psi_X$. Suppose $a \in E^*(E)$, $x \in E_*(X)$ and $\psi_X x = \sum_i e_i \otimes x_i$, where $e_i \in E_*(E)$, $x_i \in E_*(X)$. Then we have:

<u>Proposition 2</u>

$$ax = \sum_i \langle a, ce_i\rangle x_i .$$

To prove this proposition, we set up the following diagram.

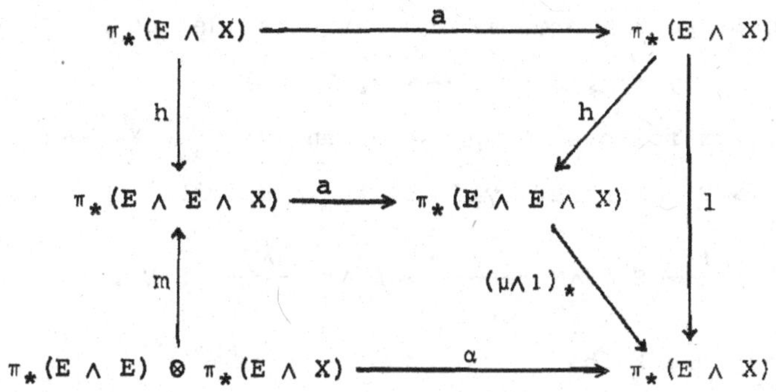


Here $\alpha$ is defined by

$$\alpha(e \otimes x) = \langle a, ce\rangle x \ .$$

It is easy to show that the diagram is commutative. This proves Proposition 2.

In the case when an element $z \in E^*(X)$ is determined by the values of $\langle z,x\rangle$ for all $x \in E_*(X)$, it is reasonable to ask for a calculation of the action of $E^*(E)$ on $E^*(X)$ in terms of $\psi_X$. There is a choice of formulae which answer this question; here I will give one which seems neater than that which I actually gave in Seattle. Suppose $a \in E^*(E)$, $y \in E^p(X)$, $x \in E_*(X)$ and $\psi_X x = \sum_i e_i \otimes x_i$, where $e_i \in E_{q(i)}(E)$, $x_i \in E_*(X)$. Then we have:

<u>Proposition 3</u>

$$\langle ay,x\rangle = \sum_i (-1)^{pq(i)} \langle a, e_i\langle y,x_i\rangle\rangle .$$

The formula on the right makes sense, because $e_i$ lies in $\pi_*(E\wedge E)$, and $\langle y,x_i\rangle$ lies in $\pi_*(E)$, which acts on the right on $\pi_*(E\wedge E)$.

To prove the proposition, we first define

$$y_*: \pi_*(F \wedge X) \longrightarrow \pi_*(F \wedge E)$$

(for any $F$) as follows. Suppose given $y: S^p \wedge X \longrightarrow E$ and $f: S^r \longrightarrow F \wedge X$; let $y_* f$ be the composite

$$S^p \wedge S^r \xrightarrow{1 \wedge f} S^p \wedge F \wedge X \xrightarrow{\tau \wedge 1} F \wedge S^p \wedge X \xrightarrow{1 \wedge y} F \wedge E .$$

Then we easily check that

$$\langle ay, x\rangle = \langle a, y_* x\rangle .$$

We now set up the following diagram.

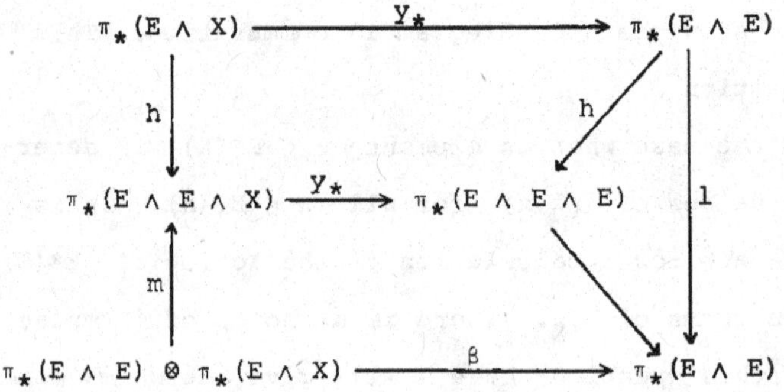


Here $\beta$ is defined by

$$\beta(e \otimes x) = (-1)^{pq} e\langle y, x\rangle$$

for $e \in E_q(E)$. It is easy to show that this diagram is commutative. This shows that

$$y_* x = \sum_i (-1)^{pq(i)} e_i \langle y, x_i\rangle ,$$

and proves Proposition 3.

We will now discuss the way in which our constructions

specialise to the case $E = K(Z_p)$. It is sufficiently clear from the definitions that $\phi$, $\eta_L$, $\eta_R$ and $\varepsilon$ specialise to their classical counterparts $\phi$, $\eta$, $\eta$ and $\varepsilon$. The right action of $\pi_*(E) = Z_p$ on $\pi_*(E \wedge E) = A_*$ coincides with the left action, because the unit acts as a unit on either side, and so the result follows for integer multiples of the unit. It follows that in Proposition 3 we can bring the factor $\langle y,x_i\rangle$ to the left of $e_i$; and after that we can bring it outside the Kronecker product, so as to obtain the following formula.

$$\langle ay,x\rangle = \sum_i (-1)^{pq(i)}\langle a,e_i\rangle\langle y,x_i\rangle .$$

It follows that $\psi_X$ is indeed the dual of the action map $A^* \otimes H^* \longrightarrow H^*$, and (specialising to the case $X = E$) that $\psi_E$ is the dual of the composition map $A^* \otimes A^* \longrightarrow A^*$. Thus $\psi_E$ and $\psi_X$ specialise to their classical counterparts.

Since we have seen that

$$\phi(1 \otimes c)\psi_E = \eta_L\varepsilon$$

and

$$\phi(c \otimes 1)\psi_E = \eta_R\varepsilon ,$$

it now follows that $c$ specialises to its classical counterpart.

It remains only to point out one difference between the classical case and the generalised case. In the generalised case we have introduced a left action of $E^*(E)$ on

$E_*(X)$. This does not specialise to the action of $A^*$ on $H_*$ which is usually considered in the classical case, since the latter is a right action, defined by

$$\langle y, xa \rangle = (-1)^{(p+q)r} \langle ay, x \rangle$$

($y \in H^p$, $x \in H_q$, $a \in A^r$) .

The connection between the two actions may be read off from Proposition 2 and 3. We have

$$xa = (-1)^{qr}(ca)x \qquad (x \in H_q,\ a \in A^r) .$$

Thus the left and right actions differ by the canonical anti-automorphism, as one might expect.

# LECTURE 4 SPLITTING GENERALISED COHOMOLOGY THEORIES WITH COEFFICIENTS

S. P. Novikov [23, 24] has emphasised the importance of the generalised cohomology theory provided by complex cobordism. This is a representable functor; if we take it "reduced", we have

$$MU^n(X) = [X, S^n MU] .$$

It has been proved by Brown and Peterson [10] that if one neglects all the primes except one prime $p$, then the MU-spectrum splits as a sum or product:

$$MU \underset{p}{\simeq} \bigvee_i S^{n(i)} BP(p) \sim \prod_i S^{n(i)} BP(p) .$$

Here $BP(p)$ means the Brown-Peterson spectrum. The sum coincides with the product since $BP(p)$ is connected and $n(i) \longrightarrow \infty$ as $i \longrightarrow \infty$. The business of neglecting all primes except one may be formalised conveniently by introducing coefficients. Let $Q_p$ be the ring of rational numbers $a/b$ with $b$ prime to $p$. Then we can form $MU^*(X;Q_p)$, and we have

$$MU^n(X;Q_p) \cong \prod_i L^{n+n(i)}(X) ,$$

where

$$L^m(X) = [X, S^m L]$$

and $L$ is a suitable version of the Brown-Peterson spectrum.

This situation has been considered by S. P. Novikov

[24]. Potentially it is very profitable. The cohomology theory $L^*$ is just as powerful as $MU^*( \;;Q_p)$; for example, it gives rise to the same "Adams spectral sequence" (see Lecture 2). However, the groups $L^*(X)$ are much smaller than the groups $MU^*(X;Q_p)$; similarly for the coefficient groups $L^*(S^0)$, the ring of operations $L^*(L)$ and the Hopf algebra $L_*(L)$ (see Lecture 3). For all these reasons, calculations with $L$ should be smaller and easier than calculations with $MU$.

Unfortunately, these benefits have not yet been fully realised in practice. The reason is that the splittings given by Brown and Peterson, and by Novikov, are not canonical; they involve large elements of choice. It is doubtless because of this that these authors have not yet given such helpful and illuminating formulae for the structure of $L^*(L)$, etc., as are available for the structure of $MU^*(MU)$, etc.

I therefore propose the following thesis. When we split a cohomology theory into summands, we should try to do so in a canonical way, issuing in helpful and enlightening formulae. To secure these ends I would even be willing to split the theory into summands larger than the irreducible ones. The method which I propose is to take a suitable ring of cohomology operations, say $A$, and construct in it canonical idempotents, say $e$. Then whenever $A$ acts on a module, say $H$, $H$ will split as the direct sum $eH \oplus (1-e)H$.

I will first show how this thesis applies to K-theory. Not only is the case of K-theory somewhat easier, but for technical reasons it is useful as a tool in attacking cobordism. For K-theory I shall give a treatment which seems tolerably complete and satisfactory (Lemma 1 to Lemma 9 below). I will then turn to cobordism (Lemma 10 to Theorem 19 below). Here the theory is somewhat less complete, but it is sufficient to show the existence of canonical summands in cobordism with suitable coefficients.

Let $R$ be a subring of the rationals. Let $K^*(X;R)$ be ordinary, complex K-theory, with coefficients in $R$. We write $K$ for $K^0$; then $K(X;R)$ is a representable functor; we write $BUR$ for the representing space. We require some information on $K^*(BUR;R)$. All that is really needed is that its $\text{Lim}^1$ subgroup [21] is zero; but our method will prove more. It is for this purpose that we introduce the first few lemmas.

Let $d$ be a positive integer, and let $f: BU \longrightarrow BU$ be the map obtained by taking the identity map of the space $BU$ and adding it to itself $d$ times, using the H-space structure of $BU$.

Lemma 1

If $d$ is invertible in $R$ then

$$f_*: H_*(BU;R) \longrightarrow H_*(BU;R)$$

and

$$f^*: H^*(BU;R) \longrightarrow H^*(BU;R)$$

are isomorphisms.

Proof. We will prove that $f^*$ is epi. Suppose, as an inductive hypothesis, that the image of $f^*$ contains the Chern classes $c_1, c_2, \ldots, c_{n-1}$. Then it contains all decomposable elements in $H^{2n}(BU;R)$. For any primitive element $p_n \in H^{2n}(BU;R)$ we have $f^*p_n = dp_n$. But we can find such a $p_n$ which is a non-zero multiple of $c_n$ mod decomposable elements. Therefore $f^*c_n = dc_n$ mod decomposables. Since $d$ is invertible in $R$, $c_n$ lies in the image of $f^*$. This completes the induction and proves that $f^*$ is epi; by duality, $f_*$ is mono.

A precisely dual argument shows that $f_*$ is epi and $f^*$ is mono. Indeed, the preceding paragraph was written so as to dualise correctly. One needs some minimal knowledge of $H_*(BU;R)$ as a ring under the Pontryagin product, and the fact that $f$ is an H-map, so that $f_*$ is a homomorphism of rings. This proves Lemma 1.

Next, let $R_1, R_2$ be two subrings of the rationals. We have an obvious map $i: BU \longrightarrow BUR_1$.

Lemma 2

If $R_1 \subset R_2$, the maps

$$i_*: H_*(BU;R_2) \longrightarrow H_*(BUR_1;R_2)$$

$$i^*: H^*(BUR_1;R_2) \longrightarrow H^*(BU;R_2)$$

$$(i\times i)_*: H_*(BU\times BU;\ R_2) \longrightarrow H_*(BUR_1\times BUR_1;\ R_2)$$

$$(i\times i)^*: H^*(BUR_1\times BUR_1;\ R_2) \longrightarrow H^*(BU\times BU;\ R_2)$$

are isomorphisms.

Proof. If $R_1 = Z$ the result is trivial, so we may assume $R_1 \neq Z$. We now construct a model for $BUR_1$. Consider the positive integers invertible in $R_1$ and arrange them in a sequence $d_1, d_2, d_3, \ldots$ . For each $d_n$ we have a map $f_n: BU \longrightarrow BU$, as in Lemma 1. Take the maps

$$BU \xrightarrow{f_1} BU \xrightarrow{f_2} BU \longrightarrow \ldots \longrightarrow BU \xrightarrow{f_n} BU \longrightarrow \ldots$$

and form a "telescope" or iterated mapping-cylinder; this gives a construction for $BUR_1$. The map $i: BU \longrightarrow BUR_1$ is the injection of the first copy of $BU$. We have

$$H_*(BUR_1;R_2) = \varinjlim(H_*(BU;R_2), f_{n*}) .$$

Now the result about $i_*$ follows from Lemma 1. The result about $(i \times i)_*$ follows from the Künneth theorem. The results about $i^*$ and $(i \times i)^*$ follow from the universal coefficient theorem. This proves Lemma 2.

Lemma 3

Suppose $R_1 \subset R_2$. Then the maps

$$i_*: K_*(BU;R_2) \longrightarrow K_*(BUR_1;R_2)$$

$$i^*: K^*(BUR_1;R_2) \longrightarrow K^*(BU;R_2)$$

$$(i\times i)^*: K^*(BUR_1\times BUR_1;\ R_2) \longrightarrow K^*(BU\times BU;\ R_2)$$

are isomorphisms. The maps $i^*$ and $(i \times i)^*$ are also homeomorphisms with respect to the filtration topology.

Proof. Let $P$ be a point. Consider the usual spectral sequence

$$H_*(X;K_*(P;R_2)) \Longrightarrow K_*(X;R_2) .$$

By Lemma 2, the map $i: BU \longrightarrow BUR_1$ induces an isomorphism between the spectral sequences for $X = BU$ and for $X = BUR_1$. This proves the result about $i_*$. The proof for $i^*$ and $(i \times i)^*$ is similar, using the spectral sequence

$$H^*(X;K^*(P;R_2)) \Longrightarrow K^*(X;R_2) .$$

The space $BUR_1$ is an H-space; let $\mu: BUR_1 \times BUR_1 \longrightarrow BUR_1$ be the product map, and let $\pi,\tilde{\omega}: BUR_1 \times BUR_1 \longrightarrow BUR_1$ be the projections onto the two factors. We retain the assumption that $R_1 \subset R_2$, and consider the set of primitive elements in $\tilde{K}(BUR_1;R_2)$, that is, the set of elements $a$ such that $\mu^*a = \pi^*a + \tilde{\omega}^*a$. This set may be identified with the set of cohomology operations

$$a: \tilde{K}(X;R_1) \longrightarrow \tilde{K}(X;R_2)$$

which are defined for all connected $X$, natural, and additive in the sense that

$$a(x + y) = a(x) + a(y) .$$

(If an operation is additive, it follows that it is $R_1$-linear.) Such operations need not be stable.

This set is to be topologised as a subset of $\widetilde{K}(BUR_1;R_2)$; in other words, an operation $\underline{a}$ is close to zero if it vanishes in all CW-complexes of dimension n.

According to Lemma 3 above, the set of operations $\underline{a}$ to be considered is essentially independent of $R_1$, so long as $R_1 \subset R_2$. (This fact would be trivial if we were dealing only with finite CW-complexes X, since then we have $\widetilde{K}(X;R_1) \cong \widetilde{K}(X) \otimes R_1$, $\widetilde{K}(X;R_2) \cong \widetilde{K}(X) \otimes R_2$.) We therefore write $\widetilde{A}(R_2)$ for the set of operations introduced above, and regard it primarily as the ring of cohomology operations on $\widetilde{K}(X;R_2)$.

We define

$$A(R) = R + \widetilde{A}(R) .$$

By making the first summand R act in the obvious way on $K(P;R)$, the set $A(R)$ may be identified with the set of cohomology operations

$$a\colon K(X;R) \longrightarrow K(X;R)$$

which are defined for all X, natural, and additive (hence R-linear).

Lemma 4

If $R_1 \subset R_2$, we have a monomorphism

$$\imath\colon A(R_1) \longrightarrow A(R_2)$$

such that for each $a \in A(R_1)$ and each X the following

diagram is commutative.

$$\begin{array}{ccc} K(X;R_1) & \xrightarrow{i_*} & K(X;R_2) \\ \downarrow a & & \downarrow a \\ K(X;R_1) & \xrightarrow{i_*} & K(X;R_2) \end{array}$$

This follows from the preceding discussion together with the fact that

$$i_*: \tilde{K}(BU;R_1) \longrightarrow \tilde{K}(BU;R_2)$$

is monomorphic.

Because of this lemma, it will be sufficient to construct idempotents in $A(Q)$ and then prove that they are defined over some suitable subring of the rationals. But over $Q$ the idempotents are obvious. The Chern character allows us to identify $K(X;Q)$ with the product

$$\prod_n H^{2n}(X;Q) .$$

Let us define $e_n$ to be projection on the $n^{th}$ factor:

$$e_n(h^0,h^2,\ldots,h^{2n-2},h^{2n},h^{2n+2},\ldots) = (0,0,\ldots,0,h^{2n},0,\ldots) .$$

Then $e_n$ is an idempotent in $A(Q)$.

I now choose a positive integer $d$, and seek to construct a "fake K-theory" with one non-zero coefficient group every $2d$ dimensions. The required idempotents are obvious. Take a residue class of integers mod $d$, say

$\alpha \in Z_d$, and define

$$E_\alpha = \sum_{n \in \alpha} e_n \in A(Q) .$$

This sum is convergent in the topology which $A(Q)$ has. If we use the Chern character to identify $K(X;Q)$ with $\prod_n H^{2n}(X;Q)$, as above, then we have

$$E_\alpha(h^0,h^2,h^4,\ldots) = (k^0,k^2,k^4,\ldots)$$

where

$$k^{2n} = \begin{cases} h^{2n} & \text{if } n \in \alpha \\ 0 & \text{if } n \notin \alpha \end{cases} .$$

<u>Theorem 5</u>

$E_\alpha$ lies in $A(R)$, where $R = R(d)$ is the ring of rationals $a/b$ such that $b$ contains no prime $p$ with $p \equiv 1 \mod d$.

For example, if $d = 2$, $R$ is the ring of fractions $a/2^f$.

For the proof, we need to work with a representation of $A(R)$. Let $\eta$ be the canonical line bundle over $CP^\infty$; then $K(CP^\infty;R)$ is the ring of formal power-series $R[[\zeta]]$, where $\zeta = \eta - 1$. We define an ($R$-linear) homomorphism

$$\theta: A(R) \longrightarrow K(CP^\infty;R)$$

by

$$\theta(a) = a(\eta) .$$

Lemma 6

$\theta$ is an isomorphism.

Proof. First we show that $\theta$ is mono. Let $a \in A(R)$ be such that $a(\eta) = 0$. Then by naturality $a(1) = 0$, so $a \in \tilde{A}(R)$. Let $\xi$ be the universal $U(n)$-bundle over $BU(n)$; then $\xi - n$ is the universal element in $\tilde{K}(BU(n))$. Since $\underline{a}$ is additive, the splitting principle shows that $a(\xi-n) = 0$ in $\tilde{K}(BU(n);R)$. Let $i: BU \longrightarrow BUR$ be as above. We have

$$\tilde{K}(BU;R) = \varprojlim_{n} \tilde{K}(BU(n);R) \ ;$$

it follows that $a(i) = 0$ in $\tilde{K}(BU;R)$. By Lemma 3 we have $a = 0$ in $\tilde{K}(BUR;R)$.

Next we show that $\theta$ is an epimorphism. For each $n$ we can find an integral linear combination $a_n$ of the operations $\psi^k$ [3, 4] such that

$$a_n \eta = \zeta^n \ ;$$

more precisely,

$$a_n = \sum_{0 \le k \le n} (-1)^{n-k} \frac{n!}{k!n-k!} \psi^k \ .$$

For any sequence of elements $r(n) \in R$, the sum

$$\sum_{n=1}^{\infty} r(n) a_n$$

is convergent in the filtration topology on $\tilde{K}(BU;R)$ and defines a primitive element $\tilde{a}$ of $\tilde{K}(BU;R)$, that is, an element $\tilde{a} \in \tilde{A}(R)$. It remains only to take

$$a = r(0)\Psi^0 + \tilde{a} \in A(R) .$$

We have

$$a(\eta) = \sum_{n=0}^{\infty} a(n)\zeta^n .$$

This proves Lemma 6.

We observe that the isomorphism $\theta$ of Lemma 6 becomes a homeomorphism if we give $K(CP^{\infty};R)$ the filtration topology. The filtration topology coincides with the usual topology on $R[[\zeta]]$: a power-series is close to zero if its first $n$ coefficients vanish.

The isomorphism $\theta$ of Lemma 6 throws the monomorphism $\iota$ of Lemma 4 onto the obvious inclusion map

$$R_1[[\zeta]] \subset R_2[[\zeta]] .$$

We now return to the proof of Theorem 5. Let $x \in H^2(CP^{\infty};Z)$ be the generator, so that

$$\mathrm{ch}\ \eta = \sum_n \frac{x^n}{n!} .$$

Consider the power-series

$$\log(1 + \zeta) = \zeta - \frac{\zeta^2}{2} + \frac{\zeta^3}{3} - \frac{\zeta^4}{4} \ldots .$$

Since ch commutes with sums, products and limits, we have

$$\begin{aligned} \mathrm{ch} \log(1 + \zeta) &= \log \mathrm{ch}(1 + \zeta) \\ &= \log \exp x \\ &= x . \end{aligned}$$

Now we have

$$\mathrm{ch}\ e_n\eta = \frac{x^n}{n!} = \mathrm{ch}\left(\frac{(\log(1+\zeta))^n}{n!}\right) .$$

Therefore

$$e_n\eta = \frac{(\log(1+\zeta))^n}{n!} .$$

We now make a formal manipulation in $Q[t][[\zeta]]$, the ring of formal power-series in $\zeta$ with coefficients which are polynomials in $t$. Namely:

$$\sum_n t^n e_n\eta = \sum_n \frac{t^n(\log(1+\zeta))^n}{n!}$$

$$= \exp(t\log(1+\zeta))$$

$$= (1+\zeta)^t$$

$$= 1 + t\zeta + \frac{t(t-1)}{1.2}\zeta^2 + \ldots .$$

This is true as a formal identity in the ring cited.

Now consider $E_\alpha\eta = \sum_{n\in\alpha} e_n\eta = \sum_r \frac{a_r}{b_r}\zeta^n$, say. We wish to show that the coefficients $\frac{a_r}{b_r}$ lie in $R = R(d)$. Take any prime $p$ such that $p \equiv 1 \mod d$; we wish to show that $a_r/b_r$ is a p-adic integer. Since $d$ divides $p - 1$, I can find in the p-adic integers a primitive $d^{th}$ root of $1$, say $\omega$. Set $\rho = \omega^m$, where the integer $m$ is fixed for the moment. Then $\rho^d = 1$ and $\rho^\alpha$ makes sense. We have

$$\sum_\alpha \rho^\alpha E_\alpha \eta = \sum_n \rho^n e_n \eta$$

$$= 1 + \rho\zeta + \frac{\rho(\rho-1)}{1.2}\zeta^2 + \ldots$$

$$= c_m(\zeta), \text{ say.}$$

Here the binomial coefficient

$$b(t) = \frac{t(t-1)\ldots(t-r+1)}{1.2 \quad \ldots \ r}$$

maps $Z$ to $Z$ and is continuous in the p-adic topology; therefore it maps p-adic integers to p-adic integers. So $c_m(\zeta)$ is a formal power-series in $\zeta$ with coefficients which are p-adic integers. Take $m = 1,2,\ldots,d$; we obtain $d$ equations for the $d$ unknowns $E_\alpha\eta$. The solution is

$$E_\alpha \eta = d^{-1} \sum_{1 \le m \le d} \omega^{-m\alpha} c_m(\zeta) .$$

Since $d^{-1}$ is a p-adic integer, this is a formal power-series whose coefficients are p-adic integers. This proves Theorem 5.

The properties of the elements $E_\alpha \in A(R)$ are as follows.

Theorem 7

(i) $E_\alpha^2 = E_\alpha$

(ii) $E_\alpha E_\beta = 0$ if $\alpha \neq \beta$

(iii) $\sum_\alpha E_\alpha = 1$

(iv) For any $x,y$ in $K(X;R)$

we have a "Cartan formula"

$$E_\alpha(xy) = \sum_{\beta+\gamma=\alpha} (E_\beta x)(E_\gamma y) .$$

<u>Proof</u>. By Lemma 4, $\iota: A(R) \longrightarrow A(Q)$ is a monomorphism. So parts (i), (ii) and (iii) follow from the corresponding equations in $A(Q)$, which are obvious. We turn to part (iv). The result is trivial when either $x$ or $y$ lies in $K(P;R)$, so it is sufficient to prove it when $x$ and $y$ lie in $\tilde{K}(X;R)$. It is sufficient to prove it for external products. Let both $x$ and $y$ be the universal elements in $\tilde{K}(BU)$; then the result holds in $\tilde{K}(BU \times BU;\ Q)$, by an obvious calculation using the Chern character. Since

$$\tilde{K}(BU \times BU;\ R) \longrightarrow \tilde{K}(BU \times BU;\ Q)$$

is monomorphic, the result holds in $\tilde{K}(BU \times BU;\ R)$. Let $x$ and $y$ be the universal element in $\tilde{K}(BUR;R)$; then the result holds in $\tilde{K}(BUR \times BUR;\ R)$ by Lemma 3. The case in which $x$ and $y$ are general follows by naturality. This proves part (iv) and completes the proof of Theorem 7.

Theorems 5 and 7 lead immediately to the results on the splitting of $K^*(X;R)$ (and indeed of $K_*(X:R)$, if required). As above, we are supposing given a positive integer $d$; $R = R(d)$ is as in Theorem 5, and $\alpha$ runs over $Z_d$.

Corollary 8

(i) We have a natural direct sum splitting

$$K(X;R) \cong \sum_{\alpha} K_{\alpha}(X),$$

where

$$K_{\alpha}(X) = E_{\alpha}K(X;R) .$$

(ii) $K_{\alpha}(X)$ is a representable functor.

(iii) If $x \in K_{\beta}(X)$ and $y \in K_{\gamma}(X)$, then $xy \in K_{\beta+\gamma}(X)$ .

(iv) We have

$$\tilde{K}_{\alpha}(S^n) = \begin{cases} R & \text{if } \frac{1}{2}n \in \alpha \\ 0 & \text{otherwise .} \end{cases}$$

(v) Define

$$\phi: \tilde{K}_{\alpha}(X) \longrightarrow \tilde{K}_{\alpha+1}(S^2 \wedge X)$$

by taking the external product with a generator of $\tilde{K}_1(S^2)$. Then $\phi$ is an isomorphism.

Proof. Part (i) follows from Theorem 7 parts (i), (ii), (iii). For part (ii), observe that a direct summand of an exact sequence is an exact sequence, and that we have no trouble about verifying the axiom about disjoint unions (for $K_{\alpha}$) or wedge-sums (for $\tilde{K}_{\alpha}$). Part (iii) follows from Theorem 7 part (iv). For part (iv), make the obvious calculation in $\tilde{K}(S^{2m};Q) \cong H^{2m}(S^{2m};Q)$. For part (v), let the representing space for $\tilde{K}_{\alpha}$ be $BUR_{\alpha}$; convert the homomorphism $\phi$ into a map $BUR_{\alpha} \longrightarrow \Omega^2 BUR_{\alpha+1}$, and check as in part (iv) that this map induces an isomorphism of homotopy groups.

It follows from part (v), iterated $d$ times, that the representable functor $K_\alpha(X)$ is periodic with period $2d$, in the same sense that standard K-theory is periodic with period 2. We therefore have no difficulty extending it to a graded cohomology theory $K^*_\alpha(X)$. Alternatively, we can first take the spectrum

$$BUR_\alpha,\ BUR_{\alpha+1},\ BUR_{\alpha+2},\ \ldots$$

and then take the resulting cohomology theory.

It follows from part (iii) that for $\alpha = 0$ the theory $K_0$ has products.

Let $BUR_\alpha$ be the representing space for $\tilde{K}_\alpha$, as above; then we have

$$BUR \simeq \prod_\alpha BUR_\alpha \ .$$

It is easy to obtain the rational cohomology of the factors $BUR_\alpha$ by inspecting their homotopy groups. In fact, $H^*(BUR_\alpha;Q)$ is a polynomial algebra on generators of dimension $2n$, where $n$ runs over the positive integers in the residue class $\alpha$.

Before moving on to cobordism, we need one more result. Given $d$, we have a map

$$E_0\colon BU \longrightarrow BUR$$

where $R = R(d)$. Let us define $E_0^!$ so that the following diagram is commutative.

$$\begin{array}{ccc} H^*(BUR;Q) & \xrightarrow{E_0^*} & H^*(BU;Q) \\ & \searrow^{\cong} \quad \nearrow_{E_0^!} & \\ & H^*(BU;Q) & \end{array}$$

We remark that in what follows, $H^*(X;Q)$ really arises as $E^0(X)$, where $E$ is the spectrum

$$\prod_{-\infty<n<+\infty} K(Q,2n) .$$

Thus $H^*$ should be interpreted as a direct product of groups $H^p$, while $H_*$ should be interpreted as a direct sum of groups $H_p$. Let

$$\text{todd} \in H^*(BU;Q)$$

be the characteristic class which has the following properties.

(i) $\text{todd}(\xi_1 \oplus \xi_2) = (\text{todd}\ \xi_1)(\text{todd}\ \xi_2)$ .

(ii) If $\eta$ is the canonical line bundle over $CP^\infty$ and $x \in H^2(CP^\infty)$ is the generator (so that $\text{ch}\ \eta = e^x$) then

$$\text{todd}\ \eta = \frac{e^x - 1}{x} .$$

Then we have the following result.

<u>Lemma 9</u>

There is a characteristic class

$$\tau \in K(BU;R)$$

such that

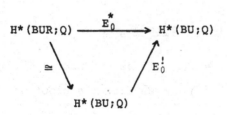

$$\frac{E_0^! \text{todd}}{\text{todd}} = \text{ch } \tau .$$

Here $R = R(d)$ is as in Theorem 5. The motivation for this result is best seen from the proof of Theorem 14.

<u>Proof</u>. Let todd' be the class in $H^*(BUR;Q)$ which maps to todd in $H^*(BU;Q)$. Then we easily see that

$$\text{todd}'(\xi_1 \oplus \xi_2) = (\text{todd}'\xi_1)(\text{todd}'\xi_2)$$

for $\xi_1$, $\xi_2$ in $K(X;R)$. We also have

$$(E_0^! \text{todd})\xi = \text{todd}' E_0 \xi$$

for $\xi$ in $K(X)$. It is now easy to see that

$$\left(\frac{E_0^! \text{todd}}{\text{todd}}\right)(\xi_1 \oplus \xi_2) = \left(\frac{E_0^! \text{todd}}{\text{todd}}\xi_1\right)\left(\frac{E_0^! \text{todd}}{\text{todd}}\xi_2\right) .$$

Now $\frac{E_0^! \text{todd}}{\text{todd}}$ is certainly equal to ch $\tau$ for some $\tau$ of augmentation 1 in $K(BU;Q)$. Using the last formula, we find that

$$\text{ch } \tau(\xi_1 \oplus \xi_2) = (\text{ch } \tau(\xi_1))(\text{ch } \tau(\xi_2)) .$$

Therefore

$$\tau(\xi_1 \oplus \xi_2) = \tau(\xi_1) \cdot \tau(\xi_2)$$

in $K(X;Q)$ for any X. We wish to show that $\tau \in K(BU;R)$. For this purpose it is now sufficient to consider $\tau(\eta)$, where $\eta$ is the canonical line bundle over $CP^\infty$; if $\tau(\eta)$ lies in $K(CP^\infty;R)$ then the splitting principle shows that $\tau$ lies in $K(BU;R)$.

Next let S be some ring containing the rationals. Let $G(CP^n;S)$ be the multiplicative group of elements of augmentation 1 in $H^*(CP^n;S)$. Then we can define a homomorphism

$$\text{todd}: K(CP^n;S) \longrightarrow G(CP^n;S)$$

by

$$\text{todd}(\xi \otimes s) = (\text{todd}\,\xi)^s$$

for $\xi \in K(CP^n)$. Here $(1 + x)^s$ is defined by the usual binomial series

$$(1 + x)^s = 1 + sx + \frac{s(s-1)}{1.2}x^2 + \ldots ;$$

in this case the series is finite. On $K(CP^n;R)$ the homomorphism agrees with todd'. Passing to inverse limits, we obtain a homomorphism

$$\text{todd}: K(CP^\infty;S) \longrightarrow G(CP^\infty;S) .$$

(Here $G(CP^\infty;S)$ is the multiplicative group of elements of augmentation 1 in $H^*(CP^\infty;S)$.) On $K(CP^\infty;R)$ this homomorphism agrees with todd'.

Take an indeterminate $t$ and take $S = Q[t]$. Consider

$$\text{todd}(1 + \zeta)^t = \text{todd}(1 + t\zeta + \frac{t(t-1)}{1.2}\zeta^2 + \ldots ) .$$

This is an element of $G(CP^\infty;Q[t])$, that is, it is a formal power-series in $x$ with coefficients which are polynomials in $t$; say

$$\text{todd}(1 + \zeta)^t = 1 + p_1(t) + p_2(t)x^2 + \ldots .$$

But for any integer $n$, $(1 + \zeta)^n$ is a line bundle, and we have

$$\mathrm{todd}(1+\zeta)^n = \frac{e^{nx}-1}{nx}$$

$$= 1 + \frac{nx}{2!} + \frac{n^2x^2}{3!} + \dots .$$

So for integer values of $t$ we have

$$p_r(n) = \frac{n^r}{(r+1)!} \quad ;$$

thus

$$p_r(t) = \frac{t^r}{(r+1)!}$$

and

$$\mathrm{todd}(1+\zeta)^t = \frac{e^{tx}-1}{tx} \quad .$$

Consider now $(\tau(\eta))^d$. A priori this is a power-series in $\zeta$ with rational coefficients. I claim that these coefficients actually lie in R. To prove this, choose a prime $p$ such that $p \equiv 1 \mod d$; we wish to prove that the coefficients of $(\tau(\eta))^d$ are p-adic integers. We work over the p-adic integers, and manipulate as follows.

$$\mathrm{ch}(\tau(\eta))^d = \mathrm{ch}\ \tau(d\eta)$$

$$= \left(\frac{E_0^!\mathrm{todd}}{\mathrm{todd}}\right)(d\eta)$$

$$= \frac{\mathrm{todd}' E_0 d\eta}{\mathrm{todd}\ d\eta}$$

$$= \frac{\mathrm{todd}\ dE_0\eta}{\mathrm{todd}\ d\eta} \quad .$$

Now the basic remark in the proof of Theorem 5 is that

$$dE_0\eta = \sum_\rho (1+\zeta)^\rho$$

where $\rho$ runs over $\rho_n = \omega^m$ for $1 \le m \le d$, and $\omega$ is a primitive $d^{th}$ root of unity as in Theorem 5. Thus we have

$$\mathrm{ch}(\tau(\eta))^d = \prod_\rho \frac{\mathrm{todd}(1+\zeta)^\rho}{\mathrm{todd}(1+\zeta)}$$

$$= \prod_\rho \frac{e^{\rho x}-1}{\rho(e^x-1)} \qquad \text{(by the remarks above)}$$

$$= \mathrm{ch} \prod_\rho \frac{(1+\zeta)^\rho - 1}{\rho\zeta}$$

Thus we have

$$(\tau(\eta))^d = \prod_\rho \frac{(1+\zeta)^\rho-1}{\rho\zeta} \quad .$$

But for each $\rho$ the coefficients of the power series

$$\frac{(1+\zeta)^\rho-1}{\zeta}$$

are p-adic integers; and the denominator

$$\prod_\rho \rho = (-1)^{d-1}$$

is invertible. Therefore the coefficients in the power-series $(\tau(\eta))^d$ are p-adic integers. This proves that these coefficients lie in $R$, as claimed.

Finally, since $d$ is invertible in $R$, we deduce that the coefficients of $\tau(\eta)$ lie in $R$. This proves Lemma 9.

We now turn to cobordism.

Let $R$ be a subring of the rationals. Let $MU^*(X;R)$ be complex cobordism with coefficients in $R$. This is a representable functor; we write $MUR$ for the representing spectrum. We require the same information as before.

<u>Lemma 10</u>

If $R_1 \subset R_2$, the maps

$$i_*: H_*(MU;R_2) \longrightarrow H_*(MUR_1;R_2)$$

$$i^*: H^*(MUR_1;R_2) \longrightarrow H^*(MU;R_2)$$

$$(i\wedge i)_*: H_*(MU\wedge MU;\ R_2) \longrightarrow H_*(MUR_1\wedge MUR_1;\ R_2)$$

$$(i\wedge i)^*: H^*(MUR_1\wedge MUR_1;\ R_2) \longrightarrow H^*(MU\wedge MU;\ R_2)$$

are iso.

<u>Proof</u>. Let $Y$ be a Moore spectrum with

$$\pi_n(Y) = 0 \quad \text{for} \quad n < 0\ ,$$

$$H_n(Y) = \begin{cases} R & \text{for} \quad n = 0 \\ 0 & \text{for} \quad n \neq 0\ . \end{cases}$$

Then we may take $MU \wedge Y$ as a construction for $MUR$. This leads immediately to the result.

Lemma 11

Suppose $R_1 \subset R_2$. Then the maps

$$i_*: K_*(MU;R_2) \longrightarrow K_*(MUR_1;R_2)$$

$$i^*: MU^*(MUR_1;R_2) \longrightarrow MU^*(MU;R_2)$$

$$(i\wedge i)^*: MU^*(MUR_1\wedge MUR_1;\ R_2) \longrightarrow MU^*(MU\wedge MU;\ R_2)$$

are iso. The maps $i^*$ and $(i \wedge i)^*$ are also homeomorphisms with respect to the filtration topology.

The proof is the same as for Lemma 3.

We now consider the set $MU^0(MUR_1;R_2)$. This set may be identified with the set of cohomology operations

$$b: MU^n(X;R_1) \longrightarrow MU^n(X;R_2)$$

which are defined for all $X$ and $n$, natural, and stable (therefore additive and $R_1$-linear). This set is topologised by the filtration topology. According to Lemma 11, the set to be considered is essentially independent of $R_1$, so long as $R_1 \subset R_2$. We therefore write $B(R_2)$ for the set of operations just introduced, and regard it primarily as the ring of stable cohomology operations of degree zero on $MU^*(X;R_2)$.

Lemma 12

If $R_1 \subset R_2$, then we have a monomorphism

$$\iota: B(R_1) \longrightarrow B(R_2)$$

such that for each $b \in B(R_1)$, each $X$ and each $n$ the the following diagram is commutative.

$$\begin{array}{ccc} MU^n(X;R_1) & \xrightarrow{i_*} & MU^n(X;R_2) \\ \downarrow{\scriptstyle b} & & \downarrow{\scriptstyle b} \\ MU^n(X;R_1) & \xrightarrow{i_*} & MU^n(X;R_2) \end{array}$$

This follows from the preceding discussion, together with the fact that

$$i_*: MU^0(MU;R_1) \longrightarrow MU^0(MU;R_2)$$

is monomorphic. (Compare Lemma 4.)

Because of this lemma, it will be sufficient to construct an idempotent in $B(Q)$. But over $Q$, stable homotopy theory becomes trivial. We will give the next construction in slightly greater generality than is needed now, for use later. Let $f: X \longrightarrow MUQ$ be a map. Then we define $f_!$ so that the following diagram is commutative.

$$\begin{array}{ccc} H_*(X;Q) & \xrightarrow{f_*} & H_*(MUQ;Q) \\ & {\scriptstyle f_!}\searrow \quad \nearrow{\scriptstyle \cong} & \\ & H_*(MU;Q) & \end{array}$$

Of course we can make a similar definition with $H_*$ replaced by $K_*$, or with $MU$, $MUQ$ replaced by $BU$, $BUQ$.

Now we define

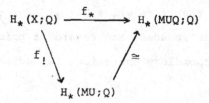

$$\theta : MU^0(X;Q) \longrightarrow \mathrm{Hom}_Q(H_*(X;Q), H_*(MU;Q))$$

as follows. If $f: X \longrightarrow MUQ$ is a map, then $\theta(f) = f_!$.

Lemma 13

$\theta$ is an isomorphism.

If we assign the obvious topology to the Hom group, then $\theta$ becomes a homeomorphism. If $X = MU$, then $\theta$ carries composition in $B(Q)$ into composition in the Hom group.

This lemma is a known consequence of Serre's C-theory [27].

I now choose a positive integer $d$, and seek to construct a "fake cobordism theory" whose coefficient groups are periodic with one multiplicative generator every $2d$ dimensions. Let $E_0 \in A(Q)$ be as above. Then we define $\varepsilon \in B(Q)$ to be the element such that the following diagram is commutative.

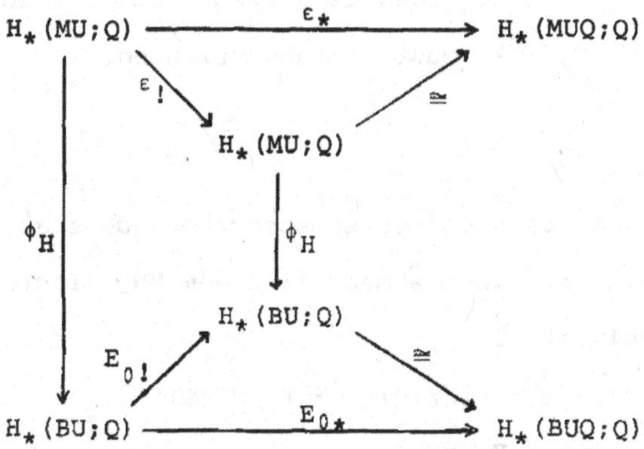

(Here $\phi_H$ is the Thom isomorphism in homology.) It is clear that $\varepsilon$ is idempotent; indeed $\varepsilon$ is the most obvious idempotent in sight.

Theorem 14

$\varepsilon$ lies in $B(R)$, where $R = R(d)$ is the ring of rationals $a/b$ such that $b$ contains no prime $p$ with $p \equiv 1 \mod d$, as in Theorem 5.

The proof will require two intermediate results.

Lemma 15

A map $f: S^p \longrightarrow MUQ$ factors through $MUR$ if and only if

$$f_!: K_*(S^p;Q) \longrightarrow K_*(MU;Q)$$

maps $K_*(S^p;R)$ into $K_*(MU;R)$.

This is the theorem of Stong and Hattori [16, 29]. Note that if $S^p$ is regarded as a space rather than as a spectrum, then $K_*(S^p)$ must be taken reduced.

Lemma 16

Let $X$ be a connected spectrum such that $H_r(X)$ is free for all $r$. Then a map $f: X \longrightarrow MUQ$ factors through $MUR$ if and only if

$$f_!: K_*(X;Q) \longrightarrow K_*(MU;Q)$$

maps $K_*(X;R)$ into $K_*(MU;R)$.

<u>Proof.</u> It is trivial that if $f$ factors, then $f_!$ maps $K_*(X;R)$ into $K_*(MU;R)$. We wish to prove the converse. First assume that $X$ is finite-dimensional, say $(n-1)$-connected and $(n+d)$-dimensional. We proceed by induction over $d$. The result is true if $X$ is a wedge of spheres, by Lemma 15. We may now assume we have a cofibering

$$A \xrightarrow{i} X \xrightarrow{j} B$$

with the following properties.

(i) For $r \leq m$ we have

$$i_*: H_r(A) \cong H_r(X), \quad H_r(B) = 0 .$$

(ii) For $r \leq m$ we have

$$H_r(A) = 0, \; j_*: H_r(X) \cong H_r(B) .$$

(iii) The result holds for $A$ and $B$.

Now suppose given a map $f: X \longrightarrow MUQ$ such that $f_!$ maps $K_*(X;R)$ into $K_*(MU;R)$. Then $fi: A \longrightarrow MUQ$ maps $K_*(A;R)$ into $K_*(MU;R)$. By (iii), we have the following commutative diagram.

$$\begin{array}{ccc} A & \xrightarrow{\;\;i\;\;} & X \\ {\scriptstyle g}\downarrow & & \downarrow{\scriptstyle f} \\ MUR & \xrightarrow{\;\;i'\;\;} & MUQ \end{array}$$

Now the spectral sequence

$$H^*(X;MUR^*(S^0)) \Longrightarrow MUR^*(X)$$

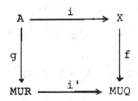

is trivial (since the differentials are zero mod torsion and the groups are torsion-free). We deduce that

$$i^*: MUR^*(X) \longrightarrow MUR^*(A)$$

is epi. So $g$ extends over $X$; say we have $h: X \longrightarrow MUR$ such that $hi = g$. Then we have

$$f = i'h + kj$$

for some $k: B \longrightarrow MUQ$. Then evidently $(kj)_!$ maps $K_*(X;R)$ into $K_*(MU;R)$. Now the spectral sequence

$$H_*(X;K_*(S^0;R)) \Longrightarrow K_*(X;R)$$

is trivial (since the differentials are zero mod torsion and the groups are torsion-free). We deduce that

$$j_*: K_*(X;R) \longrightarrow K_*(B;R)$$

is epi. Therefore $k_!$ maps $K_*(B;R)$ into $K_*(MU;R)$. By (iii), $k$ factors through $MUR$. Therefore $f$ factors through $MUR$. This completes the induction and proves the result when $X$ is finite-dimensional.

We now tackle the case of a general $X$. Approximate $X$ by $X^n$ such that $i_*: H_r(X^n) \longrightarrow H_r(X)$ is iso for $r \leq n$ and $H_r(X^n) = 0$ for $r > n$. We have

$$MUR^*(X) = \varprojlim_n MUR^*(X^n)$$

$$MUQ^*(X) = \varprojlim_n MUQ^*(X^n)$$

(since the usual spectral sequences satisfy the Mittag-Leffler condition). Take a map $f: X \longrightarrow MUQ$ such that $f_!$ maps $K_*(X;R)$ into $K_*(MU;R)$. Then the composite

$$X^n \xrightarrow{i_n} X \xrightarrow{f} MUQ$$

is such that $(fi_n)_!$ maps $K_*(X^n;R)$ into $K_*(MU;R)$. Hence $fi_n$ factors through an element $g_n \in MUR^0(X^n)$. Since $MUR^*(X^n) \longrightarrow MUQ^*(X^n)$ is mono, the elements $g_n$ define an element of $\varprojlim_n MUR^*(X^n)$ and thus give a factorisation of $f$. This proves Lemma 16.

Proof of Theorem 14. Let $\varepsilon$: $MU \longrightarrow MUQ$ be as above. We aim to apply Lemma 16 to $\varepsilon$. We equip ourselves with various formal remarks.

(i) The following diagram is not commutative.

$$\begin{array}{ccc} K_*(MU;Q) & \xrightarrow{ch} & H_*(MU;Q) \\ \downarrow{\scriptstyle \phi_K} & & \downarrow{\scriptstyle \phi_H} \\ K_*(BU;Q) & \xrightarrow{ch} & H_*(BU;Q) \end{array}$$

In fact, for a suitable choice of $\phi_K$ we have

$$ch\ \phi_K z = todd \cdot \phi_H\ chz\ .$$

(Here the product of a cohomology class and a homology class is taken in the sense of the cap product. The reader who prefers to work entirely in cohomology may write out an argument dual to the one which follows, to verify that $\varepsilon$ satisfies the analogue of Lemma 16 for $K^*$.)

(ii) The following diagrams are commutative.

$$\begin{array}{ccc} K_*(MU;Q) & \xrightarrow{\varepsilon_!} & K_*(MU;Q) \\ \downarrow{\scriptstyle ch} & & \downarrow{\scriptstyle ch} \\ H_*(MU;Q) & \xrightarrow{\varepsilon_!} & H_*(MU;Q) \end{array}$$

$$\begin{array}{ccc} K_*(BU;Q) & \xrightarrow{E_{0!}} & K_*(BU;Q) \\ \downarrow{\scriptstyle ch} & & \downarrow{\scriptstyle ch} \\ H_*(BU;Q) & \xrightarrow{E_{0!}} & H_*(BU;Q) \end{array}$$

(iii) If $u \in H^*(BU;Q)$, $v \in H_*(BU;Q)$ we have

$$E_{0!}((E_0^!u) \cdot v) = u \cdot (E_{0!}v) .$$

Now we wish to check that $\varepsilon\colon MU \longrightarrow MUQ$ satisfies the conditions of Lemma 16. So take any element $x$ in $K_*(MU;R)$; we wish to check that $\varepsilon_! x$ lies in $K_*(MU;R)$. Since $\phi_K$ is iso, it is sufficient to prove that $\phi_K \varepsilon_! x$ lies in

$$\phi_K K_*(MU;R) = K_*(BU;R) .$$

But we have

$$\begin{aligned} ch\phi_K\varepsilon_! x &= \text{todd} \cdot \phi_H ch \varepsilon_! x \\ &= \text{todd} \cdot \phi_H \varepsilon_! ch\ x \\ &= \text{todd} \cdot E_{0!}\phi_H ch\ x \\ &\qquad\qquad \text{(by definition of } \varepsilon) \\ &= E_{0!}(E_0^! \text{todd} \cdot \phi_H ch\ x) \end{aligned}$$

$$= E_{0!}\left(\frac{E_0^! \text{todd}}{\text{todd}} \cdot \text{ch}\,\phi_K x\right)$$

$$= E_{0!}(\text{ch}\ \tau \cdot \text{ch}\ \phi_K x)$$

(where $\tau$ is as in Lemma 9)

$$= E_{0!}\text{ch}(\tau \cdot \phi_K x)$$

$$= \text{ch}E_{0!}(\tau \cdot \phi_K x) .$$

Since ch is iso, we have

$$\phi_K \varepsilon_! x = E_{0!}(\tau \cdot \phi_K x) .$$

But $\tau \in K^*(BU;R)$ and $\phi_K x \in K_*(BU;R)$, so $\tau \cdot \phi_K x \in K_*(BU;R)$. Again, we have $E_0\colon BU \longrightarrow BUR$, so $E_{0!}$ maps $K_*(BU;R)$ into $K_*(BU;R)$. Thus $\phi_K \varepsilon_! x$ lies in $K_*(BU;R)$ and $\varepsilon_! x$ lies in $K_*(MU;R)$. Therefore $\varepsilon$ satisfies the conditions of Lemma 16, and $\varepsilon \in B(R)$. This proves Theorem 14.

The properties of the element $\varepsilon \in B(R)$ are as follows.

<u>Theorem 17</u>

(i) $\varepsilon^2 = \varepsilon$ in $B(R)$.

(ii) For any $x,y$ in $MU^*(X;R)$ we have

$$\varepsilon(xy) = (\varepsilon x)(\varepsilon y) .$$

<u>Proof</u>. Since $B(R) \longrightarrow B(Q)$ is mono, part (i)

follows trivially from the corresponding equation in $B(Q)$.

To prove part (ii), we have to compare the following composites.

$$MU \wedge MU \xrightarrow{\mu} MU \xrightarrow{\varepsilon} MUQ$$

$$MU \wedge MU \xrightarrow{\varepsilon \wedge \varepsilon} MUQ \wedge MUQ \xrightarrow{\mu} MUQ \ .$$

We have to compare $(\varepsilon\mu)_!$ with $(\mu(\varepsilon \wedge \varepsilon))_!$. If we compose with the map

$$H_*(MU;Q) \otimes H_*(MU;Q) \longrightarrow H_*(MU \wedge MU;Q) \ ,$$

we obtain the two ways of chasing round the following commutative diagram.

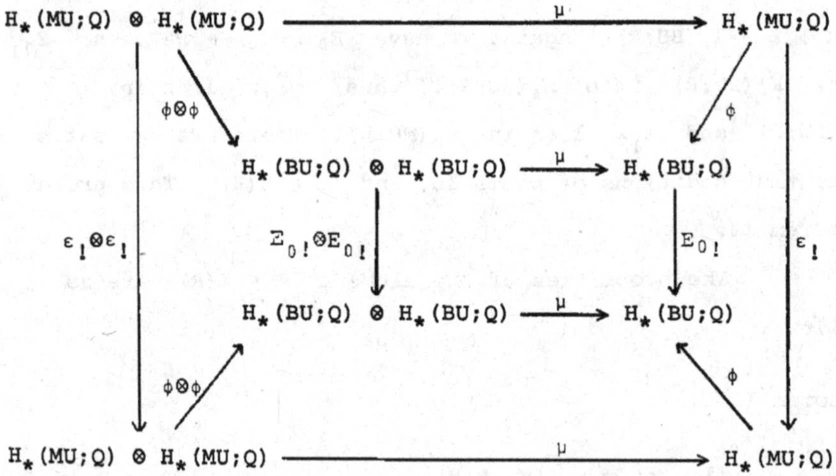


Here the commutativity of the central square arises from the fact that $E_0$ is additive; that is, the following square is commutative.

$$\begin{array}{ccc} BU \times BU & \xrightarrow{\mu} & BU \\ \Big\downarrow{\scriptstyle E_0 \times E_0} & & \Big\downarrow{\scriptstyle E_0} \\ BUQ \times BUQ & \xrightarrow{\mu} & BUQ \end{array}$$

(Here $\mu$ is the product map in BU which represents addition in $\tilde{K}$.) This proves that

$$(\varepsilon\mu)_! = (\mu(\varepsilon \wedge \varepsilon))_! \ ,$$

and (using Lemma 13) that the following square is homotopy-commutative.

$$\begin{array}{ccc} MU \wedge MU & \xrightarrow{\mu} & MU \\ \Big\downarrow{\scriptstyle \varepsilon\wedge\varepsilon} & & \Big\downarrow{\scriptstyle \varepsilon} \\ MUQ \wedge MUQ & \xrightarrow{\mu} & MUQ \end{array}$$

In other words, we have the formula

$$\varepsilon(xy) = (\varepsilon x)(\varepsilon y)$$

for the external product, when $x$ and $y$ are both the generator in $MU^*(MU)$ and the equality takes place in $MU^*(MU \wedge MU;\ Q)$. Since

$$MU^*(MU \wedge MU;\ Q) \longrightarrow MU^*(MU \wedge MU;\ R)$$

is mono, the equality holds in $MU^*(MU \wedge MU;\ R)$. Since

$$MU^*(MUR \wedge MUR;\ R) \longrightarrow MU^*(MU \wedge MU;\ R)$$

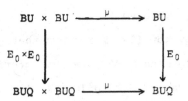

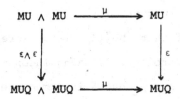

is iso, the equality holds in $MU^*(MUR \wedge MUR;\ R)$ when $x$ and $y$ are both the generator in $MU^*(MUR;R)$. Therefore it always

holds. This proves Theorem 17.

S. P. Novikov [24] has shown that multiplicative cohomology operations on $MU^*$ are characterised by their values on the generator $\omega \in MU^2(CP^\infty)$. It might perhaps be of interest to examine $\varepsilon\omega$, and to see if this provides an alternative approach to $\varepsilon$.

We now define $MU_0^*(X) = \varepsilon MU^*(X;R)$.

<u>Corollary 18</u>

$MU_0^*(X)$ is a cohomology theory with products, and is a representable functor.

The proof that $MU_0^*(X)$ is a representable functor is exactly as for Corollary 8, using Theorem 17 (i). The fact that $MU_0^*(X)$ has products is immediate from Theorem 17 (ii).

We write $MUR_0$ for the representing spectrum for $MU_0^*(X)$. In order to lend credibility to the idea that $MUR_0$ is an acceptable "Thom complex" corresponding to the space $BUR_0$, we remark that the following diagram factors to give a unique "Thom isomorphism" $\phi_0$.

$$\begin{array}{ccccc}
H_*(MUR_0;R) & \xrightarrow{i_*} & H_*(MUR;R) & \xleftarrow{\cong} & H_*(MU;R) \\
\Big\downarrow{\scriptstyle \phi_0} & & & & \Big\downarrow{\scriptstyle \phi} \\
H_*(BUR_0;R) & \xrightarrow{i_*} & H_*(BUR;R) & \xleftarrow{\cong} & H_*(BU;R)
\end{array}$$

This follows immediately from the definition of $\varepsilon$.

Theorem 19

(i) The coefficient ring $\pi_*(MUR_0)$ is a polynomial ring over $R$ with generators in dimensions $2d$, $4d$, $6d$, ... .

(ii) $MU^*(X;R)$ is a direct product of theories isomorphic to $MU_0^*(X)$.

Note. In part (ii) the splitting is not asserted to be canonical, but the injection of $MU_0^*(X)$ and the projection onto $MU_0^*(X)$ are of course canonical; this is sufficient for the applications.

Proof. For any connected algebra $A$, let $Q(A)$ be its indecomposable quotient. Then

$$\varepsilon: \pi_*(MUR) \longrightarrow \pi_*(MUR)$$

induces

$$Q(\varepsilon): Q(\pi_*(MUR)) \longrightarrow Q(\pi_*(MUR))$$

with $Q(\varepsilon) \cdot Q(\varepsilon) = Q(\varepsilon)$. We have

$$Q(\operatorname{Im}\varepsilon) \cong \operatorname{Im}(Q\varepsilon) .$$

Now $Q(\pi_*(MUR))$ is R-free with generators $x_1$, $x_2$, $x_3$,... in dimensions 2, 4, 6,... [20, 30]. For each $x_n$ we have either $Q(\varepsilon)x_n = x_n$ or $Q(\varepsilon)x_n = 0$. We may thus choose a homogeneous R-base for $\operatorname{Im}Q(\varepsilon)$ and extend it to a homogeneous R-base for $Q(\pi_*(MUR))$. Lift the basis elements in $\operatorname{Im}Q(\varepsilon)$ to elements $g_i$ in $\operatorname{Im}\varepsilon$, and lift the remaining basis elements in any way to elements $h_j$. Then $\pi_*(MUR)$ is the polynomial algebra generated by the $g_i$ and $h_j$, and $\operatorname{Im}\varepsilon$ is precisely the subalgebra generated by the $g_i$. But this subalgebra is polynomial. It remains only to find the dimensions of the generators.

We have

$$\pi_*(MUR_0) \otimes Q \cong H_*(MUR_0;Q) \cong H_*(BUR_0;Q)$$

(by the remarks above). But as remarked above, $H^*(BUR_0;Q)$ is a polynomial algebra with generators in dimension 2d, 4d, 6d,... . Now part (i) follows by counting dimensions over Q.

The preceding proof actually shows that $\pi_*(MUR)$ is free as a module over $\pi_*(MUR_0)$. Choose a $\pi_*(MUR_0)$-free base for $\pi_*(MUR)$ (beginning with the unit element 1) and represent the basis elements by maps

$$f_j: S^{n(j)} \longrightarrow MUR .$$

We now consider the map

$$g: \bigvee_j S^{n(j)} \wedge MUR_0 \longrightarrow MUR$$

which on the $j^{th}$ factor is given by

$$S^{n(j)} \wedge MUR_0 \xrightarrow{f_j \wedge i} MUR \wedge MUR \xrightarrow{\mu} MUR .$$

It is clear that g induces an isomorphism of homotopy groups. Since $MUR_0$ is connected and $n(j) \longrightarrow \infty$ as $j \longrightarrow \infty$ the infinite wedge-sum is also a product. Therefore

$$[X,MUR] \cong \prod_j [X, S^{n(j)} \wedge MUR_0] .$$

This proves part (ii).

We have now accomplished our object of splitting $MU^*(X;R)$ into a direct sum of similar functors. I believe that the functors $MU_0^*$ and $K_0^*$, together with the spectrum $MUR_0$ and the space $BUR_0$, are of some interest. I would like to give further results to prove that $MUR_0$ is related to $BUR_0$ as $MU$ is to $BU$; for lack of time in writing up these notes I offer the following in the disguise of an exercise.

Exercise 20

Show that Proposition 25 of Lecture 1 (the Conner-Floyd theorem) applies to the case $E = MUR_0$, $F = \underline{BUR_0}$.

Hints.

(a) $H^p(MUR_0;R) = 0$ unless $p \equiv 0 \mod 2d$. Therefore $K_\alpha(MUR_0) = 0$ for $\alpha \neq 0$, and $K_0(MUR_0)$ is the whole of $K(MUR_0;R)$. Take the orientation class $u$ in $K(MU)$, map it into $K(MU;R)$, lift it into $K(MUR;R)$ and restrict it to $K(MUR_0;R)$; the result must lie in $K_0(MUR_0)$. This gives the necessary orientation class.

(b) In checking Assumption 20 and 24 of Lecture 1, exercise care in approximating $MUR_0$ and $BUR_0$ by finite complexes.

## LECTURE 5. FINITENESS THEOREMS

In this lecture I want to give an exposition of certain finiteness theorems in algebra which seem useful in algebraic topology. These results are slight generalisations of known results on coherent rings; one may find the latter in Bourbaki [9, pp. 62-63]. I became interested in the subject in the course of reproving certain results of S. P. Novikov [24]. Independently, Joel M. Cohen became interested in similar results for a different topological application. I am most grateful to Cohen for sending me preprints of his two papers [12, 13]. (So far as I know these papers have not yet appeared.)

The following results 1-5 will serve as illustrations of the sort of topological application which I have in mind.

Theorem 1 (S. P. Novikov)

If $X$ is a finite CW-complex, then $MU^*(X)$ is finitely-generated as a module over the coefficient ring $MU^*(S^0)$.

The methods I will give also yield the following result, which is slightly stronger.

Theorem 2

Let $X$ be a finite CW-complex. Then $MU^*(X)$, considered as a module over the coefficient ring $MU^*(S^0)$, admits a resolution of finite length

$$0 \longrightarrow C_n \longrightarrow C_{n-1} \longrightarrow \cdots \longrightarrow C_1 \longrightarrow C_0 \longrightarrow MU^*(X) \longrightarrow 0$$

by finitely-generated free modules.

Since giving the original lecture I have heard that this result is also known to P. E. Conner and L. Smith; it may also be known to other workers in the field. I am grateful to L. Smith for sending me a preprint.

I will not quote the results of Cohen verbatim, but will reword them to suit the present lecture. I will use the words "almost all" to mean "with a finite number of exceptions".

Theorem 3 (J. M. Cohen)

Let $X$ be a spectrum whose stable homotopy groups $\pi_r(X)$ are finitely generated, and are zero for almost all $r$. Then $H^*(X;Z_p)$ is finitely-presented as a module over the mod $p$ Steenrod algebra $A$.

This result can be used to show that under mild restrictions, a space $Y$ (as distinct from a spectrum) must have infinitely many non-zero stable homotopy groups. Even better for this purpose is the variant which follows next. We will say that an abelian group $G$ is $p$-trivial if $p: G \longrightarrow G$ is iso. Spelling this out, it asks that the torsion subgroup of $G$ should contain no elements of order $p$, and that the torsion-free quotient of $G$ should be divisible by $p$.

Theorem 4 (J. M. Cohen)

Let X be a connected spectrum whose stable homotopy groups $\pi_r(X)$ are p-trivial for almost all r. Then the A-module $H^*(X;Z_p)$ can be presented by generators in only finitely many dimensions and relations in only finitely many dimensions.

In particular, of course, the theorem applies if $\pi_r(X) = 0$ for almost all r. The difference between this case and Theorem 3 is that if the groups $\pi_r(X)$ are not finitely-generated, then $H^*(X;Z_p)$ may need infinitely many generators in some dimensions.

Corollary 5 (J. M. Cohen)

Let Y be a space such that $\tilde{H}^*(Y;Z_p) \neq 0$. Then there are infinitely many values of r such that the stable homotopy group $\pi_r^S(Y)$ is not p-trivial (and therefore non-zero).

This answers a question of Serre [26, p.219].

To prove these results, we will present a slight axiomatisation of Bourbaki's results. We will first set up our assumptions, definitions and general theory. From Corollary 12 onwards we turn to the topological applications, and sketch the proof of the results given above. Topologists looking for motivation might perhaps turn to the passage beginning immediately after Example 14.

We suppose given a graded ring $R$ with unit. The word "module" will mean a graded left R-module, unless otherwise specified. We suppose given a class $\underline{C}$ of projective modules. The class $\underline{C}$ is supposed to satisfy two axioms*.

(i) If $P \cong Q$ and $P \in \underline{C}$, then $Q \in \underline{C}$.

(ii) If $P \in \underline{C}$ and $Q \in \underline{C}$, then $P \oplus Q \in \underline{C}$.

Examples.

(i) We define $\underline{F}$ to be the class of finitely-generated free modules.

(ii) We define $\underline{D}$ to be the class of free modules with generators in only a finite number of dimensions.

(iii) We define $\underline{E}$ to be the class of free modules such that for each $n$ there are only a finite number of generators in dimensions $\leq n$.

(iv) We define $\underline{O}$ to be the class containing only the zero module.

In what follows the symbols $\underline{F}$ and $\underline{D}$ will always have the meanings just given to them. In proving Theorems 1, 2 and 3 we take $\underline{C} = \underline{F}$; in proving Theorem 4 and Corollary 5 we take $\underline{C} = \underline{D}$. The axiomatisation simply saves us from giving the same proof twice over.

Definition 6

An R-module $M$ is of $\underline{C}$-type $n$ if it has a projective resolution

---

* Note added in proof. It should also be assumed that $0 \in \underline{C}$.

$$0 \leftarrow M \leftarrow C_0 \leftarrow C_1 \leftarrow \cdots \leftarrow C_r \leftarrow \cdots$$

such that $C_r \in \underline{C}$ for $0 \leq r \leq n$. (Compare Bourbaki p. 60, exercise 6.)

Examples.

(i) All modules are of $\underline{C}$-type $-1$.

(ii) A module is of $\underline{C}$-type $\infty$ if and only if it has a projective resolution by modules in $\underline{C}$.

(iii) A module of $\underline{F}$-type 0 is a finitely-generated module.

(iv) A module of $\underline{F}$-type 1 is a finitely-presented module.

(v) A module of $\underline{D}$-type 0 is one which can be generated by generators in only finitely many dimensions.

(vi) A module of $\underline{D}$-type 1 is one which can be presented using generators in only finitely many dimensions and relations in only finitely many dimensions.

Thus, the conclusion of Theorem 1 states that $MU^*(X)$ is of $\underline{F}$-type 0. The conclusion of Theorem 3 states that $H^*(X;Z_p)$ is of $\underline{F}$-type 1. The conclusion of Theorem 4 states that $H^*(X;Z_p)$ is of $\underline{D}$-type 1.

We could also say that $M$ is of $\underline{C}$-cotype $n$ if it has a projective resolution such that $C_r \in \underline{C}$ for $r > n$. With $\underline{C} = \underline{0}$, for example, we would be discussing homological dimension. It would perhaps be interesting to see if known results about homological dimension generalize to cotype (perhaps in the presence of extra assumptions on $\underline{C}$ ). In particular, is the analogue of Lemma 7 (iii) below true for cotype*? We will not pursue this further here.

If we do not need to emphasise $\underline{C}$ , we will write "type" for "$\underline{C}$-type". The basic property of Definition 6 is as follows.

Lemma 7

Suppose given an exact sequence

$$0 \longrightarrow L \xrightarrow{i} M \xrightarrow{j} N \longrightarrow 0$$

of R-modules.

(i) If L is of type (n - 1) and M is of type n , then N is of type n .

(ii) If L is of type n and N is of type n , then M is of type n .

(iii) If M is of type n and N is of type (n + 1), then L is of type n .

(Compare Bourbaki p.60, exercise 6 a, c, d. For the most significant special case see Bourbaki p. 37, Lemma 9.)

Proof. We begin with part (ii). Given resolutions

$$0 \longleftarrow L \longleftarrow C_0' \longleftarrow C_1' \longleftarrow \cdots \longleftarrow C_r' \longleftarrow \cdots$$

$$0 \longleftarrow N \longleftarrow C_0'' \longleftarrow C_1'' \longleftarrow \cdots \longleftarrow C_r'' \longleftarrow \cdots$$

of L and N , one knows how to construct a resolution of M in which $C_r = C_r' \oplus C_r''$ ; see [11, p.80]. If $C_r' \in \underline{C}$ for $r \le n$ , and $C_r'' \in \underline{C}$ for $r \le n$ , then $C_r \in \underline{C}$ for

* Note added in proof. An affirmative answer to this problem has been obtained by Mrs. S. Cormack.

$r \leq n$ . This proves part (ii).

We proceed similarly for part (i). Suppose that we are given resolutions

$$0 \longleftarrow L \longleftarrow C'_0 \longleftarrow C'_1 \longleftarrow \cdots \longleftarrow C'_r \longleftarrow \cdots$$

$$0 \longleftarrow M \longleftarrow C_0 \longleftarrow C_1 \longleftarrow \cdots \longleftarrow C_r \longleftarrow \cdots$$

of $L$ and $M$ . By constructing a chain map over $i: L \longrightarrow M$ and forming its mapping cylinder, we can construct a resolution for $N$ in which $C''_0 = C_0$ and $C''_r = C_r \oplus C'_{r-1}$ for $r \geq 1$ . If $C_r \in \underline{C}$ for $r \leq n$ and $C'_r \in \underline{C}$ for $r \leq n - 1$ , then $C''_r \in \underline{C}$ for $r \leq n$ . This proves part (i).

To prove part (iii), we begin by considering the special case in which $M$ is projective. Since $N$ is of type $(n + 1)$, we have an exact sequence

$$0 \longrightarrow K \longrightarrow F \longrightarrow N \longrightarrow 0$$

with $F \in \underline{C}$ and $K$ of type $n$ . Compare this with the exact sequence

$$0 \longrightarrow L \longrightarrow M \longrightarrow N \longrightarrow 0 \ .$$

By Schanuel's Lemma [18, p.101] we have

$$L \oplus F \cong M \oplus K \ .$$

So we have an exact sequence

$$0 \longrightarrow F \longrightarrow M \oplus K \longrightarrow L \longrightarrow 0 \ .$$

Here $F$ is of type $\infty$ and $M \oplus K$ is of type $n$ by part (ii). Therefore $L$ is of type $n$ by part (i).

We now turn to the general case. Since $M$ is of type $n$ , and the result is empty for $n = -1$ , we may suppose given an exact sequence

$$0 \longrightarrow K \longrightarrow F \xrightarrow{q} M \longrightarrow 0$$

with $F \in \underline{C}$ and $K$ of type $(n - 1)$. Let $P$ be the kernel of the composite $jq: F \longrightarrow N$; then $P$ has type $n$ by the special case already considered. We can construct the following diagram.

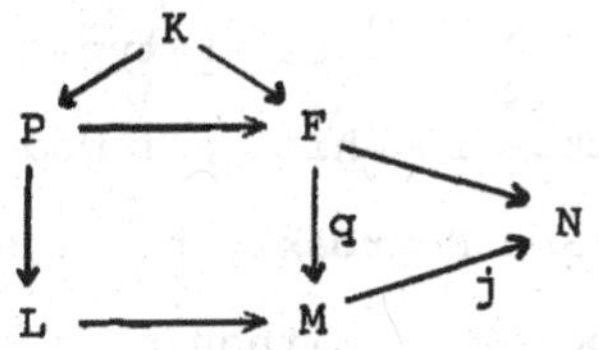


The sequence

$$0 \longrightarrow K \longrightarrow P \longrightarrow L \longrightarrow 0$$

is exact. Here $K$ has type $(n - 1)$ and $P$ has type $n$, so $L$ has type $n$ by part (i). This completes the proof of Lemma 7.

For technical reasons we need the following corollary.

<u>Corollary 8</u>

Suppose given an exact sequence

$$0 \longrightarrow K \longrightarrow C_0 \longrightarrow C_1 \longrightarrow \cdots \longrightarrow C_{n-1} \xrightarrow{d} C_n \longrightarrow M \longrightarrow 0$$

in which $C_r$ is of type $r$. Then $M$ is of type $n$.

<u>Proof</u>. The result is true for $n = 0$, by 7(i). As an inductive hypothesis, suppose it true for $(n - 1)$. Then $d(C_{n-1})$ is of type $(n - 1)$, and we have the following exact sequence.

$$0 \longrightarrow d(C_{n-1}) \longrightarrow C_n \longrightarrow M \longrightarrow 0 \ .$$

So M is of type n by 7 (i). This completes the induction and proves Corollary 8.

The next question which we consider arises as follows. The "Noetherian" case is essentially that in which all modules of $\underline{F}$-type 0 are of $\underline{F}$-type $\infty$. The "coherent" case is essentially that in which all modules of $\underline{F}$-type 1 are of $\underline{F}$-type $\infty$. (See Bourbaki, p. 61 exercise 7a and p. 63 exercise 12d, or below). Although it is not necessary for the applications, it seems worth describing a hierarchy of more subtle cases; the $n^{th}$ case is that in which all modules of type n are of type $\infty$.

Theorem 9

Suppose given $\underline{C}$ and $n \geq 0$. Then the following conditions are all equivalent.

(i) If $C \in \underline{C}$ and P is a submodule of C of type (n-1), then P is of type n.

(ii) If M is of type n and P is a submodule of M of type (n-1), then P is of type n.

(iii) Suppose given an exact sequence

$$0 \longrightarrow K \longrightarrow C_n \longrightarrow C_{n-1} \longrightarrow \cdots \longrightarrow C_1 \longrightarrow C_0 \longrightarrow M \longrightarrow 0$$

in which $C_r$ is of type n for each r. Then K is of type n.

(iv) Suppose given an exact sequence

$$C_n \longrightarrow C_{n-1} \longrightarrow \dots \longrightarrow C_1 \longrightarrow C_0 \longrightarrow M \longrightarrow 0$$

in which $C_r \in \underline{C}$ for each $r$. Then we can extend it to an exact sequence

$$C_{n+1} \longrightarrow C_n \longrightarrow C_{n-1} \longrightarrow \dots \longrightarrow C_1 \longrightarrow C_0 \longrightarrow M \longrightarrow 0$$

in which $C_{n+1} \in \underline{C}$.

(v) Every module of type $n$ is of type $\infty$.

We note that in conditions (iii) and (iv) the module $M$ at the right-hand end of the sequence is included only to avoid making an exception of the case $n = 0$. If $n \geq 1$, we can suppose given the sequence

$$C_n \longrightarrow C_{n-1} \longrightarrow \dots \longrightarrow C_1 \xrightarrow{d} C_0$$

and define $M = C_0/dC_1$.

Proof of Theorem 9. First we prove that (i) implies (ii). Suppose that $M$ is of type $n$. Then by definition, we can find a sequence

$$0 \longrightarrow K \longrightarrow C_0 \xrightarrow{j} M \longrightarrow 0$$

with $C_0 \in \underline{C}$ and $K$ of type $(n-1)$. Let $P$ be a submodule of $M$ of type $(n-1)$; then we have the following exact sequence.

$$0 \longrightarrow K \longrightarrow j^{-1}P \longrightarrow P \longrightarrow 0 .$$

Since $K$ and $P$ are of type $(n-1)$, $j^{-1}P$ is of type $(n-1)$ by 7 (ii). Since $j^{-1}P$ is a submodule of $C_0$ and $C_0 \in \underline{C}$,

$j^{-1}P$ is of type $n$ by 9 (i), which we are assuming. Hence $P$ is of type $n$ by 7 (i). This proves (ii).

We prove that (ii) implies (iii). Suppose given an exact sequence

$$0 \longrightarrow K \longrightarrow C_n \xrightarrow{d} C_{n-1} \longrightarrow \dots \longrightarrow C_1 \xrightarrow{d} C_0 \xrightarrow{\varepsilon} M \longrightarrow 0$$

in which $C_r$ is of type $n$ for each $r$. Let $Z_r \subset C_r$ be the submodule

$$\mathrm{Im}(d\colon C_{r+1} \longrightarrow C_r) = \mathrm{Ker}(d\colon C_r \longrightarrow C_{r-1}) ,$$

with the obvious interpretation for $r = 0,n$. Then by Corollary 8 (or trivially if $n = 0$) $Z_0$ is of type $(n-1)$. Since $Z_0$ is a submodule of $C_0$ and we are assuming 9 (ii), $Z_0$ is of type $n$. Assume as an inductive hypothesis that $Z_{r-1}$ is of type $n$. We have the following exact sequence.

$$0 \longrightarrow Z_r \longrightarrow C_r \longrightarrow Z_{r-1} \longrightarrow 0 .$$

So $Z_r$ is of type $(n-1)$ by 7 (iii). Since $Z_r$ is a submodule of $C_r$ and we are assuming 9 (ii), $Z_r$ is of type $n$. This completes the induction. The induction proves that $K = Z_n$ is of type $n$. This proves (iii).

We prove that (iii) implies (iv). Suppose given an exact sequence

$$C_n \xrightarrow{d} C_{n-1} \longrightarrow \dots \longrightarrow C_1 \longrightarrow C_0 \longrightarrow M \longrightarrow 0$$

in which $C_r \in \underline{C}$ for each $r$. Then certainly $C_r$ is of type $n$. Let $Z_n$ be as in the proof of (iii); then by (iii), $Z_n$ is of type $n \geq 0$. Thus we can find an epimorphism

$$C_{n+1} \longrightarrow Z_n$$

with $C_{n+1} \in \underline{C}$. This proves (iv).

We prove that (iv) implies (v). Suppose given a module $M$ of type $n$. By definition, we have an exact sequence

$$C_n \longrightarrow C_{n-1} \longrightarrow \ldots \longrightarrow C_1 \longrightarrow C_0 \longrightarrow M \longrightarrow 0$$

in which $C_r \in \underline{C}$ for each $r$. By (iv) we can extend it to an exact sequence

$$C_{n+1} \longrightarrow C_n \longrightarrow \ldots \longrightarrow C_1 \longrightarrow C_0 \longrightarrow M \longrightarrow 0$$

in which $C_{n+1} \in \underline{C}$. Now (iv) applies again to the sequence

$$C_{n+1} \longrightarrow C_n \longrightarrow \ldots \longrightarrow C_2 \longrightarrow C_1 \longrightarrow Z_0 \longrightarrow 0 .$$

Continue by induction. The induction constructs a resolution of $M$ by modules $C_r$ in $\underline{C}$ and shows that $M$ is of type $\infty$. This proves (v).

We prove that (v) implies (i). Suppose given $C \in \underline{C}$ and a submodule $P \subset C$ of type $(n-1)$. Then we have an exact sequence

$$0 \longrightarrow P \longrightarrow C \longrightarrow C/P \longrightarrow 0 .$$

Here $C/P$ is of type $n$ (by 7 (i) or direct from the definition). By 9 (v), which we are assuming, $C/P$ is of type $(n+1)$. Therefore $P$ is of type $n$ by 7 (iii). This proves (i). We have completed the proof of Theorem 9.

It now seems reasonable to make the following definition.

Definition 10

The ring $R$ is $(n,\underline{C})$-coherent if the equivalent conditions stated in Theorem 9 are satisfied.

It is clear from 9 (v) that if $R$ is $(n,\underline{C})$-coherent, it is $(m,\underline{C})$-coherent for $m \geq n$.

Examples.

(i) The ring $R$ is $(0,\underline{F})$-coherent if and only if it is (left) Noetherian.

(ii) We say that $R$ is finite-dimensional if it has non-zero components in only finitely many dimensions, so that $R = \sum_{-N}^{N} R_n$. Such a ring is $(0,\underline{D})$-coherent; the proof is trivial.

(iii) Coherence, as defined in Bourbaki, is $(1,\underline{F})$-coherence. More precisely, condition 9 (i) says in this case that every submodule $P$ of $\underline{F}$-type $0$ in $C$ is of $\underline{F}$-type $1$. This coincides with Bourbaki's condition "$C$ is pseudo-coherent". If

$$0 \longrightarrow C' \longrightarrow C \longrightarrow C'' \longrightarrow 0$$

is exact and $C'$, $C''$ satisfy this condition, then so does $C$. (This follows easily from Lemma 7; see Bourbaki p. 62 exercise 11a). So in order to check the condition for every $C$ in $\underline{F}$, it is sufficient to check it for $C = R$ (compare Bourbaki p. 63 exercise 12a). This proves the equivalence of our definition with Bourbaki's.

We will now prove that for $n \geq 1$ the property of n-coherence passes to suitable direct limits. We suppose given a (graded) ring $R$ containing subrings $R^\alpha$, and make the following assumptions. First, we assume that $\underline{C}$ is either $\underline{F}$ or $\underline{D}$, and we divide cases accordingly. If $\underline{C} = \underline{F}$, we assume that for any finite set of elements $r_1, r_2, \ldots, r_n$ in $R$ we can find an $\alpha$ such that $r_1, r_2, \ldots, r_n$ lie in $R^\alpha$. If $\underline{C} = \underline{D}$, we assume that for any finite set of dimensions $n, m, \ldots, p$ we can find an $\alpha$ such that $R_n, R_m, \ldots, R_p$ are contained in $R^\alpha$. This assumption ensures that the $R^\alpha$ approximate sufficiently closely to $R$, in a sense depending on $\underline{C}$. Secondly, we assume that, for each $\alpha$, $R$ is free as a right module over $R^\alpha$. With these assumptions we have:

<u>Theorem 11</u>

(i) For $0 < n < \infty$, the R-modules of type $n$ are precisely those of the form $R \otimes_{R^\alpha} M^\alpha$, where $R^\alpha$ runs over the subrings and $M^\alpha$ runs over the $R^\alpha$-modules of type $n$.

(ii) If $n > 0$ and $R^\alpha$ is $(n,\underline{C})$-coherent for each $\alpha$ then $R$ is $(n,\underline{C})$-coherent.

(Compare Bourbaki p. 63 exercise 12e. A check through the proof below shows that for $n = 1$ we need only assume that $R$ is flat, rather than free, as a right module over $R^\alpha$.)

Proof. For part (i), we begin by showing that the R-modules of the form $R \otimes_{R^\alpha} M^\alpha$ are of type $n$. For suppose that $M^\alpha$ is of type $n$; then there is an exact sequence of $R^\alpha$-modules

$$C_n^\alpha \longrightarrow C_{n-1}^\alpha \longrightarrow \cdots \longrightarrow C_1^\alpha \longrightarrow C_0^\alpha \longrightarrow M^\alpha \longrightarrow 0$$

with $C_t^\alpha$ in $\underline{F}$ or $\underline{D}$ as the case may be. The functor $R \otimes_{R^\alpha}$ preserves exactness, so we have the following exact sequence.

$$R \otimes_{R^\alpha} C_n^\alpha \longrightarrow \cdots \longrightarrow R \otimes_{R^\alpha} C_0^\alpha \longrightarrow R \otimes_{R^\alpha} M^\alpha \longrightarrow 0 .$$

Here the modules $R \otimes_{R^\alpha} C_t^\alpha$ are free and lie in $\underline{F}$ or $\underline{D}$ as the case may be. Thus $R \otimes_{R^\alpha} M^\alpha$ is of type $n$.

To prove the converse, suppose given an R-module $M$ of type $n$, where $0 < n < \infty$. Then we have an exact sequence of R-modules

$$C_n \xrightarrow{d} C_{n-1} \longrightarrow \cdots \longrightarrow C_1 \xrightarrow{d} C_0 \longrightarrow M \longrightarrow 0$$

with $C_t \in \underline{C}$ for each $t$. Choose R-free bases in each $C_t$; then each map $d$ can be represented by a matrix $r_{ij}$. If $\underline{C} = \underline{F}$, there are only a finite number of elements $r_{ij}$ in all. If $\underline{C} = \underline{D}$, the elements $r_{ij}$ lie in only a finite number of dimensions. In either case, we can find an $\alpha$ such that all the elements $r_{ij}$ lie in $R^\alpha$. Let $C_t^\alpha$ be the free $R^\alpha$-module generated by the R-free base of $C_t$. Then the maps $d$ restrict to give

$$C_n^\alpha \xrightarrow{d^\alpha} C_{n-1}^\alpha \longrightarrow \cdots \longrightarrow C_1^\alpha \xrightarrow{d^\alpha} C_0^\alpha .$$

The original sequence

$$C_n \xrightarrow{d} C_{n-1} \longrightarrow \cdots \longrightarrow C_1 \xrightarrow{d} C_0$$

is, up to isomorphism,

$$R \otimes_{R^\alpha} C_n^\alpha \xrightarrow{1\otimes d^\alpha} R \otimes_{R^\alpha} C_{n-1}^\alpha \longrightarrow \cdots \longrightarrow R \otimes_{R^\alpha} C_1^\alpha \xrightarrow{1\otimes d_\alpha} R \otimes_{R^\alpha} C_0^\alpha .$$

Since $R$ is free as a right module over $R^\alpha$, this sequence (as a sequence of groups) is isomorphic to a direct sum of copies of the sequence

$$C_n^\alpha \xrightarrow{d^\alpha} C_{n-1}^\alpha \longrightarrow \cdots \longrightarrow C_1^\alpha \xrightarrow{d^\alpha} C_0^\alpha .$$

Since the original sequence was exact, the sequence

$$C_n^\alpha \xrightarrow{d^\alpha} C_{n-1}^\alpha \longrightarrow \cdots \longrightarrow C_1^\alpha \xrightarrow{d^\alpha} C_0^\alpha$$

must be exact. We can define $M^\alpha = C_0^\alpha/dC_1^\alpha$, and $M^\alpha$ is an $R^\alpha$-module of type $n$, since $C_t^\alpha$ lies in $\underline{F}$ or $\underline{D}$ as the case may be. Since $R \otimes_{R^\alpha}$ preserves exactness, the sequence

$$R \otimes_{R^\alpha} C_1^\alpha \xrightarrow{1\otimes d^\alpha} R \otimes_{R^\alpha} C_0^\alpha \longrightarrow R \otimes_{R^\alpha} M^\alpha \longrightarrow 0$$

is exact, and we have

$$M \cong R \otimes_{R^\alpha} M^\alpha .$$

This proves part (i).

To prove part (ii), we assume that $R^\alpha$ is n-coherent for each $\alpha$. Let $M$ be an R-module of type $n$. By part (i) $M$ has the form $M \cong R \otimes_{R^\alpha} M^\alpha$ with $M^\alpha$ of type $n$. By 9 (v) for $R^\alpha$, $M^\alpha$ is of type $(n+1)$. By part (i), $M$

is of the type (n+1). We have shown that each R-module of type n is of type (n+1). By the proof that 9 (v) implies 9 (i), this is sufficient to show that R is (n,C)-coherent.

Corollary 12

The ring $MU^*(S^0)$ is (1,F)-coherent but not Noetherian.

In fact, $MU^*(S^0)$ is a polynomial ring $Z[x_1, x_2, \ldots, x_n, \ldots]$ on a countable set of generators [20, 30]. Each finite subset of the generators generates a Noetherian subring, and we take these subrings for the $R^\alpha$ in Theorem 11. (Compare Bourbaki p. 63 exercise 12f.)

Corollary 13

The Steenrod algebra A is both (1,F)-coherent and (1,D)-coherent, but neither Noetherian nor finite-dimensional.

In fact, any finite subset of A, and any finite-dimensional part $\sum_{r=0}^{N} A_r$ of A, is contained in a Hopf subalgebra which is finite [19], and therefore both (0,F)-coherent and (0,D)-coherent. We take such subalgebras for the $R^\alpha$ in Theorem 11; the whole algebra is free over $R^\alpha$ since $R^\alpha$ is a Hopf subalgebra [22].

Example 14

The stable homotopy groups of spheres form (under

composition) a graded ring which is neither $(1,\underline{F})$-coherent nor $(1,\underline{D})$-coherent.

We may now summarise our guiding philosophy. The most classical finiteness theorems in algebra concern finitely-generated modules over a Noetherian ring. In our applications, however, we have to use rings which are not Noetherian. The Noetherian condition gives us finiteness results on submodules. But in algebraic topology and in homological algebra we can do without information about general submodules, provided that we have information about kernels. (I mean, of course, kernels of maps from one "good" module to another.) In other words, we can use the following result.

<u>Corollary 15</u>

Suppose that $R$ is $(1,\underline{C})$-coherent, that $L$ and $M$ are modules of $\underline{C}$-type $1$ and that $f\colon L \longrightarrow M$ is an $R$-map. Then $\operatorname{Ker} f$ is of $\underline{C}$-type $1$.

This follows immediately from Theorem 9 (iii).

<u>Corollary 16</u>

Suppose that $R$ is $(1,\underline{C})$-coherent, and that

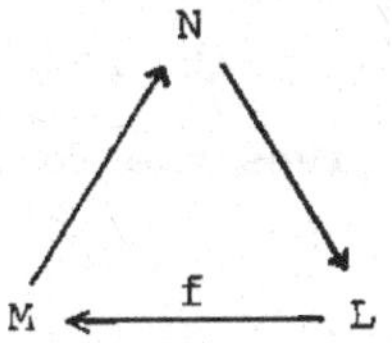

is an exact triangle of R-modules in which L and M are of $\underline{C}$-type 1. Then N is of $\underline{C}$-type 1.

Proof. Coker f is of type 1 by Corollary 8 and Ker f is of type 1 by Corollary 15. Thus N is of type 1 by Lemma 7 (ii).

For the next proposition we assume that the class $\underline{C}$ contains any free module on one generator. This is true, of course, for $\underline{C} = \underline{F}$ and $\underline{C} = \underline{D}$. We assume that $E^*$ is a (reduced) generalised cohomology theory with products, and that the coefficient ring $E^*(S^0)$ is $(1,\underline{C})$-coherent.

Proposition 17

If X is a finite CW-complex, then $E^*(X)$ is a module of $\underline{C}$-type $\infty$ over $E^*(S^0)$.

Proof. The result is true if $X = S^n$, for $E^*(S^n)$ is a free module over $E^*(S^0)$ on one generator. This serves to start an induction over the number of cells in X. If X is not a sphere, we can find a cofibering

$$A \longrightarrow X \longrightarrow B$$

in which A and B have fewer cells than X. (For example, take A to be any proper subcomplex of X.) As our inductive hypothesis, we suppose that $E^*(A)$ and $E^*(B)$ are of type 1. The cofibering gives the following exact triangle of modules over $E^*(S^0)$.

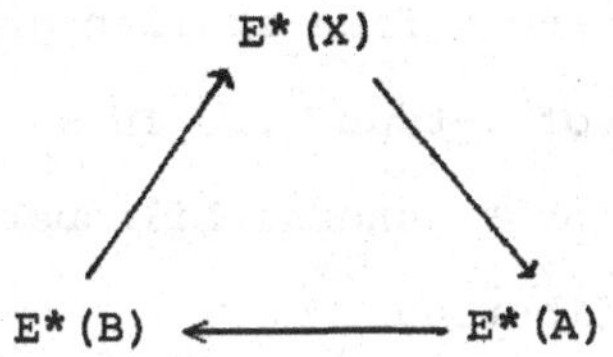


By Corollary 16, $E^*(X)$ is of type 1. This completes the induction. Of course, by Theorem 9 (v) a module of type 1 is of type $\infty$. This proves Proposition 17.

It is clear that Theorem 1 follows immediately from Corollary 12 and Proposition 17.

To prove Theorem 2, one uses Theorem 11 to reduce the problem to the study of a module $M^\alpha$ over a polynomial ring $R^\alpha$ on finitely many generators (see Corollary 12). For $M^\alpha$ we know the existence of a resolution of the sort required; take such a resolution and apply $R \otimes_{R\alpha}$, as in the proof of Theorem 11.

We will sketch the proof of Theorem 4. Let $G$ be an abelian group which is p-trivial, and let $K(G)$ be the corresponding Eilenberg-MacLane spectrum. Then $H^*(K(G);Z_p) = 0$, for $p: G \longrightarrow G$ must induce an isomorphism $p_*$ of $H_*(K(G);Z_p)$, but $p_* = 0$. Next let $X$ be a connected spectrum such that $\pi_r(X)$ is p-trivial for each $r$; then again we have $H^*(X;Z_p) = 0$. It follows that the general case of Theorem 4 can be deduced, without changing the module $H^*(X;Z_p)$, from the special case in which $\pi_r(X)$ is zero for almost all $r$.

Next let $F$ be a free abelian group; one can show that $H^*(K(F);Z_p)$ is of D-type 1. This allows us to deduce the same result for a general Eilenberg-MacLane spectrum $K(G)$; we consider a fibering

$$K(F_1) \longrightarrow K(F_2) \longrightarrow K(G)$$

and apply Corollary 16 to the resulting exact triangle of cohomology modules.

Now we can prove the result for a spectrum $X$ with just $n$ non-zero homotopy groups. This is done by induction over $n$, as for Proposition 17, but applying Corollary 16 to the exact triangle of cohomology modules arising from a suitable fibering. This completes the proof.

The proof of Theorem 3 can now safely be left to the reader.

To deduce Corollary 5, we suppose given a space $Y$ which contradicts Corollary 5, so that $\tilde{H}^*(Y;Z_p) \neq 0$ and Theorem 4 applies to the corresponding spectrum. Let $y$ be a non-zero class of lowest dimension in $\tilde{H}^*(Y;Z_p)$; then

$$P^{p^f} y = 0$$

for all sufficiently large $f$; this makes it extremely plausible that $\tilde{H}^*(Y;Z_p)$ cannot have a presentation with relations in only finitely many dimensions, and this can indeed be proved. This contradicts Theorem 4 and proves Corollary 5.

## REFERENCES

[1] Adams, J. F., "Une Relation entre Groupes d'Homotopie et Groups de Cohomologie", *Comptes Rendues de l'Acad. des Sci., Paris*, 245; 24-26 (1957).

[2] ________, "On the Structure and Applications of the Steenrod Algebra", *Comm. Math. Helv.*, 32; 180-214 (1958).

[3] ________, "Vector Fields on Spheres", *Bull. Amer. Math. Soc.*, 68; 39-41 (1962).

[4] ________, "Vector Fields on Spheres", *Annals of Math.*, 75; 603-632 (1962).

[5] Anderson, D. W., and Hodgkin, L., "The K-theory of Eilenberg-MacLane Complexes", (preprint, to appear).

[6] Atiyah, M. F., "Vector Bundles and the Künneth Formula", *Topology*, 1; 245-248 (1962).

[7] Boardman, J. M., Thesis, Cambridge, 1964.

[8] ________, "Stable Homotopy Theory", mimeographed notes, University of Warwick, 1965 onward.

[9] Bourbaki, N., "Algebre Commutative", Chapters 1 and 2 of *Éléments de Mathématique*, 27, Hermann, (1961). (Act. Sci. et Ind. 1290)

[10] Brown, E. H., and Peterson, F. P., "A Spectrum Whose $Z_p$ Cohomology is the Algebra of Reduced $p^{th}$ Powers", *Topology*, 5; 149-154 (1966).

[11] Cartan, H., and Eilenberg, S., *Homological Algebra*, Princeton University Press (Princeton Mathematical Series, no. 19) (1956).

[12] Cohen, J. M., "Coherent Graded Rings", (preprint, to appear).

[13] __________, "The Non-existence of Spaces of Finite Stable Homotopy Type", (preprint, to appear).

[14] Conner, P. E., and Floyd, E. E., *The Relation of Cobordism to K-theories*, (Lecture Notes in Mathematics, no. 28) Springer, (1966).

[15] Douady, A., Seminaire H. Cartan 11 (1958/59), exposés 18, 19.

[16] Hattori, A., "Integral Characteristic Numbers for Weakly Almost Complex Manifolds", *Topology*, 5; 259-280 (1966).

[17] Hodgkin, L., "K-theory of Eilenberg-MacLane Complexes I, II", (preprints).

[18] MacLane, S., *Homology*, (Grundlehren der Math. #114) Springer, (1963).

[19] Milnor, J., "The Steenrod Algebra and Its Dual", *Annals of Math*, 67; 150-171 (1958).

[20] __________, "On the Cobordism Ring $\Omega^*$ and a Complex Analogue", *Amer. Jour. Math.*, 82; 505-521 (1960).

[21] __________, "On Axiomatic Homology Theory", *Pacific Jour. Math.*, 12; 337-341 (1962).

[22] __________, and Moore, J. C., "On the structure of Hopf Algebras", *Annals of Math.*, 81; 211-264 (1965).

[23] Novikov, S. P., "Rings of Operations and Spectral Sequences", *Doklady Akademii Nauk S.S.S.R.*, 172; 33-36 (1967).

[24] __________, *Izvestija Akademii Nauk S.S.S.R., Serija Matematiceskaja*, 31; 855-951 (1967).

[25] Puppe, D., "Stabile Homotopietheorie I", *Math. Annalen*, 169; 243-274 (1967).

[26] Serre, J.-P., "Cohomologie Modulo 2 des Complexes d'Eilenberg-MacLane", *Comm. Math Helv.*, 27; 198-232 (1953).

[27] __________, "Groupes d'Homotopie et Classes de Groupes Abéliens", *Annals of Math.*, 58; 258-294 (1953).

[28] Stasheff, J. D., "Homotopy Associativity of H-spaces I", *Trans. Amer. Math. Soc.*, 108; 275-292 (1963).

[29] Stong, R. E., "Relations Among Characteristic Numbers I", *Topology*, 4; 267-281 (1965).

[30] Thom, R., "Travaux de Milnor sur le Cobordisme", *Séminaire Bourbaki*, 11, no. 180 (1958/59).

[31] Whitehead, G. W., "Generalized Homology Theories", *Trans. Amer. Math. Soc.*, 102; 227-283 (1962).

# On H-spaces and infinite loop spaces

by

Jon Beck

By a <u>topological category</u> I think is generally understood a category $\mathfrak{X}$ in which:

(1) For each pair of objects $X,Y \in \mathfrak{X}$ there is a hom object $(\underline{X,Y})$ which is a topological space, and composition is a continuous unitary, associative operation $(\underline{X,Y}) \times (\underline{Y,Z}) \to (\underline{X,Z})$.

(2) For every space A and object $Y \in \mathfrak{X}$ there is an object $(A,Y) \in \mathfrak{X}$ and a natural homeomorphism $\underline{(X,(A,Y))} \overset{\sim}{\to} (A,\underline{(X,Y)})$.

(3) For every space A the functor $(A,\ ):\mathfrak{X} \to \mathfrak{X}$ has a left adjoint $A \times (\ ):\mathfrak{X} \to \mathfrak{X}$.

The category of topological spaces is itself a topological category. The precaution of course is actually taken of restricting to a category of spaces or something like them for which the conversion $(X,(\underline{Y,Z}) \overset{\sim}{\to} (\underline{X\times Y,Z})$ holds. Specifically, I have compactly generated spaces in mind (E.Spanier, Annals of Math. <u>79</u> (1959), 142-197,§2), but with care and compactness assumptions everything can be pushed through in the ordinary category of topological spaces.

Another topological category is that of spaces with base points and base point preserving maps. We will practically always work in this category, which we denote simply by <u>Top</u>.

In this case pairing (3)

<u>Topological spaces</u> × <u>Top</u> → <u>Top</u>

is naturally written as $A\otimes X$. For example, if I is the unit interval and $X \in$ <u>Top</u>, then $I\otimes X$ is the reduced cylinder over the pointed space X, and maps $I\otimes X \to Y$ are base point preserving homotopies.

Of course, <u>Top</u> is also a <u>closed</u> category, that is, itself a <u>Top</u>-category. This fact gives rise to a different pairing $X\otimes Y$ where X,Y are both pointed spaces, namely $X \times Y/X \times 0 + 0 \times Y$, usually written $X \wedge Y$.

$\mathfrak{X}$ is called a <u>pointed</u> <u>topological</u> <u>category</u> if it possesses a hom functor with values in <u>Top</u> and pairings $A\otimes X, (\underline{A,Y})$ as in (2),(3) exist for $A \in$ <u>Top</u>, $X,Y \in \mathfrak{X}$.

The point is that as soon as a category has a hom functor with values in Top and adjoints as specified above, then the constructions of algebraic topology are available in that category. Let $\mathfrak{X}$ be a Top-category in this sense and let $X_0$ be a fixed object in $\mathfrak{X}$. Usually $\mathfrak{X}$ has some well known underlying-(pointed)-topological-space functor and $X_0$ is chosen as the object which represents this via $\underline{(X_0,\ )}:\mathfrak{X} \to \text{Top}$. The tensor product gives adjoint functors

$$\text{Top} \underset{\underline{(X_0,\ )}}{\overset{(\ )\otimes X_0}{\rightleftarrows}} \mathfrak{X}.$$

Since cells and spheres are in Top we have objects $e^{n+1}\otimes X_0$, $S^n\otimes X_0$ in $\mathfrak{X}$, which are the (n+1)-cell and n-sphere in $\mathfrak{X}$. Let us write $e^{n+1}\otimes\mathfrak{X}$, $S^n\otimes\mathfrak{X}$ instead. Modulo minor assumptions of completeness, CW-objects exist in $\mathfrak{X}$. Such are built up by glueing $\mathfrak{X}$-cells onto lower-dimensional skeleta via attaching maps $S^n\otimes\mathfrak{X} \to Y$ in $\mathfrak{X}$; by adjointness, these are the same as maps $S^n \to \underline{(X_0,Y)}$, the latter being the "underlying space" of Y. The usual development of CW-theory can be conducted in such a category. The essential fact to be supplied is that

$$\pi_i(S^n\otimes\mathfrak{X}) = \begin{cases} 0, & i < n, \\ Z, & i = n. \end{cases}$$

This is true in all of the tripleable or "theoretical" examples of Top-categories used in this paper.

As an example, consider the category of topological groups. The continuous homomorphisms $G \to H$ form a space $\underline{(G,H)} \in \text{Top}$. The group structure of $\underline{(A,H)}$ for $A \in \text{Top}$ is value-wise, and $A\otimes G$ is the free topological group generated by all symbols $a\otimes g$ modulo the relations $a\otimes(g_0g_1) = (a\otimes g_0)(a\otimes g_1)$. The discrete group Z plays the role of $X_0$. Given a complex X with cells $e^{n+1} \to X$, the cells $e^{n+1}\otimes Z \to X\otimes Z$ give a group-cellular decomposition of $X\otimes Z$. Homotopy theory in this category can now be carried out in the usual manner. Some of this has been done under the guise of the theory of simplicial groups.

One application: let $0 \in I$ be the base point of the unit interval. Then $I\otimes G$ is the group-theoretical cone on G. The natural map $G \to I\otimes G$ at the 1-end is an embedding of G into a contractible topological group. Under standard assumptions on the topology of G near its neutral element, the projection $I\otimes G \to I\otimes G/G$ is easily shown to be a fiber bundle. Thus $I\otimes G/G$ is a classifying space for G. Later on we shall construct classifying spaces for other types of H-spaces. Lack of a $\otimes$-product in those cases make the construction more difficult.

Another algebraic topology arises in the category of spaces <u>over</u> a fixed space X (no base points are needed). An object in this category is a map $A \to X$, a map is a commutative triangle. The maps $A \to B$ over X form a closed subspace $(\underline{A,B})_X$ of the usual $(\underline{A,B})$. The n-cell in this topological category is $e^n \times X$. Homotopy equivalence is what is usually called fiber homotopy equivalence.

Notice that differential (or PL) topology exists over X even when X is a quite arbitrary space. Euclidean space/X is $R^n \times X$ and a map $R^m \times X \to R^n \times X/X$ is differentiable if it is so with respect to the real component. For example, there should be an isomorphism $\Gamma^*(X) \to [X,PL/O]$ where $\Gamma^n(X)$ is the group of diffeomorphisms of $S^{n-1} \times X$ modulo those which can be extended to $D^n \times X$, all /X.

The category of spaces /X could be taken as a base category for algebraic topology. Pointed objects (those with zero sections), H-objects, ... can be defined and have their usual properties. When X = 1 this program reduces to ordinary topology.

However, in this paper we will adhere to the standard base category <u>Top</u> of pointed topological spaces, and concentrate on categories tripleable over <u>Top</u> (which actually counter-includes the case of spaces /X). We recall that a tripleable category is one whose objects are determined by a free-object functor (the definition follows), and for these we have:

(4) <u>Theorem</u>. Let T be a pointed topological triple. Then the category of T-spaces is a pointed topological category; more precisely, axioms (1),(2) for a topological category hold and the tensor product $A \otimes (X,\xi)$ which is asserted to exist in (3) does exist, at least when T is derivable from a topological theory.

We define a <u>pointed topological triple</u> $T = (T,\eta,\mu)$ to be a functor $T$: <u>Top</u> $\to$ <u>Top</u> with $oT = o$ and Tcontinuous, that is, effecting for all $X,Y \in$ <u>Top</u> a continuous map $(\underline{X,Y}) \to (\underline{XT,YT})$, together with natural transformations $\eta$:id. $\to T$, $\mu$:TT $\to$ T such that

commute. A T-algebra, or T-space, is a pair $(X,\xi)$ where $X \in$ Top and $\xi: XT \to X$ is a continuous map such that the unitary and associative laws hold:

$$\begin{array}{ccc} X & \xrightarrow{X\eta} & XT \\ & \searrow\!\!= & \downarrow \xi \\ & & X \end{array} \qquad \begin{array}{ccc} XTT & \xrightarrow{X\mu} & XT \\ \xi T \downarrow & & \downarrow \xi \\ XT & \xrightarrow[\xi]{} & X \end{array}$$

$\xi$ is called the T-structure of the space. With an evident definition of morphisms, T-spaces form a category $\mathrm{Top}^T$.

The usefulness of this concept arises from the fact that, by composition, adjoint functors give rise to triples T, and the corresponding categories of T-spaces consist precisely of those spaces which possess the general structure of values of the right adjoints.

As an example, consider the adjoint functors $\Sigma,\Omega$: Top $\to$ Top. Let $\eta: X \to X\Sigma\Omega$, $\epsilon: X\Omega\Sigma \to X$ be the usual adjointness maps. Then the composite functor $\Sigma\Omega$ is a triple in Top with unit and multiplication

$$\mathrm{id.} \xrightarrow{\eta} \Sigma\Omega, \qquad \Sigma\Omega\Sigma\Omega \xrightarrow{\mu = \Sigma\epsilon\Omega} \Sigma\Omega.$$

A $\Sigma\Omega$-space is then a pair $(X,\xi)$ where $X \in$ Top and $\xi: X\Sigma\Omega \to X$ is a unitary, associative structure map:

$$\begin{array}{ccc} X & \xrightarrow{\eta} & \Sigma\Omega \\ & \searrow\!\!= & \downarrow \xi \\ & & X \end{array} \qquad \begin{array}{ccc} X\Sigma\Omega\Sigma\Omega & \xrightarrow{\Sigma\epsilon\Omega} & X\Sigma\Omega \\ \xi\Sigma\Omega \downarrow & & \downarrow \xi \\ X\Sigma\Omega & \xrightarrow[\xi]{} & X \end{array}$$

Such a map $\xi$ furnishes X with all of the structure which loop spaces possess in general. Algebraically, for example, let $\theta$ be any n-variable operation on loops and $x_o,\ldots,x_{n-1}$ any n points in X. Then the value of $\theta$ in X is $[(x_o\eta,\ldots,x_{n-1}\eta)\theta]\xi$. The fact that $\xi$ is associative implies that $\theta$ satisfies all of the identities in X which it satisfies in the world of loop spaces.

In particular, every loop space has a $\Sigma\Omega$-structure, by evaluation of loops: $(B\Omega)\Sigma\Omega = (B\Omega\Sigma)\Omega \to B\Omega$. As to whether there are any $\Sigma\Omega$-spaces that are not loop spaces a priori, that question will be investigated in (16).

In the general case, if $F: \underline{A} \to \underline{B}$ is left adjoint to $U: \underline{A} \leftarrow \underline{B}$, let $T = FU$ be the corresponding triple in $\underline{A}$. There is a canonical functor

$$\begin{array}{ccc} \underline{A}^T & \xleftarrow{\Phi} & \underline{B} \\ & {\scriptstyle U^T}\searrow & \uparrow\downarrow {\scriptstyle F,\ U} \\ & & \underline{A} \end{array}$$

defined by $B\Phi = (BU, B\epsilon U)$, where $\epsilon : UF \to id.$ is the adjointness morphism. In the case of $\Sigma, \Omega$, the value of this canonical functor

$\underline{Top}^{\Sigma\Omega} \leftarrow \underline{Top}$ at a space $B$ is $B\Omega$ considered as a $\Sigma\Omega$-space.

The adjoint pair $(F,U)$ is tripleable if $\Phi$ is an equivalence of categories. The tripleableness problem is in general very difficult, and we shall not go into it. Suffice it to say that a left adjoint $\check{\Phi}$ for $\Phi$ is easily constructed as the coequalizer

$$XFUF \underset{XF\epsilon}{\overset{\xi F}{\rightrightarrows}} XF \longrightarrow (X,\xi)\check{\Phi}$$

and that the greatest difficulty ordinarily attends on showing that the adjointness map $(X,\xi) \to (X,\xi)\check{\Phi}\Phi$ is an isomorphism of T-algebras; this map is essentially the composition of $\eta : X \to XFU$ and $XFU \to (X,\xi)\check{\Phi}U^T$. The following fact is used in studying this map, and is relevant later:

(5) The augmented simplicial object

$$X(FU)^{n+1}, \qquad n \geq -1,$$

with face operators $\epsilon_i : X(FU)^{n+2} \to X(FU)^{n+1}$ given by $\epsilon_o = \xi(FU)^{n+1}$, $\epsilon_i = XF(UF)^{i-1}\epsilon(UF)^{n-i+1}U$, $1 \leq i \leq n+1$, and suitable degeneracy operators induced by $\eta : id \to FU$, has a "contraction"

$$X(FU)^{n+2} \xleftarrow{h_n} X(FU)^{n+1}, \qquad n \geq -1$$

obeying $h_n\epsilon_i = \epsilon_i h_n$, $0 \leq i \leq n$, $h_n\epsilon_{n+1} = id.$, namely $h_n = X(FU)^{n+1}\eta$, and is therefore homotopy equivalent, as a simplicial object, to the constant or "discrete" simplicial object $X$.

Finally, as to the topological nature of $\underline{Top}^T$, if $(X,\xi)$, $(Y,\theta)$ are T-spaces, their T-space maps $f : X \to Y$ form a closed subspace of the space of all maps $(\underline{X,Y})$, namely the equalizer

$$(\underline{X,Y})_T \longrightarrow (\underline{X,Y}) \underset{fT.\theta}{\overset{\xi f}{\rightrightarrows}} (\underline{XT,Y}).$$

If $A$ is a space, $(\underline{A,Y})$ is a T-space by means of the composition $(\underline{A,Y})T \to (\underline{A,YT}) \to (\underline{A,Y})$, where the first map is adjoint to $A \otimes (\underline{A,Y})T \to YT$ and the second is induced by the T-structure of Y. The T-space $A \otimes (X,\xi)$ should be produced as a quotient of $(A\otimes X)T$ (cf. the example of topological groups). But for topological triples in general it is not known whether a T-structure can be defined on the quotient. This problem is open, in particular, for the triple $\Sigma\Omega$. Parenthetically, the same results and difficulties carry over to any suitable notion of "enriched" category and triple thereon.

On the other hand, when T is a triple arising from a topological theory, which we shall shortly define, the above quotient problem is easily disposed of.

Another problem also leads us to introduce topological theories. That the continuous triple morphisms $S \to T$ form a topological space is evident. But for other constructions such as the function space $(\underline{A,T})$, the product $A \otimes T$ and the rest of the algebraic topology of continuous triples, it is necessary to reveal the internal structure of triples, and restrict to those for which this is a topological theory.

Since there will be a lot of deliberate confusion between topological theories and triples, the same letter T will be used to refer to both concepts. The original notion of theory, over the category of sets, is due to F.W. Lawvere (Proc. NAS USA 50 (1963), 869-872).

(6) By a (finitary) pointed topological theory is meant a pointed topological category T whose objects are the natural numbers 0, 1, 2, ... and in which m is the coproduct 1+1+...+1 (m times). Thus the hom object $(m,n)T$ is a topological space with base point and is the cartesian power $(n)T^m$, where $(n)T$ is the space of n-ary operations $(1,n)T$. The composition in T is an associative family of continuous base point preserving maps

$$(m,n)T \otimes (n,p)T \to (m,p)T.$$

Probably with a more advanced concept the objects $A \otimes n$ could also be attributed to the theory, but we shall not bother with that.

A map $T \to T'$ is a continuous, pointed, 1- and coproduct-preserving functor.

An algebra over a topological theory T, or for greater clarity, a T-space, is a pointed continuous product-preserving functor

$$T^* \xrightarrow{X} \underline{Top}.$$

$T^*$ is the dual or opposite of the topological category T. By the product-preserving property, such a functor is determined by the image of $1 \in T$ which is also denoted by X (and then $n \longmapsto X^n$). Thus the functor is equivalent to a family of continuous maps

$$(n)T \to (\underline{X^n,X})$$

subject to various identities, that is, the n-ary operations of T are continuously represented by actual maps $X^n \to X$.

A map $f:X \to Y$ of T-spaces is a natural transformation of X,Y thought of as functors; equivalently, a continuous map such that all diagrams

$$\begin{array}{ccc} (n)T \otimes X^n & \to & (n)T \otimes Y^n \\ \downarrow & & \downarrow \\ X & \to & Y \end{array}$$

commute. The resulting category of T-spaces is denoted by $\underline{Top}^T$.

(7) Examples of topological theories. The easiest theories are those which arise from "algebraic" theories in the category of sets. Let Alg(Top) be the category of algebras of any specified discrete type, but interpreted in the category of topological spaces, for example, topological groups, topological rings, topological Lie algebras, ... . A free-algebra functor Top → Alg(Top) manifestly exists and is left adjoint to the forgetful Alg(Top) → Top. The values of the resulting triple T on sums of 0-spheres, $(nS^0)T$, $n > 0$, defines a theory T. The space of maps $m \to n$ in the theory is then the cartesian power $(nS^0)T^m$. The category of models for the theory, i.e., spaces X equipped with maps $(n)T \to (\underline{X^n,X})$, is exactly the category Alg(Top).

For topological groups, $(nS^0)T$ is just the free topological group generated by $S^0 + \ldots + S^0$ (n times), and this is the free discrete group on n generators. (As the group is free relative to pointed spaces, the apparent generator furnished by the base point is suppressed). Thus the elements of (n)T are exactly all of the n-variable operations in the theory of topological groups, and these are the same as in the discrete theory of groups. Maps $(n)T \to (\underline{X^n,X})$ as above clearly make X into a topological group.

Topological monoids, that is, strictly associative H-spaces, arise similarly.

More significant and indicative of the reason for introducing topologies into theories are categories of H-spaces in which the defining equations hold only up to specified homotopies.

Consider the theory of homotopy-associative H-spaces (with strict unit). This theory T has $(o)T = 0$, and its operations of higher power are generated by a binary operation $x,y \to xy$ which is a 0-cell in (2)T, and 1-cell in (3)T whose vertices are $(x\ y)z$ and $x(yz)$. This results in considerable complexity, (1)T for example containing 0-cells corresponding to the unary operations $x \to x^n$, 1-cells corresponding to the various ways of introducing parentheses into these homotopy-associative "powers", 2-cells as products of these 1-cells, ... in fact (1)T is an infinite-dimensional CW complex; (2)T, (3)T, ... are more complicated. Maps in this category $\underline{\text{Top}}^T$ of canonically homotopy-associative H-spaces are, of course, required to preserve the generating 1-simplex in (3)T, and all of its consequences.

More complicated is Stasheff's theory $A_\infty$ (cf. Trans. AMS <u>108</u> (1963), 275-292). This is the theory of H-spaces which have (unnecessarily) strict multiplicative units and homotopy associativities as above, as well as many "higher associativities". For example, in the space $(4)A_\infty$ the homotopy associativity generates an $S^1$:

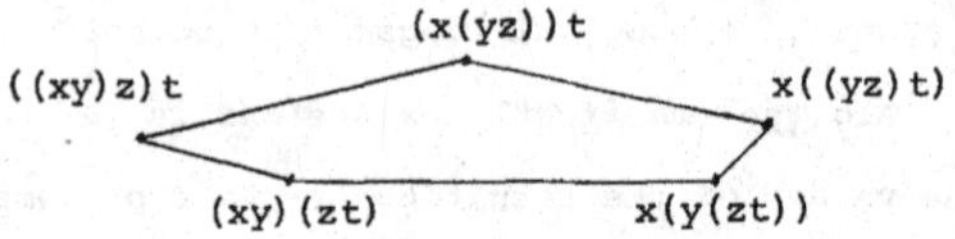


The cell structure of $(4)A_\infty$ then includes a 2-cell with this $S^1$ as its boundary, and so on.

In order to be able to manipulate these constructs with confidence, it is essential to know that every graded topological space (i.e. sequence of spaces) generates a free topological theory, or even more, that topological theories are tripleable over graded spaces, and that arbitrary theories can therefore be constructed as coequalizers of maps between free theories.

If T is a topological theory, the adjoint pair $\underline{\text{Top}} \to \underline{\text{Top}}^T \to \underline{\text{Top}}$ is easily seen to be tripleable. In fact, from now on we confuse topological theories with the triples in <u>Top</u> which they generate via their free algebra functors. We will find it useful to have the following formula for the triple in terms of the spaces of operations in the theory:

$$AT = \sum_{n \geq 0} A^n \otimes (n)T \,/\, (\otimes\text{-identities}).$$

The precise identifications made are generated by maps of finite sets. If $\alpha : m \to n$, then $a \otimes \theta.\alpha T$ is identified with $a.A^\alpha \otimes \theta$ for $a \in A^n$, $\theta \in (m)T$, much as in Milnor's geometrical realization of a s. s. complex.

(8) <u>Theorem</u>. Let $f: T \to T'$ be a map of finitary pointed topological theories such that

$(n)f:(n)T \to (n)T'$ is a homotopy equivalence for each $n \geq 0$. Then $Xf:XT \to XT'$ is a homotopy equivalence for every CW complex X.

To prove this, note that the above formula for AT makes sense for any functor $T:\underline{S}_{fin} \to \underline{Top}$ and in fact is the $\otimes$-product of functors $A^* \otimes T$ where $A^*:(\underline{S}_{fin})^* \to \underline{Top}$ is the powers of A ($\underline{S}_{fin}$ = finite sets). Let $T \to T' \to T_f$ be a mapping cone sequence in the topological category of functors $\underline{S}_{fin} \to \underline{Top}$. Then $AT \to AT' \to A^* \otimes T_f$ is a mapping cone sequence in $\underline{Top}$. One demonstrates that the space $A^* \otimes T_f$ is contractible by induction on p applied to the spaces $\Sigma\ A^n \otimes (n)T_f\ /\ (\otimes - \text{id.})$, $n<p$.

The concept of discreteness gives rise to certain operations on topological triples. Let the two adjoints to the inclusion of discrete spaces be written $X_d \to X$ (discrete topology on X) and $X \to \pi_o X$ (for good X). Similar functors exist for topological theories. Actually, we have no use for the atomization $T_d \to T$. The other discretization $T \to \pi_o T$ gives us the theory which has $(n)(\pi_o T) = \pi_o(nT)$ with the obvious composition of operations. For example, the discretization $A_\infty \to \pi_o A_\infty$ yields A, the theory of monoids (see (7)). For the similar theory of groups up to compatible homotopies, $G_\infty$, we have $G_\infty \to \pi_o G_\infty = G$, the theory of groups.

(9) Proposition. In the diagram of natural transformations

$$\begin{array}{ccc} A_\infty & \xrightarrow{\ \sim\ } & A \\ \downarrow & & \downarrow \\ G_\infty & \xrightarrow{\ \sim\ } & G \end{array}$$

the horizontal arrows are homotopy equivalence of CW theories, or of CW triples (by (8)). The vertical arrows have the property that they are equivalences when evaluated on any connected CW complex X.

For the horizontals, the spaces $(n)A_\infty$ are unions of contractible components, the $(n)G_\infty$ as well. For the verticals let

$$\underline{Top} \overset{F}{\to} \text{Mon}(\underline{Top}) \overset{U}{\to} \underline{Top}$$

be the usual free topological monoid and underlying space functors. Both of these functors "preserve" the subcategory $\underline{W}_o \to \underline{Top}$ of connected CW complexes. Moreover, by use of the homotopy extension theorem, F,U remain adjoint modulo homotopy, i.e.

$$[\underline{W}_o] \overset{F}{\to} \text{Mon}[\underline{W}_o] \overset{U}{\to} [\underline{W}_o]$$

is an adjoint pair of functors, where $[\underline{W}_o]$ denotes the category modulo homotopy or weak homotopy. The same holds for groups. The natural forgetful functor from groups to monoids

$$\begin{array}{ccc} \mathrm{Mon}[\underline{W}_o] & \longleftarrow & \mathrm{Gp}[\underline{W}_o] \\ {\scriptstyle A}\ \nwarrow\!\searrow & & \swarrow\!\nearrow\ {\scriptstyle G} \\ & [\underline{W}_o] & \end{array}$$

is actually an isomorphism by the well known fact that a connected CW H-space always possesses a multiplicative inverse up to homotopy. Since both categories are tripleable over $[\underline{W}_o]$, this implies that the triple map $A \to G$ is an isomorphism when restricted to the category $[\underline{W}_o]$.

(10) Discretization of the powers of operations of topological triples is an important process. Let $T = (T,\eta,\mu)$ be a topological triple with $OT = O$. Let $T_{fin}$ be the pointed topological theory $(m,n)T_{fin} = ((nS^o)T)^m$, with composition law $(m,n)T \otimes (n,p)T \to (m,p)T$ given by $\alpha \otimes \beta \to \alpha.\beta T.(pS^o)\mu$.

The $T_{fin}$ construction reflects triples into the subcategory of topological theories, and is about the same process as was applied to an "algebraic triple" in <u>Top</u> in (7). Of course, $T_{fin}$ can be considered as a triple itself, and this <u>finitary reflection</u> or <u>truncation</u> of T is an injection

$$T_{fin} \to T$$

(cf. Linton, La Jolla Conference on Categorical Algebra, Springer Verlag, 1966).

As an example, consider the suspension-loops triple $\Sigma\Omega$. The finitary theory $(\Sigma\Omega)_{fin}$ has as its space of n-ary operations the space of loops on the sum $S^1 + \ldots + S^1$ of n circles.

(11) <u>Proposition</u>. The inclusion

$$X(\Sigma\Omega)_{fin} \to X\Sigma\Omega$$

is a weak homotopy equivalence (homotopy equivalence if X is a CW complex).

For the proof we use the fact that the operation $(\ )_{fin}$ can actually be applied to any endofunctor of <u>Top</u>:

$$X.F_{fin} = \varinjlim_{nS^o \to X} (nS^o)F$$

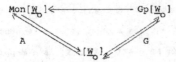

where the $\varinjlim$ has to be understood in the right closed-category, i.e. topological, sense, namely as a quotient space of $\Sigma X^n \otimes (nS^o)F$. We then have the diagram

$$\begin{array}{ccc} X(\Sigma\Omega)_{fin} & \longrightarrow & X\Sigma\Omega \\ \downarrow & & \downarrow \\ X(\Sigma E)_{fin} & \longrightarrow & X\Sigma E \\ \downarrow & & \downarrow \\ X\Sigma_{fin} & = & X\Sigma \end{array}$$

where E is the contractible path space functor. The left column is essentially a fibration by means of a path lifting function which shifts the terminal segments of paths from "generator to generator" of the suspension. From the homotopy exact sequences,

$$\pi_n X(\Sigma\Omega)_{fin} \stackrel{\simeq}{\to} \pi_n X\Sigma\Omega.$$

Since loop spaces are $G_\infty$-spaces there is a topological theory map $G_\infty \to (\Sigma\Omega)_{fin}$. By (8), this is a homotopy equivalence of CW theories. Thus in the following diagram of triples in Top, all of the arrows are homotopy equivalences, at least when evaluated on connected CW complexes.

$$\begin{array}{ccc} A_\infty & \longrightarrow & A \\ \downarrow & & \downarrow \\ G_\infty & \longrightarrow & G \\ \downarrow & & \\ (\Sigma\Omega)_{fin} & & \\ \downarrow & & \\ \Sigma\Omega & & \end{array}$$

(12) Theorem. The triples A and $\Sigma\Omega$ are naturally equivalent on the category of connected CW complexes, that is, if X is such a space there is a natural homotopy equivalence of the "reduced product space"

$$X_\infty = XA \stackrel{\simeq}{\to} X\Sigma\Omega.$$

(I.M. James, Ann. Math. 62 (1955), 170-197).

The above fibration argument can be iterated to obtain the same result about the triples $\Sigma^k\Omega^k$, $k>0$. The inductive step in the proof of the following theorem has the form

$$\begin{array}{ccc} X(\Sigma^{k+1}\Omega^{k+1})_{fin} & \longrightarrow & X\,\Sigma^{k+1}\Omega^{k+1} \\ \downarrow & & \downarrow \\ X(\Sigma^{k+1}\Sigma^{k}E)_{fin} & \longrightarrow & X\,\Sigma^{k+1}\Omega^{k}E \\ \downarrow & & \downarrow \\ X(\Sigma^{k+1}\Omega^{k})_{fin} = X\,\Sigma(\Sigma^k\Omega^k)_{fin} & \longrightarrow & X\,\Sigma^{k+1}\Omega^{k} \end{array}$$

(13) Theorem. The horizontal inclusions in the diagram

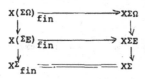

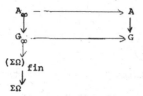

$$\begin{array}{ccc} X(\Sigma^k\Omega^k)_{fin} & \xrightarrow{\sim} & X\,\Sigma^k\Omega^k \\ \downarrow & & \downarrow \\ X\,Q_{fin} & \xrightarrow{\sim} & X\,Q \end{array}$$

are weak homotopy equivalences (homotopy equivalences if X is a CW complex); here $Q = \lim_{k\to\infty} \Sigma^k\Omega^k$.

The functor Q was introduced by Dyer-Lashof (Am. J. Math. 84 (1962), 35-88). As a direct limit of direct limit preserving triples (the circle is compact), Q is itself a direct limit preserving triple. The discretization $Q \to \pi_o Q$ is the natural map $Q \to AG$, the latter the free abelian group triple, and is not a homotopy equivalence. Indeed, $Q \to AG$ induces the Hurewicz homomorphism stably, and Q-spaces generally have non-trivial k-invariants.

Here Q-space means an algebra over the Q triple:

$$\begin{array}{ccc} X & \xrightarrow{\eta} & XQ \\ & \searrow{=} & \downarrow \xi \\ & & X \end{array} \qquad\qquad \begin{array}{ccc} XQQ & \xrightarrow{X\mu} & XQ \\ \xi Q \downarrow & & \downarrow \xi \\ XQ & \xrightarrow[\xi]{} & X \end{array}$$

Q-spaces more or less coincide with the homotopy-everything H-spaces of Boardman-Vogt (Bull. AMS 74 (1968), 1117-1122). At least the functor h.e.-spaces $\leftarrow$ $\underline{\text{Top}}^Q$ is evident, and we do demonstrate that Q-spaces are infinite loop spaces ((17) below; we mean h.e.-spaces X with $\pi_o X$ abelian groups). It would be desirable to have direct demonstrations that the infinite objects of algebraic and differential topology, O,U,BO,PL,... are Q-spaces.

The homotopy-finitary character of a topological triple has various consequences. Restricting to a suitable class of "linear" finitary triples, it is possible to demonstrate theorems of the type

(14) $\qquad H_*(XT) \simeq (H_*X)(H_*T)$ ,

that is, existence of triples $H_*T$ on the category of graded vector spaces over a field rendering the diagram

$$\begin{array}{ccc} \underline{\text{Top}} & \xrightarrow{T} & \underline{\text{Top}} \\ H_* \downarrow & & \downarrow H_* \\ \underline{\text{Vect}} & \xrightarrow{H_*T} & \underline{\text{Vect}} \end{array}$$

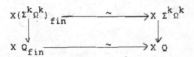

commutative (non-naturally, for coefficients in a hereditary ring). I hope to carry this out in a later paper for the triples $\Sigma^k\Omega^k$, Q (which are not themselves linear but have linear approximations). It should be possible to press triple-theoretic techniques far enough to obtain the calculations of Milgram (Ann. of Math. <u>84</u> (1966), 386-403) and Dyer-Lashof (op. cit., their Theorem 5.1 gives $H_*(Q,Z/pZ)$ explicitly). For triples with "simplicial" bases it is possible to demonstrate formulas like (14), replacing vector spaces with coalgebras over a field and constructing $H_*T$ from theories which have hom objects in the cartesian-closed category of coalgebras.

(15) The purely categorical question of whether the $\Sigma,\Omega$ adjoint pair is tripleable leads to the construction of universal base spaces. Tripleableness would mean that the functor

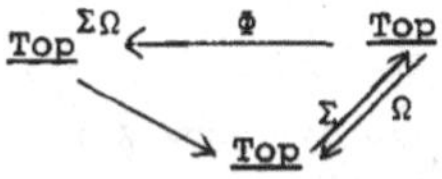

is an equivalence. The question can be examined in several parts. If X is a connected CW complex, it is of "descent type" for this adjoint pair, that is,

$$X\Omega\Sigma\Omega\Sigma \underset{\Omega\Sigma\epsilon}{\overset{\epsilon\Omega\Sigma}{\rightrightarrows}} X\Omega\Sigma \xrightarrow{\ \epsilon\ } X$$

is a coequalizer diagram. Restricted to spaces for which this diagram is a coequalizer, $\Phi$ is full. For a $\Sigma\Omega$-space to be "effective", letting $B = (X,\xi) \otimes_{\Sigma\Omega}\Sigma$ be the coequalizer

$$X\Sigma\Omega\Sigma \underset{\Sigma\epsilon}{\overset{\xi\Sigma}{\rightrightarrows}} X\Sigma \longrightarrow B,$$

the composition $X \to X\Sigma\Omega \to B\Omega$ would have to be a homeomorphism. Yet from experience it is unreasonable to expect this map to be better than a homotopy equivalence. If even this were so, B would be a <u>classifying space</u> for the $\Sigma\Omega$-algebra $(X,\xi)$ in the usual sense of homotopy theory. But as a caution: if X is a discrete group and $\xi: X\Sigma\Omega \to X$ is the structure induced by the group law, then B = 0. Perhaps for connected X the result is better, but connectivity would be an awkward assumption later on.

Before resolving the difficulty, worsen it by considering the general case of a $\Sigma^k\Omega^k$-space $(X,\xi)$ where k is any integer > 0; thus $\xi$ is a unitary, associative structure map $X\Sigma^k\Omega^k \to X$.

Regard the simplicial space

$$X\,\Sigma^k(\Omega^k\Sigma^k)^n, \qquad n\geq 0,$$

with continuous face operators

$$X\,\Sigma^k(\Omega^k\Sigma^k)^{n+1} \xrightarrow{\ \epsilon_i\ } X\,\Sigma^k(\Omega^k\Sigma^k)^n, \qquad 0\leq n$$

$$\epsilon_i = \begin{cases} \xi\Sigma^k(\Omega^k\Sigma^k)^{n-1}, & i=0,\\ X\,\Sigma^k(\Omega^k\Sigma^k)^{i-1}\epsilon(\Omega^k\Omega^k)^{n+i-1}, & 1\leq i\leq n,\end{cases}$$

where $\epsilon$ is the adjointness map $\Omega^k\Sigma^k \to$ id., and degeneracy operators $\eta_i$ similarly defined in terms of $\eta$: id. $\to \Sigma^k\Omega^k$. It is intuitively reasonable to replace the coequalizer above with

$$B_k = \text{geometrical realization of } X\,\Sigma^k(\Omega^k\Sigma^k)^*$$
$$= \sum_{n\geq 0} \Delta_n \times X\,\Sigma^k(\Omega^k\Sigma^k)^n \;/\; \otimes\text{-identities},$$

exactly as defined originally by Milnor (Ann. of Math. <u>65</u> (1957), 357-362). For example, $B_0 = X$.

Using the distributivity of $\Omega^k$ over the realization identifications, we have a natural map

$$\text{geom. realiz. } (X\,\Sigma^k(\Omega^k\Sigma^k)^*\Omega^k) \to B_k\Omega^k.$$

By an elaboration of (5), there is also a natural homotopy equivalence of X into the above geometrical realization, hence by composition a map $X \to B_k\Omega^k$.

(16) <u>Theorem</u>. Every $\Sigma^k\Omega^k$-space $(X,\xi)$ has a k-classifying space. Precisely, the above map $X \to B_k\Omega^k$ is a $\Sigma^k\Omega^k$-map and a weak homotopy equivalence (homotopy equivalence if X has the homotopy type of a CW complex).

It suffices to prove $X \to B_k\Omega^k$ is a homology equivalence. When k=0 this is X=X, and when k>0 iterated cobar constructions are applied.

We can also de-loop in the limit:

(17) <u>Theorem</u>. Every Q-space $(X,\xi)$ has a classifying space B which is also a Q-space. Precisely, there exist a Q-space B which is a functor of X and a natural map $(X,\xi)\to B\Omega$ which is a Q-homomorphism relative to the induced Q-structure on $B\Omega$ and is a homotopy equivalence.

Each $\Sigma^k\Omega^k \to Q$ is a triple map, so X has induced $\Sigma^k\Omega^k$-structures $\xi_k$ for k>0. Let $B_k$ be the "classifying spaces" for these (16). Maps $B_k \to B_{k+\ell}\,\Omega^\ell$ are easily obtained such that the diagrams

$$\begin{array}{ccc} & X & \\ \swarrow & & \searrow \\ B_k\Omega^k & \longrightarrow & B_{k+\ell}\Omega^{\ell}\Omega^k \end{array} \qquad \ell > 0$$

commute. As $B_k$ contains no cells of dimensions $< k$, $B_k \to B_{k+\ell}\Omega^{\ell}$ is a homotopy equivalence.

For the proof of (17), let $B = \underrightarrow{\lim}(B_{k+1}\Omega^k)$ as $k \to \infty$. B has compatible $\Sigma^k\Omega^k$-structures for all $k>0$ by the direct limit preserving property of $\Sigma^k\Omega^k$, hence B has a natural Q-structure. BΩ also has a natural Q-structure via the transposition $\Omega\Sigma \to \Sigma\Omega$ which gives rise to compatible $\Sigma^k\Omega^k$-structures:

$$B\Omega\Sigma^k\Omega^k \to B\Sigma^k\Omega^k\Omega \to B\Omega.$$

These $\Sigma^k\Omega^k$-structures coincide with those defined "internally" by the fact that $B\Omega = \underrightarrow{\lim}(B_{k+\ell}\Omega^{k+\ell})$. Using the "internal" point of view, $X \to B\Omega$ is seen to be a Q-map, and it is obviously a homotopy equivalence.

We could have de-looped k times at once by using $B = \underrightarrow{\lim}(B_{k+\ell}\Omega^{\ell})$. If B is erroneously defined as the "telescope" of

$$B_1 \to B_2\Omega \to B_3\Omega^2 \to \dots ,$$

an example of a "Q-space up to canonical homotopies" results. This might prove to be a useful concept for the triple Q, and for other topological triples. In contrast to "homotopy-everything" structures, Q-structures are not transportable along homotopy equivalences.

(18) Dualizing the foregoing produces a rather striking phenomenon; we mean dualizing in both the categorical and Eckmann-Hilton senses. The composition $\Omega\Sigma$ is a cotriple in Top, and an $\Omega\Sigma$-costructure on a space X is a counitary, coassociative map $\sigma: X \to X\Omega\Sigma$. By adjointness every suspension canonically has such a costructure. Does the existence of a costructure imply that X is a suspension? Although in the case of loop spaces the general "tripleableness" theorem was more or less useless, in this instance the dual "cotripleableness" theorem, or a simple manual approach, shows that every X with an $\Omega\Sigma$-structure is canonically homeomorphic to a suspension. This fact, which also holds for the cotriples $\Omega^k\Sigma^k$, was pointed out by Luke Hodgkin.

# FUNCTORS BETWEEN CATEGORIES OF VECTOR SPACES

by

D. B. A. Epstein*

and

M. Kneser**

Let K and L be fields. We consider the problem of classifying functors from $\underline{\underline{V}}(K)$, the category of finite dimensional K-vector spaces and K-linear maps, to $\underline{\underline{V}}(L)$. For any two categories $\underline{\underline{A}}$ and $\underline{\underline{B}}$ and for any object B of $\underline{\underline{B}}$, we have a functor from $\underline{\underline{A}}$ to $\underline{\underline{B}}$, which assigns to each object of $\underline{\underline{A}}$ the object B, and to each morphism of $\underline{\underline{A}}$ the identity map $1_B$. Any functor isomorphic to such a functor will be called a <u>constant</u> functor.

The following results are substantial improvements on the results in [2].

<u>Theorem 1</u>

Let $F: \underline{\underline{V}}(K) \longrightarrow \underline{\underline{V}}(L)$ be a non-constant functor, and let K be infinite. Then K and L have the same characteristic and L is infinite.

If K is finite then Theorem 1 is false. For we

* University of Warwick, Coventry, England

** Göttingen, Germany

have the forgetful functor from $\underline{\underline{V}}(K)$ to the category of finite sets, and for any field L there are many functors from the category of finite sets to $\underline{\underline{V}}(L)$. This remark is due to A. Borel. We shall therefore assume throughout this paper that K is infinite.

<u>Theorem 2</u>

Let K and L have characteristic zero and let $F: V(K) \longrightarrow V(L)$ be a non-constant functor. Then for each integer $n \geq 0$, there exists a functor

$$G_n: \underline{\underline{V}}(K) \times \ldots \times \underline{\underline{V}}(K) = \underline{\underline{V}}(K)^n \longrightarrow \underline{\underline{V}}(L)$$

and a functor $F_n: \underline{\underline{V}}(K) \longrightarrow \underline{\underline{V}}(L)$, such that

i) $G_n$ is additive in each of its n variables

ii) $F_n$ is a subfunctor of $G_n \cdot \Delta_n$ where

$\Delta_n: \underline{\underline{V}}(K) \longrightarrow \underline{\underline{V}}(K)^n$

is the diagonal functor

iii) $F = \oplus F_n$ .

<u>Note</u>. When $n = 0$, $\underline{\underline{V}}(K)^n$ is defined as the category with one object and one morphism. So $G_0$, $\Delta_0$ and $F_0$ are constant functors.

Corollary 3

K can be embedded in a finite field extension of L.

Corollary 4

If $K = \mathbb{Q}$, the field of rationals, then F is a polynomial functor. Such functors are completely classified in [2].

Corollary 3 is a consequence of Theorem 2. The easy proof is given in Lemma 5. Corollary 4 follows from [2] Lemma 7.2.

Since FV is naturally isomorphic to $F0 \otimes \ker(FV \longrightarrow F0)$, we may assume without loss of generality that $F0 = 0$. (The constant functor $V \longmapsto F0$ corresponds to the case $n = 0$ in Theorem 2. That is $F_0V = F0$ .)

We now commence the proofs of the theorems.

Lemma 5

Let G be an abelian group and W a finite dimensional vector space over L. Let $\theta: G \longrightarrow \mathrm{Aut}_L W$ be a homomorphism. Then the smallest extension field $L_1$ of L, which contains all eigenvalues of $\theta g$ for all $g \in G$, is a finite extension. Moreover there is a basis of $W \otimes_L L_1$ with respect to which all elements of $\theta G$ are upper triangular.

Proof. We use induction on the dimension of W. Suppose that for some element $g \in G$, $\theta g$ is not scalar multiplication. Let $\lambda$ be an eigenvalue of $\theta g$. Without loss of generality, we may suppose that $\lambda \in L$. Then

$$W \neq W_1 = \{w \in W \mid (\theta g)w = \lambda w\} \neq 0$$

is a representation space for G. We apply the induction hypothesis to $W_1$ and $W/W_1$. This completes the proof.

We use this lemma to show that Theorem 2 implies Corollary 3. Since F is not constant, $G_n \neq 0$ for some $n \geq 1$. Therefore $M = G_n(K,\ldots,K) \neq 0$. Now M is an L-vector space which is also a vector space over K (via the action on the first variable, for example). By Lemma 5 we may find a finite field extension $L_1$ of L and an $L_1$-basis for $M \otimes_L L_1$, with respect to which the action of any element $k \in K$ is upper triangular. Each diagonal position then gives rise to a field embedding of K in $L_1$.

Lemma 6

Let G be a nilpotent group and let L be an algebraically closed field. Let W be a finite dimensional vector space and let $\theta: G \longrightarrow \mathrm{Aut}_L W$ be a homomorphism. Suppose that G has no finite cyclic quotient group. Then it is possible to choose a basis for W, so that each element $g \in G$ acts as an upper triangular matrix.

Proof. We need only show that $G$ acts by scalar multiplication for every simple G-module $W$ which is finite dimensional over $L$. Let $\{1\} = G_0 < G_1 < \ldots < G_r = G$ be a central series for $G$, and suppose we have shown by induction on $i$, that $G_i$ acts on $W$ by scalar multiplication. Let $\theta: G \longrightarrow \mathrm{Aut}_L W$ be the representation.

Let $x \in G_{i+1}$ and $g \in G$. Then $xgx^{-1}g^{-1} \in G_i$ and so we can define $\lambda(x,g) \in L^*$ by $\lambda(x,g) = \theta(xgx^{-1}g^{-1})$ or

$$\theta x \cdot \theta g = \theta g \cdot \theta x \lambda(x,g) .$$

Taking determinants, we see that $\lambda(x,g)$ is a $k^{th}$ root of unity, where $k = \dim W$. It is obvious that for fixed $x$, $\lambda(x,\ )$ gives a homomorphism of $G$ into $L^*$, and hence into the group of $k^{th}$ roots of unity in $L$. But the $k^{th}$ roots of unity form a finite cyclic group, and so $\lambda(x,g) = 1$ for all $x \in G_{i+1}$ and all $g \in G$. Hence $\theta x$ commutes with the action of $G$ on $W$. By Schur's Lemma, $\theta x$ is therefore scalar multiplication. This completes the proof of the lemma.

We recall that a linear map $A: W \longrightarrow W$ is called unipotent if $(A - 1)$ is nilpotent, i.e. if $(A - 1)^r = 0$ for large enough $r$. This is equivalent to being able to find a basis for $W$, with respect to which $A$ is unitriangular (upper triangular with ones down the diagonal).

Theorem 7

Let $K$ be an infinite field and let $\dim_K V > 2$. Let $SL(V,K)$ be the group of automorphisms with determinant one. Let $\theta: SL(V,K) \longrightarrow Aut_L W$ be a non-trivial homomorphism. Then

i) $L$ is infinite;

ii) $K$ and $L$ have the same characteristic;

iii) $\theta$ maps unipotent elements to unipotent elements.

iv) In fact, if we fix a basis for $V$, then there exists a basis for $W$ such that $\theta$ maps unitriangular matrices to unitriangular matrices, but we do not prove this.

Proof. Every normal subgroup of $SL(V,K)$ is contained in the group of scalar multiplications by the $r^{th}$ roots of unity $(\dim V = r)$ [1] p. 38. So $SL(V,K)$ has only trivial homomorphisms into finite groups. In particular $L$ must be infinite.

Let $K$ have characteristic $p$. Then the unipotent elements are exactly those whose order is a power of $p$. So if $L$ has characteristic $p$, then unipotent elements are mapped to unipotent elements. The matrices of the form $1 + xE_{12} (x \in K)$ form an infinite abelian group $H$ of exponent $p$. By Lemma 5, we may choose a basis for $W \otimes_L \bar{L}$, so that $\theta H$ consists of upper triangular matrices. The diagonal entries of an element of $\theta H$ are $p^{th}$ roots of unity. Hence the subgroup $S$ of $H$ consisting of elements mapping under $\theta$ to unitriangular elements, is non-trivial. On the other hand, if the characteristic of $L$ is not $p$, $S = 1$. It follows, as in the first paragraph, that $\theta$ is trivial. So we have proved that if $K$ has characteristic $p$, then so has $L$, and unipotent elements are mapped to unipotent elements.

Now let $K$ have characteristic zero. Without loss of generality, we may suppose that $L$ is algebraically closed (which would not be legitimate if we were proving iv). We now apply Lemma 6, to deduce that unitriangular matrices in $SL(V,K)$ are sent to upper triangular matrices in $\mathrm{Aut}_L W$. Let $i < j < k$. Then the commutator

$$(1 + \lambda E_{ij})(1 + E_{jk})(1 - \lambda E_{ij})(1 - E_{jk}) = 1 + \lambda E_{ik} .$$

It follows that $\theta(1 + \lambda E_{ik})$ is a commutator of upper triangular matrices and is therefore unitriangular. So $\theta(1 + \lambda E_{ik})$ is unipotent.

Changing the basis of $V$, we see that $\theta(1 + \lambda E_{ij})$ is unipotent for all $i \neq j$. It follows that for $i < j$, $\theta(1 + \lambda E_{ij})$ is unitriangular. Since the elements $1 + \lambda E_{ij}$ generate the group of unitriangular matrices in $SL(V,K)$, we see that $\theta$ maps unitriangular matrices to unitriangular matrices, and hence unipotent elements to unipotent elements.

We therefore have $\theta(1 + xE_{12}) = 1 + \Sigma_{i<j}\alpha_{ij}(x)E_{ij}$, where $\alpha_{ij}: K \longrightarrow L$. Since $\theta$ is non-trivial on $SL(V,K)$, some $\alpha_{ij}$ must be non-zero. We pick a pair of integers $i < j$, with $\alpha_{ij}$ non-zero and $j - i$ minimal. Then $\alpha_{ij}$ is an additive homomorphism of the divisible abelian group $K$ into $L$. It follows that $L$ has characteristic zero, and the proof of Theorem 7 is complete.

We can now deduce Theorem 1. If Theorem 1 is false, then by Theorem 7, each homomorphism $SL(V,K) \longrightarrow Aut_L FV$ induced by $F$ is trivial. Let $V$ be a vector space of even dimension, such that $FV \neq 0$. Let $i,j: V \longrightarrow V \oplus V$ be the canonical injections and $p,q: V \oplus V \longrightarrow V$ be the canonical projections. Let $\alpha = jp + iq$. Then $\alpha$ has determinant one and $\alpha^2 = 1$. We have $0_{V\oplus V} = ip\,\alpha\, ip\,\alpha$. Applying $F$ and remembering that $F\alpha = 1_{V\oplus V}$, we have $0 = F(ip)^2 = F(ip)$. Now $1_V = pi\,pi$. Therefore

$$1_{FV} = F(1_V) = F(p)\,F(ip)\,F(i) = 0 ,$$

which is a contradiction.

We assume from now on that $K$ and $L$ have characteristic zero. For any nilpotent endomorphism $N$ of $W$, we can define $\exp N$ and $\log(1 + N)$ with the usual power series. The functions exp and log give inverse bijections between the set of unipotent endomorphisms and the set of all nilpotent endomorphisms.

Lemma 8

Let $G$ be the group of automorphisms of $V$ of the form $(1 + xN)$, where $N$ is a fixed endomorphism of $V$ with $N^2 = 0$ and $x \in K$. Let $\theta: G \longrightarrow GL(m,L)$ be a homomorphism which maps $G$ into unipotent matrices. Let $\theta_{ij}: K \longrightarrow L (1 \leq i,j \leq m)$ be the function defined by $\theta_{ij}(x) = \theta(1 + xN)_{ij}$. Then $\theta_{ij}$ is a sum of products of additive homomorphisms from $K$ to $L$.

Proof. $\theta(1 + xN) = \exp \log \theta(1 + xN)$. Now $x \longrightarrow \log \theta(1 + xN)$ is an additive homomorphism into the additive group of $(m \times m)$ matrices over $K$. The result follows by expanding the exponential series (which is zero after a finite number of terms).

Lemma 9

Let $x_V: V \longrightarrow V$ be scalar multiplication by $x \in K$. There exist endomorphisms $A_i$ of $FV (1 \leq i \leq r)$ and functions $\alpha_i: K \longrightarrow L$ such that $\alpha_i$ is a sum of products of additive homomorphisms and $F(x_V) = \Sigma_{i=1}^{r} \alpha_i(x) A_i$ .

Proof. Let $i,j: V \longrightarrow V\oplus V$ be the canonical injections and $p,q: V\oplus V \longrightarrow V$ be the canonical projections. Then, $F(V\oplus V)$ has $FV\oplus FV$ as a direct summand. The first factor has as its canonical injection and projection the maps $Fi$ and $Fp$, and the second factor the maps $Fj$ and $Fq$.

We consider the subgroup of $\mathrm{Aut}_K(V\oplus V)$ consisting of elements of the form $1 + xiq (x \in K)$. By Theorem 7, $F(1 + xiq)$ is unipotent for all $x \in K$. By Lemma 8, if we choose a basis for $F(V\oplus V)$, then each entry in the matrix of $F(1 + xiq)$ is a sum of products of additive homomorphisms of $K$ into $L$.

Now $Fp \cdot F(1 + xiq) \cdot Fj = F(x_V)$. The lemma follows.

We now apply Lemma 5, with $G = \mathbb{Q}^* \subset K^*$. Here $\mathbb{Q}$ is the field of rational numbers. $G$ acts on $FV$ by $\lambda \longmapsto F(\lambda_V)$. We can choose a basis for $FV\otimes_L L_1$, such that $F(\lambda_V)$ is upper triangular for each $\lambda \in \mathbb{Q}$. Now any additive homomorphism $\mathbb{Q} \longmapsto L$ has the form $\lambda \longmapsto a\lambda$ for some $a \in L$. By Lemma 9 each entry in the matrix of $F(\lambda_V)$ is a polynomial function in $\lambda$, with coefficients in $L$ and zero constant term. The diagonal entries are multiplicative homomorphisms $\mathbb{Q} \longmapsto L$. But every multiplicative homomorphism, which is polynomial, has the form $\lambda \longmapsto \lambda^i$ for some $i \geq 0$. Hence the diagonal entries of $F(\lambda_V)$ are all of the form $\lambda^i$, where the value of $i$ may depend on the position of the entry. It follows that $L_1$ (the extension field of $L$ described in Lemma 5), is in fact equal to $L$.

For each integer $i > 0$, we define

$$F_i V = \{w \in FV \mid (F\lambda_V - \lambda^i)^N w = 0 \text{ all } \lambda \in \mathbb{Q} \text{ and } N = \dim FV\}$$

It is easy to see that if $\alpha: V \longrightarrow W$ is linear, then $F_\alpha$ carries $F_i V$ into $F_i W$. We have $F \cong \oplus_i F_i$.

For the sake of completeness, we repeat some material contained in [2] concerning deviation functors, which is a notion due to Eilenberg and MacLane [3].

Let $\underline{C}$ be an arbitrary category with finite products and a zero object, and let $\underline{A}$ be an abelian category. Let $F: \underline{C} \longrightarrow \underline{A}$ be a functor such that $F0 = 0$. If $C$ and $D$ are two objects in $\underline{C}$, we have canonical injections and projections

$$i: C \longrightarrow C \times D,\ j: D \longrightarrow C \times D,\ p: C \times D \longrightarrow C,\ q: C \times D \longrightarrow D,$$

such that $pi = 1_C$, $qj = 1_D$, $pj = 0_{DC}$, $qi = 0_{CD}$. Let $F^1(C,D) = \ker Fp \cap \ker Fq$. We have a direct sum decomposition, which is natural for morphisms $C \longrightarrow C^1$ $D \longrightarrow D^1$.

Lemma 10

$F(C \times D) \cong FC \oplus FD \oplus F^1(C,D)$. The projections on to the three factors are $FP$, $Fq$ and $1 - Fj \cdot Fq - Fi \cdot Fp$.

The injections are Fi, Fj and inclusion.

Lemma 11

If $F_1 \longrightarrow F_2$ is a morphism of functors, then we obtain an induced morphism $(F_1)^1 \longrightarrow (F_2)^1$ of functors from $\underline{\underline{C}}^2 = \underline{\underline{C}} \times \underline{\underline{C}}$ to $\underline{\underline{A}}$, which respects the direct sum decomposition 10.

If $G: \underline{\underline{C}}^k \longrightarrow \underline{\underline{A}}$ is a functor of $k$ variables, one can perform the above process on the $i^{th}$ variable for some fixed $i$, to obtain a functor $G^i: \underline{\underline{C}}^{k+1} \longrightarrow \underline{\underline{A}}$. (We use Lemma 11 for this.) Suppose $F$ is as above, and we have defined $F^I: \underline{\underline{C}}^{k+1} \longrightarrow \underline{\underline{A}}$, where $I = \{i_1,\ldots,i_k\}$ is a k-tuple of integers such that $1 \leq i_j \leq j$ for each $j$. Then we define

$$F^J = (F^I)^j: \underline{\underline{C}}^{k+2} \longrightarrow \underline{\underline{A}}$$

where $J = \{i\ ,\ldots,i_k,j\}$ and $1 \leq j \leq k+1$ .

If $C = C_1 \times \ldots \times C_n$ and $K$ is a subset of $\{1,2,\ldots,n\}$, we denote by $\psi_K: C \longrightarrow C$ the morphism such that $p_i\psi_K = p_i$ for $i \in K$ and $p_i\psi_K = 0$ for $i \notin K$. Let $|K|$ be the number of elements of $K$. It is easy to prove, by induction on the length $n$ of n-tuple $I = \{i_1,\ldots,i_n\}$ such that $1 \leq i_j \leq j$ for each $j$, that

$$F^I(C_1,\ldots,C_n) = \mathrm{Im}(\Sigma_K(-1)^{|K|}F(\psi_K)) \subset FC ,$$

where $K$ runs over all the subsets of $\{1,\ldots,n\}$. Hence $F^I$ depends only on the length of $I$ and is independent of the order of the variables $C_1,\ldots,C_n$. We define

$$F^{(n)}(C_1,\ldots,C_n) = \mathrm{Im}(\Sigma_K(-1)^{|K|}F(\psi_K)) .$$

$F^{(n)}$ is called the <u>$n^{th}$ deviation</u> of $F$. We obviously have:

<u>Lemma 12</u>

$F^{(n)}$ is additive in each variable if and only if $F^{(n+1)} = 0$ .

Now we turn to functors $F: \underline{V}(K) \longrightarrow \underline{V}(L)$, and we suppose that $F0 = 0$, as we can do (see just before Lemma 5). A functor $F$ will be called <u>homogeneous of degree i</u> if for each vector space $V$ over $K$, and each $\lambda \in \mathbb{Q}$, the eigenvalues of $F\lambda_V$ are all equal to $\lambda^i$ . We have shown above (just before the section on deviation functors), that if $K$ and $L$ have characteristic zero, then every functor is the direct sum of homogeneous functors. In order to prove Theorem 2, we may therefore assume $F$ is homogeneous.

<u>Lemma 13</u>

If $F$ is homogeneous of degree n, then $F^{(n)} \neq 0$ and $F^{(n+1)} = 0$ .

<u>Proof</u>. If $A$ and $B$ are commuting endomorphisms of a vector space, then every eigenvalue of $AB$ is the

product of an eigenvalue of $A$ and an eigenvalue of $B$. Regarding $F^{(k)}$ as a functor of the $i^{th}$ variable only, we have shown that $F^{(k)}(1,\ldots,1,\lambda,1,\ldots,1)\,(\lambda \in \mathbb{Q})$ has each of its eigenvalues of the form $\lambda^{n_i}$. It follows that the eigenvalues of $F^{(k)}(\lambda,\ldots,\lambda)$ are of the form $\lambda^{n_1+n_2+\ldots+n_k}$. Since $F^{(k)}(V,\ldots,V)$ is a subfunctor of $F(V\oplus \ldots \oplus V)$, it follows that $n_1 + \ldots + n_k = n$. Since $n_i \geq 1$ for $1 \leq i \leq k$, we deduce that $F^{(n+1)} = 0$. Let $k$ be the largest integer such that $F^{(k)} \neq 0$. Then $F^{(k)}$ is additive in each variable by Lemma 12 and so each $n_i$ is equal to one.

Lemma 14

Let $F\colon \underline{\underline{V}}(K) \longrightarrow \underline{\underline{V}}(L)$ be homogeneous of degree $n$. Then $F$ can be embedded in the direct sum $G$ of $(n - 1)!$ copies of $F^{(n)} \cdot \Delta_n$, where $\Delta_n\colon \underline{\underline{V}}(K) \longrightarrow \underline{\underline{V}}(K)^n$ is the diagonal functor. This embedding is natural in $F$ - that is, if $\alpha\colon F_1 \longrightarrow F_2$ is a morphism of functors of degree $n$, then we have a commutative diagram

$$\begin{array}{ccc} F_1 & \xrightarrow{\ \alpha\ } & F_2 \\ \downarrow & & \downarrow \\ G_1 & \xrightarrow{\ \beta\ } & G_2 \end{array}$$

where $\beta$ is induced by $\alpha^{(n)}$.

Proof. Consider the composition

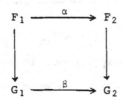

$$\begin{array}{ccc} & V \oplus V & \\ {}^{\Delta}\nearrow & & \searrow^{m} \\ V & \xrightarrow{\;2\;} & V \end{array}$$

where $\Delta$ is the diagonal and $m$ is addition. We have $F(2_V)$ expressed as the composition

$$FV \xrightarrow{F\Delta} FV \oplus FV \oplus F^{(2)}(V,V) \xrightarrow{Fm} FV$$

which is equal to $F1_V + F1_V + Fm \cdot \gamma_V$ where $\gamma_V \colon FV \longrightarrow F^{(2)}(V,V)$ is equal to $(1 - FiFp - FjFq)F\Delta = F\Delta - Fi - Fj$. So $\gamma_V$ gives rise to a morphism of functors $\gamma$ from $\underline{V}(K)$ to $\underline{V}(L)$. Moreover, if $\alpha\colon F_1 \longrightarrow F_2$ is a morphism of functors, then we have a commutative diagram

$$\begin{array}{ccc} F_1V & \xrightarrow{\;\alpha\;} & F_2V \\ \downarrow{\scriptstyle\gamma} & & \downarrow{\scriptstyle\gamma} \\ F_1^{(2)}(V,V) & \xrightarrow{\;\alpha^{(2)}\;} & F_2^{(2)}(V,V) \end{array}$$

We know that $F(2_V) = 1_{FV} + 1_{FV} + Fm \cdot \gamma_V$. Without loss of generality, we may suppose that $n > 1$. Then all the eigenvalues of $Fm \cdot \gamma_V$ are equal to $2^n - 2$ and so $\gamma_V$ is a monomorphism for each $V$. Hence $F$ is embedded in $F^{(2)} \cdot \Delta_2$.

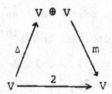

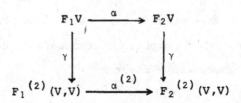

The lemma is now proved by induction on $n$. We can write $F^{(2)} = \oplus_{r=1}^{n-1} T_{r,n-r}$, where

$$T_{r,n-r}: \underline{\underline{V}}(K) \longrightarrow \underline{\underline{V}}(L)$$

has degree $r$ in the first variable and degree $n - r$ in the second variable. This is done by regarding $F^{(2)}$ as a functor of the first variable only and writing it as a sum of homogeneous functors. (We recall that $F^{(2)}(V,V) \subset F(V \oplus V)$, so that the degrees in the two variables must add up to $n$.)

Let $G: \underline{\underline{V}}(K)^2 = \underline{\underline{V}}(K) \times \underline{\underline{V}}(K) \longrightarrow \underline{\underline{V}}(L)$. We define $G^{(r,s)}: \underline{\underline{V}}(K)^{r+s} \longrightarrow \underline{\underline{V}}(L)$ by taking the $r^{th}$ deviation with respect to the first variable and the $s^{th}$ deviation with respect to the second variable. By induction on $n$, using the naturality of the embeddings for lower values of $n$, we embed $T_{r,n-r}$ in $(r - 1)!(n - r - 1)!$ copies of $(T_{r,n-r})^{(r,n-r)} \cdot (\Delta_r \times \Delta_{n-r})$. From Lemma 13, it follows that $T_{r,n-r})^{(r,n-r)} = (F^{(2)})^{(r,n-r)} = F^{(n)}$. Hence $T_{r,n-r}$ may be embedded in $(r - 1)!(n - r - 1)!(\leq (n - 2)!)$ copies of $F^{(n)} \cdot \Delta_r \times \Delta_{n-r}$. It follows that $T_{r,n-r} \cdot \Delta_2$ may be embedded in the direct sum of $(r - 1)!(n - 1 - r)!$ copies of $F^{(n)} \cdot \Delta_n$. Therefore $F \subset F^{(2)} \cdot \Delta_2$ may be embedded in $(n-1)!$ copies of $F^{(n)} \cdot \Delta_n$. This completes the proof of Theorem 2.

## REFERENCES

[1] Dieudonné, J., "Sur la géométrie des groupes classiques", Springer, 1955.

[2] Epstein, D. B. A., "Group representations and functors", to appear in *Am. J. Math.*

[3] Eilenberg, S., and MacLane, S., "On the groups $H(\pi,n)$, II", *Annals of Math.*, 60; 75-83, (1954).

# NATURAL VECTOR BUNDLES

by

D. B. A. Epstein

## 1. DEFINITIONS AND RESULTS

Let $0 \leq r \leq s \leq \infty$ be integers. An <u>(r,s) natural vector bundle</u> is a functor $V$ which assigns to each $C^s$ manifold $M$ a $C^r$ vector bundle $\pi_M: VM \longrightarrow M$ and to each $C^s$ map $f: M \longrightarrow N$ a $C^r$ map $Vf: VM \longrightarrow VN$, which is linear on fibres and makes the diagram

$$\begin{array}{ccc} VM & \xrightarrow{Vf} & VN \\ \downarrow{\scriptstyle \pi_M} & & \downarrow{\scriptstyle \pi_N} \\ M & \xrightarrow{f} & N \end{array}$$

commutative. (In this paper, a manifold will have no boundary, will have a countable basis, but need not be connected or compact. The dimension of the fibre of a vector bundle will be assumed not to vary from one component to another.)

Examples are the tangent bundle $(r = s - 1)$ and the bundle of $k^{th}$ order differential operators $(r = s - k)$. New examples can be generated by taking tensor products, direct sums etc. of existing examples. Given a vector bundle $E \longrightarrow M$, $J^kE$ is the bundle of $k$-jets of sections of $E$. Given a natural vector bundle $V$, we obtain a new natural vector bundle $(J^k(V^*))^*$.

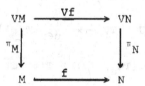

1.1. Let VLin be the category of finite dimensional real vector spaces and linear maps. Let

$$T: \underline{\text{VLin}} \longrightarrow \underline{\text{VLin}}$$

be continuous. By [1] 1.7, $T$ is polynomial. Then, given any $C^r$ vector bundle $\pi: E \longrightarrow M$, we can construct a new vector bundle $TE \longrightarrow M$ as follows. The underlying point set of $TE$ is $\cup_{x \in M} T(\pi^{-1}x)$, and $T(\pi^{-1}x)$ is the fibre over $x$. Given any open set $U$ of $M$, over which $E$ is trivial, we have an isomorphism of vector bundles $G: U \times V \longrightarrow \pi^{-1}U$ over $U$. We give $TE$ a topology and differential structure, by insisting that

$$TG: U \times TV \longrightarrow \cup_{u \in U} T(\pi^{-1}u)$$

defined by $TG(u,v) = (TG_u)(v) \in T(\pi^{-1}u)$, be a $C^r$ isomorphism. If $V$ is a natural vector bundle, then we can define a new natural vector bundle $TV$ by $M \longrightarrow T(VM)$.

Definition 1.2

We say that an (r,s) natural vector bundle is continuous, if, for any $C^s$ manifolds $P$, $M$ and $N$ and for any $C^s$ map $f: P \times M \longrightarrow N$, the induced map $F: P \times VM \longrightarrow VN$, defined by

$$F(p,w) = V(f_p)w$$

is $C^r$.

Definition 1.3

A natural vector bundle $V$ is said to be myopic if $\dim M = \dim N$ implies that $\dim VM = \dim VN$. (This name is due to P. Freyd.)

Theorem 1.4

Every natural vector bundle is continuous and myopic.

Theorem 1.5

Let $V$ be an $(r,r)$ natural vector bundle, where $r < \infty$. Then there is a vector space $W$, wuch that $V$ is isomorphic to the constant natural vector bundle $M \longrightarrow M \times W$.

Theorem 1.6

Let $V$ be an $(r,s)$ natural vector bundle. Then $V$, restricted to $C^\infty$ manifolds and maps, is isomorphic to a unique $(\infty,\infty)$ natural vector bundle.

Theorem 1.7

Let $V$ be an $(r,s)$ natural vector bundle. Then $V$ is filtered by $(r,s)$ natural subbundles

$$0 = V_{-1} \subseteq V_0 \subseteq V_1 \subseteq$$

such that

i) For a fixed manifold $M$, $V_r M = VM$ for $r$ large.

ii) For each integer $i \geq 0$, $V_i/V_{i-1} \cong F_i\tau$, where

$\tau$ is the tangent bundle and $F_i$ is a homogeneous continuous functor of degree $i$ from VLin to VLin. (To say $F_i$ has degree $i$ means that $F_i$ sends scalar multiplication by $\lambda$ to scalar multiplication by $\lambda^i$. See [1] for the classification of such functors.)

iii) Let $f,g: M,m \longrightarrow N,n$ have the same $i$-jet at $m$. Then $Vf$ and $Vg$ induce the same map $(V_jM)_m \longrightarrow (V_jN)_n$ if either $j \leq i$ or if $i = s$.

## 2. CONTINUITY

In this section we begin the proof that a natural vector bundle is continuous (see 1.2). The following lemma provides the necessary point set topology.

Lemma 2.1 Let X be a complete metric space and let Y be a Hausdorff topological space. Let $y \in Y$ be a point with a countable basis of neighbourhoods. Let $f: X \to Y$ be a function with the following property. For each $x \in X$ and each neighbourhood W of $\{y, fx\}$ there is a neighbourhood V of x such that $fV \subseteq W$. Then f is continuous at a dense set of points in X.

Proof: Let $U_1, U_2, U_3 \ldots$ be a countable base of open neighbourhoods of y. Let

$$X_i = \{x \mid fx \in U_1 \text{ or } f \text{ is continuous at } x\}$$

We claim that $X_i$ is open. In fact it is obvious that $\{x \mid fx \in U_i\}$ is open, by the conditions stated. Let f be continuous at x, let $W_1$ and $W_2$ be disjoint open neighbourhoods of fx and y respectively and let V be an open neighbourhood of x such that $f V \subseteq W_1$. Then f is continuous at each point x' of V. For let $W_3$ be any neighbourhood of fx'. By the hypothesis we can find an open neighbourhood V' of x' such that $f V' \subseteq W_2 \cup (W_3 \cap W_1)$. Then we obviously have $f(V \cap V') \subseteq W_3 \cap W_1$.

We also claim that $X_i$ is dense. For suppose $x \notin X_i$. Then $fx \neq y$. There are disjoint open neighbourhoods $W_1$ and $W_2$ of fx and y respectively, such that $W_2 \subseteq U_i$ and for each neighbourhood V of x, $f V \not\subseteq W_1$. By the hypothesis $W_2$ meets fV for each neighbourhood V of x. Hence V meets $f^{-1} U_i \subseteq X_i$.

Now let $X' = \cap_i X_i$. By the Baire Category Theorem X' is dense in X. If $x \in X'$ then f must be continuous at x. For suppose not. Then $f x \in U_i$ for each i and so fx = y. The continuity of f then follows from the hypothesis and this is a contradiction. This completes the proof of the lemma.

We wish to show that V is a continuous (r,s) natural vector bundle. We need only show that for all $C^s$ manifolds P and M, the map

$$\Phi_{P,M}\colon P \times VM \to V(P \times M)$$

defined by $\Phi(p,w) = V(i_p)w$, is $C^r$. Here $i_p\colon M \to P \times M$ is defined by $i_p(m) = (p,m)$. By restricting our attention to a neighbourhood of P, we may in fact assume that $P = R^n$. The proof that $\Phi_{P,M}$ is $C^r$ will take place in a number of steps.

Lemma 2.2

$\Phi_{\mathbb{R},M}: \mathbb{R} \times VM \longrightarrow V(\mathbb{R} \times M)$ is continuous in the first variable.

Proof. Let $w \in VM$ lie over $m \in M$. Let $\pi: V(\mathbb{R} \times M) \longrightarrow \mathbb{R} \times M$ be the projection of the vector bundle and let $Y$ be the one point compactification of the Euclidean space $\pi^{-1}(\mathbb{R} \times m)$. $Y$ is homeomorphic to a sphere, and the usual metric on the sphere makes $Y$ into a uniform space.

Let $f: \mathbb{R} \longrightarrow Y$ be defined by $f(t) = \Phi(t,w)$. We must show $f$ is continuous. We first verify the hypotheses of Lemma 2.1. We shall prove that as $t$ tends to $t_0$, $f(t)$ tends either to $f(t_0)$ or to $\infty \in Y$. For suppose not. Let $\{t_i\}_{i>0}$ be a sequence such that

$$0 < | t_{i+1} - t_0 | < | t_i - t_0 |$$

and $f(t_i)$ tends to $w_0 \neq f(t_0)$.

We choose a strictly increasing sequence of integers $n(i)\,(i > 0)$, which increases sufficiently rapidly so that there is a $C^\infty$ function $\varphi: \mathbb{R} \longrightarrow \mathbb{R}$ with the following properties:

i) $\varphi(t_{2i}) = t_0$,

ii) $\varphi(t_{2i+1}) = t_{n(i)}$.

Then

$$V(\varphi \times 1_M)\ \Phi(t_i,w) = \Phi(\varphi t_i,w) \quad .$$

The left hand side tends to the limit $V(\varphi \times 1_M)w_0$, and so the right hand side tends to a limit. But this means

$$
\begin{aligned}
w_0 &= \lim_i \Phi(t_{n(i)},w) \\
&= \lim_i \Phi(\varphi t_{2i+1},w) \\
&= \lim_i \Phi(\varphi t_{2i},w) \\
&= f(t_0)
\end{aligned}
$$

which is a contradiction.

Hence, by Lemma 2.1, $f$ is continuous at some point $t_0 \in \mathbb{R}$. Let $\gamma_t\colon \mathbb{R} \longrightarrow \mathbb{R}$ be translation by $t$. Then $V(\gamma_t \times 1_M)$ is a homeomorphism, since it has an inverse $V(\gamma_{-t} \times 1_M)$. Now

$$V(\gamma_s \times 1_M)\Phi(t,w) = \Phi(s+t,w) \quad .$$

Taking $s = t_1 - t_0$, we see that $\Phi$ is a continuous function of $t$ at all points in $\mathbb{R}$.

<u>Proposition 2.3</u>

Let $w_1,\cdots,w_k$ be a basis for the fibre of $V\mathbb{R}^n$ over $0 \in \mathbb{R}^n$. Then we have a $C^r$ isomorphism $\Psi$ of vector bundles over $\mathbb{R}^n$

$$
\begin{array}{ccc}
\mathbb{R}^n \times \mathbb{R}^k & \xrightarrow{\ \Psi\ } & V\mathbb{R}^n \\
\downarrow & & \downarrow \\
\mathbb{R}^n & = & \mathbb{R}^n
\end{array}
$$

given by $\Psi(x,y) = \Sigma_{i=1}^k y_i[V(\gamma_x)w_i]$,

where $\gamma_x: \mathbb{R}^n \longrightarrow \mathbb{R}^n$ is translation by $x$ .

Proof. By induction on $n$ , we know that for each $w \in VM$ , the composite

$$\mathbb{R}^n \xrightarrow{i_w} \mathbb{R}^n \times VM \xrightarrow{\Phi} V(\mathbb{R}^n \times M)$$

is continuous. Hence the composite

$$\mathbb{R}^n \xrightarrow{i_w} \mathbb{R}^n \times VM \xrightarrow{\Phi} V(\mathbb{R}^n \times M) \xrightarrow{V(+)} V(\mathbb{R}^n)$$

is continuous. This composite sends $x$ to $V(\gamma_x)w$ . It follows that the map $\Psi$ of the theorem is a continuous isomorphism. $\mathbb{R}^n$ acts as a transformation group on the manifold $V\mathbb{R}^n$ , by $p(x,w) = V(\gamma_x)w$ . The action is continuous, since

$$p(s,\Psi(x_1,y)\ ) = \Psi(x + x_1,y)$$

and $\Psi$ is a homeomorphism. Moreover, for fixed $x$ , $V(\gamma_x)$ is a $C^r$ isomorphism of $V\mathbb{R}^n$ . By [3] p. 212, the map

$$p: \mathbb{R}^n \times V\mathbb{R}^n \longrightarrow V\mathbb{R}^n$$

is $C^r$ . (The author wishes to thank R. Palais for drawing his attention to the relevance of this result.) It follows that $\Psi$ is a $C^r$ isomorphism.

Proposition 2.4

$$\Phi_{U,W}: U \times VW \longrightarrow V(U \times W)$$

is $C^r$ for finite dimensional vector spaces $U$ and $W$ .

<u>Proof</u>. Let $i: W \longrightarrow U \times W$ and $p: U \times W \longrightarrow W$ be the canonical injection and projection. Then the composite

$$VW \xrightarrow{V(i)} V(U \times W) \xrightarrow{V(p)} VW$$

is the identity. Let $w_1, \cdots, w_r$ be a basis for the fibre of $VW$ over $0$. Let $u_j = V(i)w_j (1 \leq j \leq r)$ and extend this to a basis $u_1, \cdots, u_k$ for the fibre of $V(U \times W)$ over $0$. By Proposition 2.3, we have $C^r$ isomorphisms

$$\Psi_1 : W \times \mathbb{R}^r \longrightarrow VW \quad \text{and} \quad \Psi_2 : U \times W \times \mathbb{R}^k \longrightarrow V(U \times W) \quad .$$

Let $i' : \mathbb{R}^r \longrightarrow \mathbb{R}^k$ be defined by

$$i'(x_1, \cdots, x_r) = (x_1, \cdots, x_r, 0, \cdots, 0) \quad .$$

The proposition follows from the commutative diagram

$$\begin{array}{ccc} U \times W \times \mathbb{R}^r & \xrightarrow{1_{U\times W} \times i'} & U \times W \times \mathbb{R}^k \\ \Big\downarrow 1_U \times \Psi_1 & & \Big\downarrow \Psi_2 \\ U \times VW & \xrightarrow{\quad \Phi \quad} & V(U \times W) \end{array}$$

## 3. BASED MANIFOLDS

Let $\underline{\underline{M}}_*(\underline{s})$ be the category of $C^s$ manifolds with base point, satisfying the second axiom of countability, not necessarily compact or connected, and which have no boundary. Morphisms in $\underline{\underline{M}}_*(\underline{s})$ preserve the base point and are $C^s$ maps. If $(M,m)$ and $(N,n)$ are $C^s$ based manifolds, we usually suppress the base points and write $C^s_*(M,N)$ for the space of $C^s$ maps from $M$ to $N$, preserving the base points. We give $C^s_*(M,N)$ the coarse $C^s$ topology (on some compact subset of $M$ the first $s$ derivatives should be close). An $(r,s)$ natural vector bundle $V$ obviously gives rise to a functor $T(V): \underline{M_*(s)} \longrightarrow \underline{VLin}$, by putting $TM$ equal to the fibre over the base point. In due course we shall show that the study of such functors is equivalent to the study of natural vector bundles.

Definition 3.1

We say that $T: \underline{M_*(s)} \longrightarrow \underline{VLin}$ is $C^r$ if for each pair of based $C^s$ manifolds $(M,*)$ and $(N,*)$, for each $C^s$ manifold $P$ and for each $C^s$ map $P \times M, P \times * \longrightarrow N,*$, the induced map $P \times TM \longrightarrow TN$ is $C^r$.

Let $\underline{VDiff(s)}$ be the category of real vector spaces and $C^s$ maps preserving the origin. We can talk of $C^r$ functors $T: \underline{VDiff(s)} \longrightarrow \underline{VLin}$.

We shall in fact prove that every $C^0$ functor is automatically $C^s$. We remark that $T$ is $C^0$ if and only if it is continuous in the sense that for each pair of based manifolds $M$ and $N$, the map

$$C^s_*(M,N) \longrightarrow Lin(TM,TN)$$

is continuous.

Definition 3.2

We say $T: \underline{M_*(s)} \longrightarrow \underline{VLin}$ is myopic if $\dim M = \dim N$ implies that $\dim TM = \dim TN$. We say that $T$ is local if for any based manifold $(M,m)$ and for any open neighbourhood U of m , the inclusion of U in M induces an isomorphism TU ⟶ TM .

We can also talk of a functor T: VDiff(s) ⟶ VLin being local. If T is local and f,g: M,* ⟶ N,* agree on some neighbourhood of the base point, then Tf = Tg: TM ⟶ TN .

Theorem 3.3

a) Let T: VDiff(s) ⟶ VLin be continuous. Then T is local.

b) Let T: $M_*(s)$ ⟶ VLin be myopic and continuous. Then T is local.

c) If T: $M_*(s)$ ⟶ VLin is local, then it is myopic.

d) Every local functor T: VDiff(s) ⟶ VLin has a unique extension to a local myopic functor T: $M_*(s)$ ⟶ VLin .

Conjecture 3.4

Every functor T: VDiff(s) ⟶ VLin is local and every functor T: $M_*(s)$ ⟶ VLin is both myopic and local.

Note that it is easy to construct functors which are not continuous by composing some given functor with a non-continuous functor VLin ⟶ VLin [1] 1.2. c).

Proof of 3.3. During this proof we shall regard the objects of VDiff(s) as being the open unit disks $D^n$ in $R^n$ , with base point at the centre.

Let $T$ be continuous. For $0 < t \leq 1$, let $m_t\colon D^n \longrightarrow D^n = D$ be defined by $m_t x = tx$. For $t$ near $1$, $m_t$ is near $1_D$ and so $Tm_t\colon TD \longrightarrow TD$ is an isomorphism. Fix such an $m_t$.

Let $\varphi\colon D,0 \longrightarrow M,*$ be a diffeomorphism on to a neighbourhood of the base point. Let $h\colon M,* \longrightarrow D,0$ be a $C^s$ map such that $h\varphi = \mathrm{id}$ on $m_t D$. Then we have a commutative diagram

$$\begin{array}{ccc} D,0 & \xrightarrow{m_t} & D,0 \\ {\scriptstyle m_t}\downarrow & & \uparrow{\scriptstyle h} \\ D,0 & \xrightarrow{\varphi} & M,* \end{array}$$

In case a) above $M,* = D,0$ so that $TD \cong TM$. In case b) this follows since $T$ is myopic. We deduce that $T\varphi$ and $Th$ are isomorphisms. It follows that $T$ is local.

Part c) needs no proof. To prove d), we fix for each $C^s$ based manifold $(M,*)$ a diffeomorphism $\varphi_M\colon D,0 \longrightarrow M,*$ on to a neighbourhood of the base point. Given a local functor $T\colon \underline{\mathrm{VDiff}(s)} \longrightarrow \underline{\mathrm{VLin}}$, we define $T\colon \underline{M_*(s)} \longrightarrow \underline{\mathrm{VLin}}$ as follows. We put $TM = T\mathbb{R}^m$. If $f\colon M,* \longrightarrow N,*$ is a $C^s$ map, we define $Tf = T(\varphi_N^{-1} f \varphi_M \gamma)(T\gamma)^{-1}$, where $\gamma\colon D,0 \longrightarrow D,0$ is equal to $m_\varepsilon$ with $\varepsilon$ small. It is easy to see that $Tf$ is well-defined and that $T$ defines a functor.

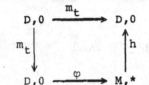

To show that the extension of $T$ to $\underline{M_*(s)}$ is unique, let $T_1$ and $T_2$ be two extensions. We define an isomorphism $\alpha\colon T_1 \longrightarrow T_2$ by putting $\alpha_M\colon T_1M \longrightarrow T_2M$ equal to $(T_2\varphi_M)(T_1\varphi_M)^{-1}$ . This completes the proof of the theorem.

Corollary 3.5

Let V be a myopic (r,s) natural vector bundle. Then V is continuous (see 1.2). If M is an open subset of N , then the inclusion induces an isomorphism of VM with $VN|M$ . If $T = T(V)$ , then T is $C^r$ myopic and local.

*Proof*. The fact that T is myopic is obvious. That T is $C^r$ follows from Proposition 2.4. By Theorem 3.3 b), T is local. The fact that VN is isomorphic to $VN|M$ now follows by letting the base point vary over all points of M . This isomorphism, together with Proposition 2.4, proves that V is continuous.

## 4. MYOPIA

In this section we prove

Theorem 4.1

Every natural vector bundle V is myopic. If M is an open and closed subset of the non-connected manifold N , then M is a retract of N . Hence VM is a retract of VN . As M varies over all manifolds of dimension m , dim VM is bounded. To see this, let $M_1, M_2, \cdots,$ be a collection of manifolds of the same dimension, and let M be their disjoint union. Then $\dim VM$ is an upper bound for $\dim VM_1$. For each integer $n > 0$, let $P(n)$ be a manifold of dimension $n$ with $\dim VP(n)$ maximal.

Lemma 4.2

Let $\dim M_1 = n_1$, $\dim M_2 = n_2$. Let

$$f,g\colon M_1 \cup P(n_1),\ M_1,\ P(n_1) \longrightarrow M_2 \cup P(n_2),\ M_2,\ P(n_2)$$

and let $f|M_1 = g|M_1$. Then $Vf$ and $Vg$ agree over $M_1$.

Proof. We extend $f$ and $g$ to

$$f',g'\colon M_1 \cup P(n_1) \cup P(n_1) \longrightarrow M_2 \cup P(n_2)$$

(disjoint unions) by fixing a point in $P(n_2)$ and sending the third summand to this point. Then we have the commutative diagram

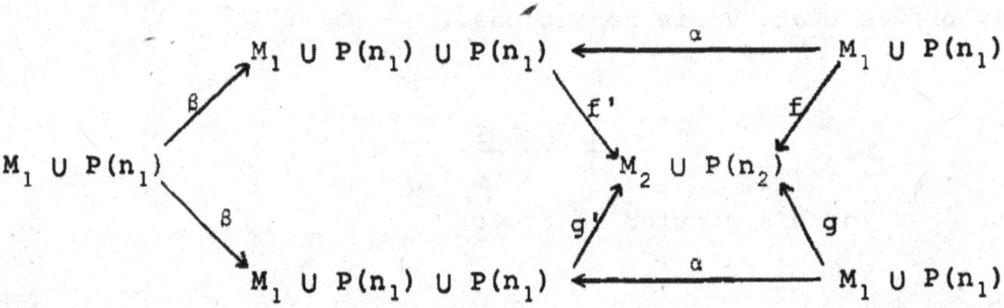


where $\alpha$ is the canonical inclusion avoiding the third summand and $\beta$ is the canonical inclusion avoiding the second summand. Both $\alpha$ and $\beta$ have one-sided inverses so $V\alpha$ and $V\beta$ are isomorphisms over $M_1$. Hence $Vf = Vg$ over $M_1$.

From the lemma, we see that we can define a new

natural vector bundle $V'$ by $V'M = V(M \cup P(n))|M$. Moreover $V'$ will be myopic and so Corollary 3.5 applies. Let $T = T(V): \underline{M}_*(\underline{s}) \longrightarrow \underline{VLin}$ and let $T' = T(V')$. With the notation of the proof of Theorem 3.3, we have a commutative diagram

$$\begin{array}{ccccc} V(D \cup P(n)) & \xrightarrow{V(\varphi \cup 1)} & V(M \cup P(n)) & \xrightarrow{V(h \cup 1)} & V(D \cup P(n)) \\ \uparrow & & \uparrow & & \uparrow \\ VD & \xrightarrow{V\varphi} & VM & \xrightarrow{Vh} & VD \end{array}$$

and hence a commutative diagram

$$\begin{array}{ccccc} T'D & \xrightarrow{T'\varphi} & T'M & \xrightarrow{T'h} & T'D \\ \uparrow & & \uparrow & & \uparrow \\ TD & \xrightarrow{T\varphi} & TM & \xrightarrow{Th} & TD \end{array}$$

In both diagrams the vertical maps are injective. By Corollary 3.5 $T'$ is $C^r$, myopic and local. Hence the composite in the top row is the identity and $T'\varphi$ and $T'h$ are inverse isomorphisms. It follows that $T\varphi$ and $Th$ are both injective and hence they are isomorphisms. This shows that $\dim TD = \dim TM$, and so $V$ is myopic, which proves Theorem 4.1.

## 5. CONSTRUCTING NATURAL VECTOR BUNDLES

Let $T: \underline{M}_*(\underline{s}) \longrightarrow \underline{VLin}$ be a myopic $C^r$ functor. We construct an $(r,s)$ natural vector bundle $V$ using $T$.

If $M$ is an n-dimensional manifold, the underlying point set and vector space structure on $VM$ is $\cup_{m\in M} T(M,m)$ , where $T(M,m)$ is the fibre over $m$. We give $VM$ a differential structure, by insisting that for each $C^s$ diffeomorphism $\varphi: \mathbb{R}^n \longrightarrow M$ on to an open subset of $M$, we have a $C^r$ diffeomorphism

$$\Phi: \mathbb{R}^n \times T(\mathbb{R}^n,0) \longrightarrow VM|\varphi\mathbb{R}^n ,$$

given by $\Phi(x,w) = T\varphi \cdot T\gamma_x \cdot w$ , where $\gamma_x: \mathbb{R}^n,0 \longrightarrow \mathbb{R}^n,x$ is translation by $x$.

To check that the differential structure is well-defined, we let $\psi: \mathbb{R}^n \longrightarrow M$ be a diffeomorphism on to an open subset of $\varphi\mathbb{R}^n$. We must show that $\Phi^{-1}\Psi$ is a $C^r$ isomorphism of vector bundles over $\varphi^{-1}\psi$ (where $\Psi$ is defined analogously to $\Phi$).

Now

$$\Phi^{-1}\Psi(x,w) = (\varphi^{-1}\psi x, T\gamma_y \cdot T(\varphi^{-1}\psi) \cdot T\gamma_x \cdot w)$$

where $y = -\varphi^{-1}\psi x$. The map $\mathbb{R}^n \times \mathbb{R}^n \longrightarrow \mathbb{R}^n$ , given by $(x,x_1) \longrightarrow (\gamma_y\varphi^{-1}\psi\gamma_x)x_1$ is $C^s$. Since $T$ is $C^r$ , we see from the definition 3.1 that $\Phi^{-1}\Psi$ is $C^r$.

If $f: M \longrightarrow N$ is $C^s$ , we define

$$Vf: \cup_{m\in M} T(M,m) \longrightarrow \cup_{n\in N} T(N,n)$$

by $Vf|T(M,m) = Tf: T(M,m) \longrightarrow T(N,fm)$ . The proof that $Vf$ is $C^r$ is the same as the proof above that $\Phi^{-1}\Psi$ is $C^r$.

We now have a map (in fact a functor)

$\alpha$: {myopic $C^r$ functors $\underline{\underline{M}}_*(\underline{s}) \longrightarrow \underline{\underline{VLin}}$}

$\longrightarrow$ {(r,s) natural vector bundles}

We also have the obvious map (also a functor)

$\beta$: {(r,s) natural vector bundles}

$\longrightarrow$ {myopic $C^r$ functors $\underline{\underline{M}}_*(\underline{s}) \longrightarrow \underline{\underline{VLin}}$}

defined by taking the fibre over the base point.

<u>Theorem 5.1</u>

These maps are inverse bijections.

<u>Proof</u>. It is obvious that $\beta\alpha = 1$.

Let us start with an (r,s) natural vector bundle $V$. Then $\alpha\beta V$ and $V$ have the same underlying point set and vector space structure in each fibre. To see that $\alpha\beta V$ and $V$ have the same differential structure, we apply Proposition 2.3.

## <u>6. THE FINE STRUCTURE</u>

Let $T: \underline{\underline{M}}_*(\underline{s}) \longrightarrow \underline{\underline{VLin}}$ be a myopic $C^r$ functor. For each based manifold $(M,*)$ we have the factorization

$$* \longrightarrow M \longrightarrow * \ .$$

This factorization is natural and so $TM$ is naturally isomorphic to $T* \oplus \ker(TM \longrightarrow T*)$.

<u>6.1</u>. We may therefore assume, whenever it is convenient, that $T* = 0$.

Lemma 6.2

If $s = 0$ , then $T$ is a constant functor.

Proof. We suppose $T^* = 0$ . By 3.3 b), $T$ is local. Let $f: \mathbb{R}^n,0 \longrightarrow \mathbb{R}^m,0$ be such that $Tf \neq 0$ . We can find $g$ arbitrarily near in the $C^0$ topology, such that there is a neighbourhood of zero on which $g$ is zero. By taking $g$ near enough to $f$ and using the continuity of $T$ , we may assume $Tg \neq 0$ . But this contradicts the fact that $T$ is local.

Theorem 6.3

There is a filtration $T^0 \subset T^1 \subset T^2 \subset \cdots$ of $T$ such that:

a) For each based manifold M, $T^r M = TM$ for some $r$ depending only on the dimension of $M$ ;

b) $T^i/T^{i-1}$ is isomorphic to $F_i \circ \tau$ , where $\tau$ is the tangent space and $F_i$: VLin $\longrightarrow$ VLin has degree $i$ ;

c) If $f,g: M,* \longrightarrow N,*$ have the same k-jet at the base point and if $t \leq k$ or if $k = s$ then $T^t f = T^t g$ .

Proof. By Theorem 3.3 d), we may take $T$ to be a $C^r$ functor from VDiff(s) to VLin . Let $I$: VLin $\longrightarrow$ VDiff(s) be the inclusion. $T \circ I$: VLin $\longrightarrow$ VLin is obviously a continuous functor. By 6.1, we may assume $T0 = 0$ . By the results of [1] or [2], we have

$T \circ I = \oplus_{i>0} F_i$ , where $F_i$: <u>VLin</u> $\longrightarrow$ <u>VLin</u> has the property that if $\lambda_W: W \longrightarrow W$ is scalar multiplication by $\lambda$ , then $F_i(\lambda_W)$ is scalar multiplication by $\lambda^i$ . We have $TW = \oplus_{i>0} F_iW$ , but the direct sum decomposition will in general not be preserved by $Tf$ , where $f$ is differentiable.

<u>Lemma 6.4</u>

Let $f: W_1,0 \longrightarrow W_2,0$ be a $C^s$ map between vector spaces, with zero k-jet at zero. Then $Tf$ sends $\oplus_{i=1}^{j} F_iW_1 \subset TW_1$ into $\oplus_{i=1}^{t} F_iW_2 \subset TW_2$ where $t = [j/(k + 1)]$ if $k < s$ and $t = 0$ if $k = s$ .

<u>Proof</u>. If $x \in TW = \oplus_{i>0} F_iW$ , we write $x_i$ for the component in $F_iW$ . Suppose that the lemma is false. Then there exists $x \in F_iW_1$ and $f: W_1,0 \longrightarrow W_2,0$ with zero k-jet, such that $[Tf(x)]_1 \neq 0$ , where $1(k + 1) > i$ if $k < s$ .

We replace $W_1$ by $\mathbb{R}^n$ and $W_2$ by $\mathbb{R}^m$ . We can write $f$ as a finite sum $f = \Sigma\alpha_i\beta_i$ , where $\alpha_i$ is a monomial of degree $k$ in the co-ordinates of $\mathbb{R}^n$ and $\beta_i: \mathbb{R}^n,0 \longrightarrow \mathbb{R},0$ is $C^{s-k}$ . Since $T$ is continuous, we can approximate $\beta_i$ by a polynomial which vanishes at zero, and hence $f$ by a polynomial $g$ , such that $g$ vanishes to order $(k + 1)$ at zero, at $[(Tg)(x)]_1 \neq 0$ . If $k = s$ , we may assume without loss of generality that $g$ vanishes to order $(i + 1)$ .

We factorize $g$ into

$$\mathbb{R}^n,0 \xrightarrow{h} \mathbb{R}^N,0 \xrightarrow{\alpha} \mathbb{R}^m,0$$

where $h(x_1,\cdots,x_n)$ is a new vector, each entry of which is a monomial in the $x_j$'s of degree greater than $k$ (greater than $i$, if $k = s$), and where $\alpha$ is a linear map. Now $T\alpha$ sends $F_j\mathbb{R}^N$ to $F_j\mathbb{R}^m$ for each $j$. Hence $[Th(x)]_1 \neq 0$.

We have a commutative diagram

$$\begin{array}{ccccc} \mathbb{R}^n,0 & & \xrightarrow{\quad\lambda\quad} & & \mathbb{R}^n,0 \\ \downarrow h & & & & \downarrow h \\ \mathbb{R}^N,0 & \xrightarrow{\lambda^q} & \mathbb{R}^N,0 & \xrightarrow{\beta} & \mathbb{R}^N,0 \end{array}$$

where $q = k + 1$ if $k < s$ and $q = i + 1$ if $k = s$, and where $\beta$ is a linear map, represented by a scalar matrix, each entry being a power (possibly $\lambda^0$) of $\lambda$.

Since each $F_i$ is a polynomial functor [1] 1.7, we know that $T\beta\colon T\mathbb{R}^N \longrightarrow T\mathbb{R}^N$ depends in a polynomial fashion on $\lambda$. Applying $T$ to the above diagram, we see that

$$\lambda^i Th \cdot x = Th \cdot T\lambda \cdot x$$
$$= T\beta \cdot T(\lambda^q) \cdot Th \cdot x \ .$$

Therefore, taking components,

$$\lambda^i (Th \cdot x)_1 = \lambda^{1q} T\beta \cdot (Th \cdot x)_1 \ .$$

Since $1q > i$, and $T\beta$ is polynomial in $\lambda$, we must have $(Th \cdot x)_1 = 0$ and this is a contradiction.

Corollary 6.5

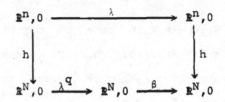

Defining $T^j W = \oplus_{i=1}^{j} F_i W$, we obtain a subfunctor of $T$ .

To complete the proof of Theorem 6.3, we need only prove part c), because b) follows from c) by putting $k = 1$ . We may assume that $f,h\colon U,0 \longrightarrow W,0$ and that the k-jet of $h$ is zero. We must show that $T^t f = T^t(f + h)$ if $t \leq k$ or if $k = s$ .

Now $f + h$ factors as

$$U \xrightarrow{\Delta} U \times U \xrightarrow{f\times h} W \times W \xrightarrow{+} W \quad .$$

Writing $S = T^t$ , we see that we need only show that $S(f \times h)\colon S(U \times U) \longrightarrow S(W \times W)$ is independent of $h$ . Now by Lemma 6.4, $Sh\colon SU \longrightarrow SW$ is zero. According to [2] 9.1, we need only show that $S^{(2)}(1_U,h)\colon S^{(2)}(U,U) \longrightarrow S^{(2)}(U,W)$ is zero. This follows by applying Lemma 6.4 to the functor $W \longrightarrow S^{(2)}(U,W)$ , since we know that this functor is the sum of functors of degree less than $t$ (see [2] 9.4).

## 7. MISCELLANEOUS

By now we have completed the proofs of most of the results stated in 1. The only outstanding points are Theorems 1.5 and 1.6.

We first prove Theorem 1.5. According to Theorem 5.1 and Lemma 6.2, we need only consider $C^r$ functors $T\colon$ <u>VDiff(r)</u> $\longrightarrow$ <u>VLin</u> with $r > 0$ . By 6.1, we may assume that $T0 = 0$ . We wish to show that $T = 0$ . Without loss of generality, we may suppose (in the notation of Theorem 6.3) that $T = T^i$ and $T^{i-1} = 0$. As in [2] 9.4, we can now reduce $i$, by replacing $T$ with $T^{(2)}(W, \ )$ for some fixed $W$. There is therefore no loss of generality in supposing $T = T^1$. But then, according to Theorem 6.3 and [1] 7.1, $T$ is simply the direct sum of a certain number of copies (say $N$) of the tangent space at the origin.

$T$ corresponds to an $(r,r)$ natural vector bundle $V$ (see 5.1). There is a natural isomorphism $\Phi$ between $V$, considered as an $(r-1,r)$ natural vector bundle, and $N\tau$, where $N$ is an integer and $\tau$ is the tangent bundle. For each $C^r$ map $f\colon \mathbb{R} \longrightarrow \mathbb{R}$, we have a commutative diagram

$$\begin{array}{ccc} V\mathbb{R} & \xrightarrow{\ \Phi\ } & N\mathbb{R} \\ {\scriptstyle Vf}\downarrow & & \downarrow{\scriptstyle N\tau f} \\ V\mathbb{R} & \xrightarrow{\ \Phi\ } & N\mathbb{R} \end{array} \quad .$$

We choose trivialisations $V\mathbb{R} \cong \mathbb{R} \times \mathbb{R}^N$ and $N\tau\mathbb{R} \cong \mathbb{R} \times \mathbb{R}^N$. In these terms $\Phi(x,y) = (x,\varphi x \cdot y)$ for $x \in \mathbb{R}$, $y \in \mathbb{R}^N$, where $\varphi x$ is a non-singular $(N \times N)$ matrix. We know that $\varphi$ is $C^{r-1}$. We have $Vf(x,y) = (fx, gx \cdot y)$ where $g$ is $C^r$ and $gx$ is an $(N \times N)$ matrix, and $N\tau f(x,y) = (fx, f'(x)y)$.

The commutative diagram above leads to the equation, which exists for all $C^r$ maps $f\colon \mathbb{R} \longrightarrow \mathbb{R}$,

$$f'(x)\varphi(x) = \varphi(f(x))g(x)$$

where $\varphi$ is independent of $f$ and $g$ depends on $f$.

We choose $i$ and $j$ so that the $(i,j)$ entry of

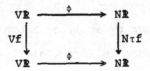

$\varphi(f(x))^{-1}\varphi(x)$ is non-zero for $x = 0$. We call this entry $k(x)$. Let $h(x) = x^r$ for $x \geq 0$ and $h(x) = -x^r$ for $x \leq 0$. Then $h$ is $C^{r-1}$ but not $C^r$. Let $f\colon \mathbb{R},0 \longrightarrow \mathbb{R},0$ be a $C^r$ function such that for $x$ near zero $f'(x)k(x) = h(x)$. But then the matrix $g(x)$, where $g$ is the function corresponding to $f$, has in its $(i,j)$ entry the function $h(x)$, which is not $C^r$, and this is a contradiction.

We now prove Theorem 1.6. In view of Theorem 5.1, we need only show that a functor $T$ satisfying Theorem 6.3 is $C^\infty$ (in the sense of 3.1), when restricted to VDiff($\infty$). Without loss of generality, we assume that $T = T^i$.

If $M$ and $N$ are vector spaces, we denote by $J^i(M,N)$ the space of i-jets at the origin, of differentiable maps preserving the origin. $J^i(M,N)$ is a finite dimensional manifold (in fact a Euclidean space). We denote by $\operatorname{Inv} J^iM$ the space of invertible i-jets. This is a finite dimensional Lie group (under composition of jets).

Let $F\colon P \times M, P \times 0 \longrightarrow N,0$ be $C^\infty$. We have to show that the composite

$$P \longrightarrow C^\infty_*(M,N) \longrightarrow \operatorname{Lin}(TM,TN)$$

is $C^\infty$. We need only prove this when both $M$ and $N$ are replaced by $M \times N$, for the general case can then be deduced by composing with the injection $M \longrightarrow M \times N$ and the projection $M \times N \longrightarrow N$. We can therefore assume $M = M = W$, without loss of generality.

We factor $C^\infty_*(W,W) \longrightarrow \mathrm{Lin}(TW,TW)$:

$$\begin{array}{ccccccc}
C^\infty_*(W,W) & \xrightarrow{\alpha} & \mathrm{Diffeo}(W \times W) & \longrightarrow & \mathrm{Aut}\, T(W \times W) & \xrightarrow{\gamma} & \mathrm{Lin}(TW,TW). \\
\downarrow & & \downarrow & \nearrow_{\beta} & & & \\
J^i(W,W) & \xrightarrow{\alpha'} & \mathrm{Inv}\, J^i(W \times W) & & & &
\end{array}$$

This diagram is commutative, and $\alpha, \beta$ and $\gamma$ are defined as follows: $(\alpha f)(x,y) = (x, y + f(x))$, $\beta$ is induced by $T$ (according to Theorem 6.3 c)) and $\gamma$ is defined by composing on the right with $Tj$ and on the left with $Tq$, where $j: W \longrightarrow W \times W$ is the injection of the first factor and $q: W \times W \longrightarrow W$ is projection on to the second factor.

We know that $\beta$ is analytic since it is a map of Lie groups. The maps $\gamma$ and $\alpha'$ are also analytic. If $P \times W, P \times 0 \longrightarrow W, 0$ is $C^\infty$, then the composite

$$P \longrightarrow C^\infty_*(W,W) \longrightarrow J^i(W,W)$$

is obviously a $C^\infty$ map between finite dimensional manifolds. The result follows.

## REFERENCES

[1] Epstein, D. B. A., "Group Representations and Functors", (to appear in *Am. J. Math.*).

[2] Epstein, D. B. A., and Kneser, M., "Functors between Categories of Vector Spaces", (to appear).

[3] Montgomery, D., and Zippin, L., "Topological Transformation Groups", *Interscience* (1955).

# SEVERAL NEW CONCEPTS:
# LUCID AND CONCORDANT FUNCTORS, PRE-LIMITS, PRE-COMPLETENESS, THE CONTINUOUS AND CONCORDANT COMPLETIONS OF CATEGORIES

by

Peter Freyd

## I LUCID FUNCTORS

Given a set-valued functor $T: \underline{A} \longrightarrow \underline{S}$ we define the category $\underline{El(T)}$, the category of elements of $T$, as the special comma category $(1,T)$, i.e.,

Objects are pairs $\langle x,A\rangle$, $A \in \underline{A}$, $x \in TA$.

Maps from $\langle x,A\rangle$ to $\langle y,B\rangle$ are maps $f: A \longrightarrow B$ such that $(Tf)(x) = y$.

A functor $T: \underline{A} \longrightarrow \underline{S}$ is called a PETTY functor (sometimes "proper", "bounded,") if $\underline{El(T)}$ has a pre-initial set, i.e., if there is a set of $\mathcal{S}$ of objects in $\underline{El(T)}$ such that for every $X \in \underline{El(T)}$ there exists $S \in \mathcal{S}$ and a map $S \longrightarrow X$. By translation, then, there exists a generating set $\{\langle x_i,A_i\rangle\}_{i \in I}$, $x_i \in TA_i$ such that for any $B \in \underline{A}$, $y \in TB$ there is a map $f: A_i \longrightarrow B$ such that $(Tf)(x_i) = y$. The subfunctor generated by $\{\langle x_i,A_i\rangle\}$ is all of $T$. (In general, given any class $\{\langle x_i,A_i\rangle\}$ we may define

$T' \subset T$ by $T'(B) = \{y \in TB \mid \exists i,\ A_i \longrightarrow B$ such that $(Tf)(x_i) = y\}$.) Note that the class of natural transformations $(T,V)$, $V$ any functor, is embedded in

$\Pi_I V(A_i)$ where the $i^{th}$ projection map is defined by $p_i(\eta) = \eta_{A_i}(x_i)$. Consequently the class of transformations may be replaced by a set. The category of petty functors from $\underline{A}$ to $\underline{S}$ is locally small. For petty $T$ the functor $\Sigma_I H^{A_i} \longrightarrow T$ defined by $H^{A_i} \xrightarrow{u_i} \Sigma H^{A_i} \longrightarrow T = \eta_i$ where $\eta_{i,B}(f\colon A_i \longrightarrow B) = (Tf)(x_i)$, is epimorphic. Conversely, any quotient functor of a sum of representable functors is petty.

$T\colon \underline{A} \longrightarrow \underline{S}$ is petty iff all chains of subfunctors $\{T_\alpha\}$, $\alpha$ ranging through the ordinals, such that $\cup\, T_\alpha = T$ are in fact eventually stationary, i.e., $T_\alpha = T$ for large $\alpha$. Clearly, petty functors have this property because for each $i \in I$ we let $\delta(i)$ be the first ordinal such that $s_i \in T_{\delta(i)}A_i$. Let $\delta = \sup \alpha(i)$. Then for $\delta \le \alpha$, $T_\alpha = T$. Conversely, if $T$ is not petty we may well-order the objects of $\underline{El(T)}$ and construct a non-terminating chain.

An example of a naturally arising functor that is not petty is the covariant power set functor $P\colon \underline{S} \longrightarrow \underline{S}$. $P(A)$ is the family of subsets of $A$. Given $f\colon A \longrightarrow B$, $(Pf)(A') = \mathrm{Im}(f|A')$. We may construct a chain of subfunctors of $\underline{P}$ as follows: for each cardinal $\underline{K}$ define $P_K$ to be such that $P_K(A) = \{A' \subset A|$ cardinality $A' < K\}$. In fact, these are the only subfunctors of $P$, and all proper subfunctors of $P$ are petty.

A functor $T\colon \underline{A} \longrightarrow \underline{B}$ is called petty if for every petty set-valued functor $S: \underline{B} \longrightarrow \underline{S}$, it is the case that $\underline{A} \xrightarrow{T} \underline{B} \xrightarrow{S} \underline{S}$ is petty.

It is easy to see that it suffices to check for representable $S$. If one replaces this last condition with primitive terms the "solution set conditions" as used to be stated for the general adjoint functor theorem is obtained. Of course, nowadays, we replace that condition with the requirement that the functor be petty.

Returning to set-valued functors, we say that $T: \underline{A} \longrightarrow \underline{S}$ is LUCID if:

(1) $T$ is petty.

(2) For every $P: \underline{A} \longrightarrow \underline{S}$ and pair of transformations $\alpha, \beta: P \longrightarrow T$ it is the case that the equalizer $\operatorname{Ker}(\alpha, \beta)$ is petty.

It should be noticed that lucid is to petty as coherent is to finitely generated. The next proposition allows us to test for lucidity by restricting $P$ to representable functors.

Proposition 1.1

A functor $T: \underline{A} \longrightarrow \underline{S}$ is lucid iff it is petty and for every representable $H^A$ and transformations $x, y: H^A \longrightarrow T$ the equalizer $\operatorname{Ker}(x, y)$ is petty.

Proof. Given petty $P$ and transformations $\alpha, \beta: P \longrightarrow T$ we first use the pettiness of $P$ to obtain an epimorphism

$\Sigma H^{A_i} \longrightarrow P$ . If

$$\begin{array}{ccc} K' & \longrightarrow & \Sigma H^{A_i} \\ \downarrow & & \downarrow \\ \mathrm{Ker}(\alpha,\beta) & \longrightarrow & P \end{array}$$

is a pullback then $K' \longrightarrow \mathrm{Ker}(\alpha,\beta)$ is epi and $K' \longrightarrow \Sigma H^{A_i}$ is the equalizer of $\Sigma H^{A_i} \longrightarrow P \xrightarrow{\alpha,\beta} T$ . It suffices to show that $K'$ is petty. But $K' = \Sigma K_i'$ where $K_i'$ is the equalizer of $H^{A_i} \longrightarrow P \xrightarrow{\alpha,\beta} T$ . ∎

We might add here that for ⊕'ive functors from an ⊕'ive category to the category of abelian groups $\underline{G}$, the theorem is true, but the proof is different. It is not the case that $\Sigma K_i' \longrightarrow K'$ is epi. It is the case that $\Sigma K'_{(i, \ldots, i_n)} \longrightarrow K'$ is epi where $K'_{i_1 \ldots i_n}$ is the kernel arising from $H^{A_{i_1} \oplus \cdots \oplus A_{i_n}}$ .

<u>Proposition 1.2</u>

<u>Arbitrary sums of lucid functors are lucid.</u>

<u>Proof</u>. Given a pair of maps $x, y: H^A \longrightarrow \Sigma T_i$, if $x(1_A)$ is not in the same component as $y(1_A)$ the equalizer of $x$ and $y$ is empty. If, on the other hand, $x(1_A) \in T_i A$ , $y(1_A) \in T_i A$ then the equalizer of $x$ and $y$ is the same as the equalizer a pair of transformations from $H^A$ to $T_i$ . ∎

Again this proposition is true in the ⊕'ive case,

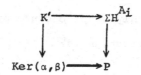

but not the proof. We shall wait until after 1.7 to describe the correction. The only other closure property on lucid functors true without restriction is:

Proposition 1.3

The full subcategory of lucid functors on A is closed under the formation of equalizers.

Proof. Given $\alpha, \beta: T_1 \longrightarrow T_2$ since $T_1$ satisfies $L_1$ and $T_2$ satisfies $L_2$ it follows that $\mathrm{Ker}(\alpha,\beta)$ satisfies $L_1$. Since $\mathrm{Ker}(\alpha,\beta)$ is a subfunctor of $T_1$ and $T_1$ satisfies $L_2$ it follows that $\mathrm{Ker}(\alpha,\beta)$ satisfies $L_2$. ∎

A category is RIGHT PRE-COMPLETE if the solution set condition holds for right limits. This may be rephrased in several ways, the easiest being that left-limits of representable functors be petty. For example:

Proposition 1.4

PRE-COEQUALIZERS exist if representable functors are lucid.

Proof. We mean by the phrase "pre-coequalizers exist" that for any pair of maps $x, y: A \longrightarrow B$ there exists a set $\{B \longrightarrow C_i\}$ such that

PCE (1) $A \xrightarrow{x} B \longrightarrow C_i = A \xrightarrow{y} B \longrightarrow C_i$, all i.

PCE (2) For any $B \longrightarrow X$ such that

$A \xrightarrow{x} B \longrightarrow X = A \xrightarrow{y} B \longrightarrow X$ there exists a triangle

$$\begin{array}{ccc} B & \longrightarrow & C_i \\ & \searrow & \downarrow \\ & & X \end{array} \quad \text{for some } i.$$

The equalizer of the transformation $H^x, H^y: H^B \longrightarrow H^A$ is a subfunctor of $H^B$, generated by the set $\{B \longrightarrow C_i\}$.

Conversely, given that the equalizer of $H^x, H^y$ is petty we may reverse the argument to obtain a pre-coequalizer for $x, y$. ∎

**Proposition 1.5**

(Finite) PRE-COPRODUCTS exist in $\underline{A}$ iff (finite) products of representable functors are petty.

*Proof*. We mean by the phrase "(Finite) pre-coproducts exist" that for any (finite) family $\{A_i\}_I$ that there exists a set $\{A_i \longrightarrow P_j\}_{I\times S}$ such that for any $\{A_i \longrightarrow X\}$ there exists a $P_j \longrightarrow X$ some $j$ such that

$$\begin{array}{ccc} A_i & \longrightarrow & P_j \\ & \searrow & \downarrow \\ & & X \end{array}$$

all $i$. Such a set is clearly seen to be nothing more nor less than a generating set for $\Pi H^{A_i}$. ∎

**Corollary 1.6**

(Finite) pre-coproducts exist in $\underline{A}$ iff products of petty functors are petty.

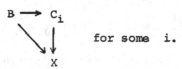

Proof. Let $\{T_i\}_{i\in I}$ be a family of non-empty petty functors. For each $i$ let $\Sigma_{J_i} H^{A_j} \longrightarrow T_i$ be epi. Then $\Pi_I \Sigma_{J_i} H^{A_j} \longrightarrow \Pi T_i$ is epi. Let $K$ be the set of choice functions $\{k: I \longrightarrow \cup_I J_i\}$ where $k(i) \in J_i$. Then there is an epi $\Sigma_K \Pi_I H^{A_{k(i)}} \longrightarrow \Pi_I \Sigma_{J_i} H^{A_j}$. ∎

Proposition 1.7

*If finite pre-coproducts exist in* $\underline{A}$ *then finite left-limits of lucid functors are lucid.*

Proof. It suffices in light of 1.3 to prove that finite products of lucid functors are lucid. The last corollary demonstrated their pettyness. It remains to verify $L_2$. Let $T_1, T_2$ be lucid, $P$ petty and $\langle \alpha_1, \alpha_2 \rangle$, $\langle \beta_1, \beta_2 \rangle : P \longrightarrow T_1 \times T_2$ transformations. The equalizer $K_1$ of the maps $\alpha_1, \beta_2 : P \longrightarrow T_1$ is petty. The equalizer of $\langle \alpha_1, \alpha_2 \rangle$ and $\langle \beta_1, \beta_2 \rangle$ is the equalizer of the pair $K_1 \longrightarrow P_{\alpha_2} \longrightarrow T_2$ and $K_1 \longrightarrow P_{\beta_2} \longrightarrow T_2$. ∎

In the ⊕'ive case finite pre-products exist and hence finite products of lucid functors are lucid. The ⊕'ive version of 1.2 is proved by observing that any map $H^A \longrightarrow \Sigma T_i$ factors through a finite subsum.

An observation, though entirely formal is worth making here. We'll say $\underline{A}$ is PRE-COREFLECTIVE in $\underline{B}$, if for every $B \in \underline{B}$ there exists a set $\{A_i \longrightarrow B\}$ $A_i \in \underline{A}$ such that for every $A \in \underline{A}$, $A \longrightarrow B$ there exists a triangle

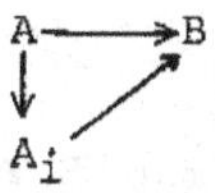

.

Proposition 1.8

$\underline{A}$ may be fully embedded as a pre-coreflective subcategory of a finitely left complete category iff $\underline{A}$ is finitely left pre-complete iff $\underline{A}$ has pre-equalizers and finite preproducts.

Proof. $\Longrightarrow$ clear

$\Longleftarrow$ By the last propositions the category of lucid functors from $\underline{A}^{op}$ is finitely left-complete and the Yoneda embedding embeds $\underline{A}$ into it by 1.4. ∎

To remove the "finitely"'s we must do more work. Particularly, we must know when arbitrary products of lucid functors are lucid. First, however, we consider the right-hand side. We know that arbitrary sums of lucid functors are lucid (1.2). It remains to show that coequalizers of lucid functors are lucid. In the ⊕'ive case this is easy. Given $S \longrightarrow T \longrightarrow C \longrightarrow 0$ exact, $S,T$ lucid, let $H^A \longrightarrow C$ be any transformation, $H^A$ is projective and we may lift to obtain a triangle

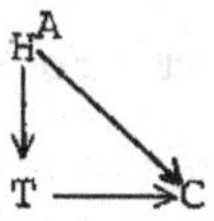

.

Let $\begin{array}{ccc} K & \longrightarrow & H^A \\ \downarrow & & \downarrow \\ S & \longrightarrow & T \end{array}$ be a pullback. $K \longrightarrow H^A$ is the kernel of $H^A \longrightarrow C$.

But $K$ is also the kernel of a map $S \oplus H^A \longrightarrow T$ and hence $K$ is petty.

For the set-valued case we must do more.

<u>Proposition 1.9</u>

<u>If $\underline{A}$ has finite pre-coproducts then coequalizers of lucid functors are lucid.</u>

<u>Proof</u>. Let $S,T$ be lucid, $\alpha,\beta\colon S \longrightarrow T$ transformations, $T \longrightarrow C$ the coequalizer of $\alpha$ and $\beta$ . $C$ is easily petty. To verify $L_2$ for $C$ we first display $C$ as the coequalizer of a reflexive, symmetric and <u>transitive</u> pair of transformations into $T$. The construction is familiar to topos-people. For each word on the symbols "$\alpha\beta$" and "$\beta\alpha$" we consider the diagram typified by:

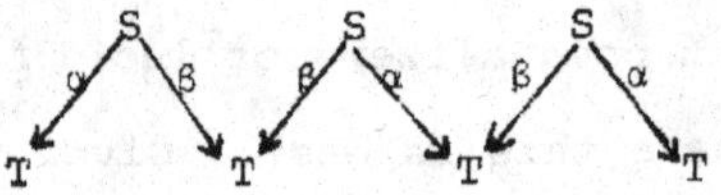


where the word is $(\alpha\beta)(\beta\alpha)(\beta\alpha)$. We let $K_w$ be the left-limit of the diagram for the word $w$, $K_w \xrightarrow{p_1} T$ the stipulated map to the left-most copy of $T$, $K_w \xrightarrow{p_2} T$ to the right-most copy. $(K_\emptyset = T)$ . Note that under the hypothesis of the propositions $K_w$ is petty. If we sum over all words $\bar{K} = \Sigma K_w$ and consider the pair of induced maps into $T$, $\bar{K} \xrightarrow{p_1} T$, $\bar{K} \xrightarrow{p_2} T$ it becomes clear that $\bar{K} \xrightarrow{p_1} T \longrightarrow C = \bar{K} \xrightarrow{p_2} T \longrightarrow C$ and that for elements $x, y \in TB$ which are sent to the same element in $CB$ that there exists $z \in \bar{K}B$ such that $(Kp_1)(z) = x$, $K(p_2)(z) = y$. Hence the image $K$ of $\bar{K} \longrightarrow T \times T$ is the kernel-pair of $T \longrightarrow C$, i.e., we obtain a pullback

$$\begin{array}{ccc} K & \xrightarrow{p_2} & T \\ {\scriptstyle p_1}\downarrow & & \downarrow \\ T & \longrightarrow & C \end{array} ,$$

and $K$ is petty (in fact, lucid). Now let $\bar{x}$, $\bar{y}$ be transformations from $H^A$ to $C$. We wish to show that the equalizer of $\bar{x}$, $\bar{y}$ is petty. We may lift each to a transformation into $T$:

$$\begin{array}{ccc} H^A & & \\ {\scriptstyle x}\downarrow & \searrow^{\bar{x}} & \\ T & \longrightarrow & C \end{array} \qquad\qquad \begin{array}{ccc} H^A & & \\ {\scriptstyle y}\downarrow & \searrow^{\bar{y}} & \\ T & \longrightarrow & C \end{array}$$

let $E$ be the left-limit of

$$\begin{array}{ccc} & & H^A \\ & & {\scriptstyle x}\downarrow\;\downarrow{\scriptstyle y} \\ K & \overset{p_1}{\underset{p_2}{\rightrightarrows}} & T \end{array} .$$

By the hypothesis of the proposition, $E$ is petty. But $E \longrightarrow H^A$ is the desired equalizer. ∎

We may collect:

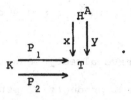

Proposition 1.(10)

If $\underline{A}$ has finite pre-products then the full category of contravariant lucid-functors is closed under finite left-limits and arbitrary right-limits.

If $\underline{A}$ furthermore has pre-equalizers then representable functors are lucid and the yoneda embedding displays $\underline{A}$ as a full pre-coreflective subcategory of the category of contravariant lucid functors. ∎

We shall now find the conditions under which lucid functors are closed under arbitrary left-limits and under which the yoneda embedding displays $\underline{A}$ as a pre-reflective subcategory.

The first is the problem of finding when arbitrary products of lucid functors are lucid.

Proposition 1.(11)

If products of representables are petty and powers of representables are lucid then products of lucid functors are lucid.

Proof. We have already seen that products of representable's petty implies products of petty being petty. Let $\{T_i\}_I$ be a set of lucid functors $x, y: H^A \longrightarrow \Pi T_i$ transformations. We wish to show that the equalizer of $x, y$ is petty. For each $i \in I$ let $u: K_i \longrightarrow H^A$ be the equalizer of $p_i x, p_i y: H^A \longrightarrow T_i$. Then the desired equalizer is the intersection of the K's. Let $P = \Pi_I K_i$ and $r_i \colon P \longrightarrow H^A$ be defined by $r_i = u_i p_i$ . The common equalizer of the $r_i$'s is the desired equalizer. Consider the maps $f, g \colon P \longrightarrow \Pi_{I \times I} H^A$ where $p_{\langle i,j \rangle} f = r_i$ , $p_{\langle i,j \rangle} g = r_j$ . Then the equalizer of f, g is the desired common equalizer. P is petty. Hence if the power $\Pi_{I \times I} H^A$ is lucid then every product of lucids is lucid. ∎

We have seen that the pettyness of products of representables is directly equivalent to the existence of precoproducts. Powers of representables are lucid if we have the further condition that for any sequence $\{\langle f_i, g_i \rangle \colon B \longrightarrow A\}_{i \in I}$ of pairs of maps that there exists a solution set $\{A \longrightarrow C_j\}_J$ as follows:

(1) $B \xrightarrow{f_i} A \longrightarrow C_j = B \xrightarrow{g_i} A \longrightarrow C_j$ all i, j.

(2) For $A \longrightarrow X$ such that $B \xrightarrow{f_i} A \longrightarrow X$ $= B \xrightarrow{j_i} A \longrightarrow X$ all i there exist

$$\begin{array}{ccc} A & \longrightarrow & C_j \\ & \searrow & \downarrow \\ & & X \end{array} \qquad \text{some } j.$$

We can further reduce the condition so that instead of sets of pairs of maps we have sets of maps. Given a set $\{f_i \colon B \longrightarrow A\}$ we define an EXTENDED PRE-COEQUALIZER to be a set $\{A \longrightarrow C_j\}$ such that

(1) $B \xrightarrow{f_i} A \longrightarrow C_j = B \xrightarrow{f_k} A \longrightarrow C_j$ all i, k, j

(2) For $A \longrightarrow X$ such that $B \xrightarrow{f_i} A \longrightarrow X$
$= B \xrightarrow{f_k} A \longrightarrow X$ all i,k there exists

$$\begin{array}{ccc} A & \longrightarrow & C_j \\ & \searrow & \downarrow \\ & & X \end{array} \qquad \text{some } j .$$

It should be clear that an extended pre-coequalizer is precisely a generating set for the common equalizer for the set of transformations $\{H^{f_i}\}$.

*Theorem 1.(12)*

*The following are equivalent.*

(1) *A is right pre-complete.*

(2) *Left-limits of representable functors are petty.*

(3) *Products of representable functors are lucid.*

(4) *Representable functors are lucid and left-limits of lucid functors are lucid.*

(5) *A has pre-coproducts and extended pre-coequalizers.*

*Proof.*

(1) <==> (2) by definition.

(2) ==> (3) one need only remember the test for lucidity (1.1).

(3) $\Longrightarrow$ (4) last propostion.

(4) $\Longrightarrow$ (2) by forgetting.

Now for 5. That (4) $\Longrightarrow$ (5) is easy. For the converse we know that simple pre-coequalizers imply that representables are lucid. (1.4) Pre-coproducts imply that products of lucid functors are petty. (1.6) Since equalizers of lucid functors are lucid it suffices to show that products of lucids are lucid and by the last proposition it suffices to show that powers of representables are lucid. By our comments after the last proof it suffices to construct for any set of pairs

$$\{\langle f_i, g_i\rangle : B \longrightarrow A\}_I$$

a solution set using the existence of extended precoequalizers and pre-coproducts.

For each $i$, let $\{A \longrightarrow C_j\}_{j\in J_i}$ be a pre-coequalizer of $f_i$, $g_i$. Let $K$ be the set of choice functions $\{r: I \longrightarrow \cup J_i \mid r(i) \in J_i\}$. For each $r \in K$ let $\{C_{r(i)} \longrightarrow P_f\}_{I\times L_r}$ be a pre-coproduct of $\{C_{r(i)}\}_I$. For each $r$, and $f \in L_r$ let $\{P_f \longrightarrow D_m\}_{Mf}$ be an extended precoequalizer of $\{A \longrightarrow C_{r(i)} \longrightarrow P_f\}_I$. There exists, then, for each $m \in M_f$, $f \in L_r$, $r \in K$ a map $A \longrightarrow D_m = A \longrightarrow C_{r(i)} \longrightarrow P_f \longrightarrow D_m$ (independent of $i$). The set $\{\{\{\{A \longrightarrow D_m\}_{m\in M_f}\}_{f\in L_r}\}_{r\in K}\}$ is the desired solution set, as follows:

(1) Choose $i, r \in K$, $f \in L_r$, $m \in M_f$. To show:

$$B \xrightarrow{f_i} A \longrightarrow D_m = B \xrightarrow{g_i} A \longrightarrow D_m .$$

But $A \longrightarrow D_m = A \longrightarrow C_{r(i)} \longrightarrow P_f \longrightarrow D_m$

and $B \xrightarrow{f_i} A \longrightarrow C_{r(i)} = B \xrightarrow{g_i} A \longrightarrow C_{r(i)}$ .

(2) Given $A \longrightarrow X$ such that

$$B \xrightarrow{f_i} A \longrightarrow X = B \xrightarrow{g_i} A \longrightarrow X \qquad \text{all } i ,$$

to find $r \in K$, $f \in L_r$, $m \in M_f$ such that there exists

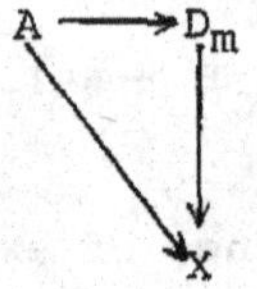

.

We know that for each $i \in I$ there exists $j \in J_i$ and

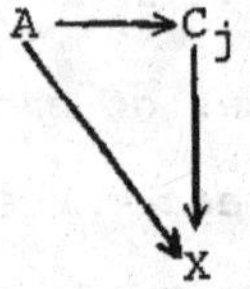

.

Choosing a single $j$ for each $i$ we define $r \in K$ such that there exists

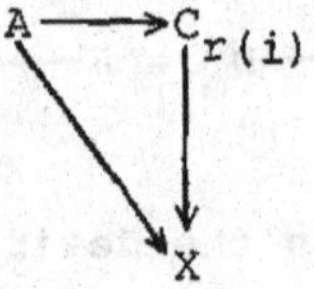

all i.

For the collection $\{C_{r(i)} \longrightarrow X\}$ there exists $f \in L_r$ and triangles

$$\begin{array}{ccc} C_{r(i)} & \longrightarrow & P_f \\ & \searrow & \downarrow \\ & & X \end{array} .$$

Remembering that there is only one map under discussion from $A$ to $X$ we note that $A \longrightarrow C_{r(i)} \longrightarrow P_f \longrightarrow X$ is independent of $i$. Hence by the definition of $\{P_f \longrightarrow D_m\}$ there exists $m$ and

$$\begin{array}{ccc} P_f & \longrightarrow & D_m \\ & \searrow & \downarrow \\ & & X \end{array} . \quad ∎$$

<u>Theorem 1.(13)</u>

<u>$\underline{A}$ may be fully embedded as a pre-coreflective subcategory (simultaneously pre-reflective) of a complete category iff $\underline{A}$ is left (and right) pre-complete.</u>

<u>Proof.</u>

$\Longleftarrow$ clear.

$\Longrightarrow$ The embedding is, of course, into the category of contravariant lucid functors, <u>Lucid</u> $(\underline{A})^{op}$. The last theorem showed that the left pre-completeness of $\underline{A}$ implies the left-completeness of <u>Lucid</u> $(\underline{A}^{op})$. 1.2 and 1.9 demonstrated its right completeness. For the parenthetically stated theorem we suppose that $T$ is a lucid functor.

We seek a set $\{T \longrightarrow H_{A_i}\}$ such that for any $T \longrightarrow H_B$ there exists $i$ and a triangle

$$\begin{array}{ccc} T & \longrightarrow & H_{A_i} \\ & \searrow & \downarrow \\ & & H_B \end{array}$$

. A transformation $T \longrightarrow H_B$ is an element of the Isbell conjugate $T^*$ when evaluated at $B$. And $\{T \longrightarrow H_{A_i}\}$ would be nothing more nor less than a generating set of elements for $T^*$. Thus we wish to show precisely that the Isbell conjugate of a lucid functor is petty. Every petty functor is a $\underrightarrow{\lim}$ of representables, conjugating carries $\underrightarrow{\lim}$'s into $\underleftarrow{\lim}$'s, hence $T^*$ is a $\underleftarrow{\lim}$ of representables. ($T$ was contravariant, $T^*$ is a $\underleftarrow{\lim}$ of covariant representables.) Now the right pre-completeness and the last theorem finish the proof. $T^*$ is lucid, a fortiore, it is petty. ∎

## II THE RIGHT-COMPLETION

Let $\underline{B}$ be a right-complete category and $\mathbb{F}$ any set, class or super-class of right-continuous functors from $\underline{B}$ (range not fixed).

Let $\underline{C} \subset \underline{B}^{\to}$ be the class of maps carried into isomorphisms by every functor in $\mathbb{F}$. Let $\overline{\underline{B}}$ be the category of fractions obtained by formally adjoining inverses of the maps in $\underline{C}$. We recall that objects of $\overline{\underline{B}}$ are the same as those of $\underline{B}$, the maps from $B$ to $A$ are represented by pairs $\langle B \longrightarrow A_1, A \longrightarrow A_1 \rangle$ where $A \longrightarrow A_1 \in \underline{C}$ subject to the congruence $\langle B \longrightarrow A_1, A \longrightarrow A_1 \rangle \equiv \langle B \longrightarrow A_2, A \longrightarrow A_2 \rangle$ if there exists

$$\begin{array}{ccc} A & \longrightarrow & A_1 \\ \downarrow & & \downarrow \\ A_2 & \longrightarrow & A_3 \end{array}$$ all sides of which are in $\underline{C}$ such that

$B \longrightarrow A_1 \longrightarrow A_3 = B \longrightarrow A_2 \longrightarrow A_3$. Composition is defined as follows: given $\langle C \longrightarrow B_1,\ B \longrightarrow B_1 \rangle$ and $\langle B \longrightarrow A_1,\ A \longrightarrow A_1 \rangle$ we let

$$\begin{array}{ccc} B & \longrightarrow & B_1 \\ \downarrow & & \downarrow \\ A_1 & \longrightarrow & A_2 \end{array}$$ be a pushout, noting that $A_1 \longrightarrow A_2$ is in $\underline{C}$ and define

the composition as $\langle C \longrightarrow B \longrightarrow A_2,\ A \longrightarrow A_1 \longrightarrow A_2 \rangle$.

Proposition 2.1

$\bar{\underline{B}}$ is right-complete, $\underline{B} \longrightarrow \bar{\underline{B}}$ right-continuous. Given any $F\colon \underline{B} \longrightarrow \underline{E}$ in $\mathbb{F}$ there exists unique $\bar{F}\colon \bar{\underline{B}} \longrightarrow \underline{E}$ such that $\underline{B} \longrightarrow \bar{\underline{B}} \longrightarrow \underline{E} = F$, $F$ is right-continuous and given any $G\colon \bar{\underline{B}} \longrightarrow \underline{C}$ and natural transformation $\dot{\eta}\colon F \longrightarrow G|\underline{B}$ there exists unique extension $\bar{F} \longrightarrow G$.

Proof. We observe first that $\underline{B} \longrightarrow \bar{\underline{B}}$ is right-continuous. Given $T\colon \underline{D} \longrightarrow \underline{B}$, $\underline{D}$ small we wish to show that $\varinjlim T$ remains the right-limit of $T\colon \underline{D} \longrightarrow \underline{B} \longrightarrow \bar{\underline{B}}$. Given any collection $\{\langle TD \longrightarrow X_D,\ X \longrightarrow X_D \rangle\}_{D \in \underline{D}}$ with $X \longrightarrow X_D$ in $\underline{C}$ all $D$, such that $\langle TD' \longrightarrow TD \longrightarrow X_D,\ X \longrightarrow X_D \rangle \equiv \langle TD' \longrightarrow X_{D'},\ X \longrightarrow X_{D'} \rangle$ all $D' \longrightarrow D$ in $X\underline{CD}$, we wish to find $\langle \varinjlim T \longrightarrow X_L,\ X \longrightarrow X_L \rangle$, $X \longrightarrow X_L$ in $\hat{\underline{C}}$ such that $\langle TD \longrightarrow \lim T \longrightarrow X_L,\ X \longrightarrow X_L \rangle \equiv \langle TD \longrightarrow X_D,\ X \longrightarrow X_D \rangle$

all D. We define first $X'$ as the right-limit of the family $\{X \longrightarrow X_D\}$. Because each functor in $\mathbb{F}$ is right-continuous it is the case that $X_D \longrightarrow X'$ and $X \longrightarrow X'$ are in $\underline{C}$. For each $f\colon D' \longrightarrow D$ let $\begin{array}{ccc} X & \longrightarrow & X_D \\ \downarrow & & \downarrow \\ X_{D'} & \longrightarrow & X_f \end{array}$ be such that

$TD' \longrightarrow TD \longrightarrow X_D \longrightarrow X_f = TD' \longrightarrow X_{D'} \longrightarrow X_f$ (as must exist because of the given equivalences). Let $\begin{array}{ccc} X & \longrightarrow & X' \\ \downarrow & & \downarrow \\ X_f & \longrightarrow & X'_f \end{array}$ be a pushout, note that $X \longrightarrow X'_f$ is in $\underline{C}$ and define $X_L$ to be the right-limit of the family $\{X \longrightarrow X'_f\}$. We now have a family $\{TD \longrightarrow X \longrightarrow X_L\}_{\underline{D}}$ such that $TD' \longrightarrow TD \longrightarrow X_L$ $= TD' \longrightarrow X_L$ all $D' \longrightarrow D$ in $\underline{D}$. Hence there exists $\lim T \longrightarrow X_L$. The pair $\langle \lim T \longrightarrow X_L, X \longrightarrow X_L \rangle$ is the pair we seek. To see that $\langle TD \longrightarrow \lim T \longrightarrow X_L, X \longrightarrow X_L \rangle$ $\equiv \langle TD \longrightarrow X_D, X \longrightarrow X_D \rangle$ recall that we have for each D a map $X_D \longrightarrow X_L$ in $\underline{C}$ such that $TD \longrightarrow X_L = TD \longrightarrow X \longrightarrow X_D \longrightarrow X_L$.

We have shown that $\varinjlim T$ is a weak-limit in $\underline{B}$.

Any functor F from a right-complete category which carries right limits into weak-right-limits is, in fact, right-continuous. We need only show that a jointly epimorphic family $\{A_i \longrightarrow B\}$ is carried into such. Let L be the $\varinjlim$ of the family $\left\{\begin{array}{ccc} A_i & \longrightarrow & B \\ \downarrow & & \\ B & & \end{array}\right\}$, $B \xrightarrow{u} L$, $B \xrightarrow{v} L$ the associated maps. FL is the weak-right-limit of the family $\left\{\begin{array}{ccc} FA_i & \longrightarrow & FB \\ \downarrow & & \\ FB & & \end{array}\right\}$. Given a pair $f,g\colon FB \longrightarrow X$ such that $FA_i \longrightarrow FB \xrightarrow{f} X = FA_i \longrightarrow FB \xrightarrow{g} X$ all $i$ there exists a map $FL \longrightarrow X$ such that $FB \xrightarrow{Fu} FL \longrightarrow X = f$ and $FB \xrightarrow{Fv} FL \longrightarrow X = g$. The joint epimorphism of the original family $\{A_i \longrightarrow B\}$ implies (and is equivalent to) $u = v$. Hence $Fu = Fv$ and $f = g$.

The right-completeness of $\underline{B}$ is proved by first observing that sums exist easily (because $\underline{B} \longrightarrow \bar{\underline{B}}$ is right-continuous and every object in $\underline{B}$ comes from $\underline{B}$). For the existence of co-equalizers let $\langle B \longrightarrow A_1, A \longrightarrow A_1\rangle$ and $\langle B \longrightarrow A_2, A \longrightarrow A_2\rangle$ be given. Let $\begin{array}{ccc} A & \longrightarrow & A_1 \\ \downarrow & & \downarrow \\ A_2 & \longrightarrow & A_3 \end{array}$ be a pushout in $\underline{B}$, noting that $A \longrightarrow A_3$ is in $\underline{C}$. We may rename the maps of the given pair as $\langle B \xrightarrow{f} A_3, A \longrightarrow A_3\rangle$, $\langle B \xrightarrow{g} A_3, A \longrightarrow A_3\rangle$. A co-equalizer of $\langle B \xrightarrow{f} A_3, A_3 \xrightarrow{1} A_3\rangle$ and $\langle B \xrightarrow{g} A_3, A_3 \xrightarrow{1} A_3\rangle$ would serve as a co-equalizer for the given pair since $\langle A \longrightarrow A_3, A_3 \longrightarrow A_3\rangle$ is an isomorphism. We know that the $\underline{B}$-co-equalizer of $\langle f,g\rangle$ remains a co-equalizer in $\underline{B}$.

Given $F\colon \underline{B} \longrightarrow \underline{E}$ in $\mathbf{F}$ its definition on the objects of $\underline{B}$ is clearly forced. Given $\langle B \xrightarrow{f} A_1, A \xrightarrow{g} A_1\rangle$, $g$ in $\underline{C}$ we define $F(f/g) = FB \xrightarrow{Ff} FA_1 \xrightarrow{(Fg)^{-1}} FA$. The uniqueness is clear. That F preserves sums and co-equalizers is easily proved along the lines of the last paragraph. ∎

Given a category $\underline{A}$ let $\underline{\text{Small}}(\underline{A}^{op})$ be the category of functors which appear as $\varinjlim$'s of contravariant representables. As Ulmer has observed, $\underline{\text{Small}}(\underline{A}^{op})$ is a free right-completion. If we define $\mathbb{F}$ to be the class of all right-continuous extensions of right-continous functors on $\underline{A}$, define $\underline{C}$ as those maps carried into isomorphisms by functors in $\mathbb{F}$ and the category $\underline{\text{Cont}}(\underline{A}^{op})$ as the category of fractions obtained from $\underline{\text{Small}}(\underline{A}^{op})$ by inverting the maps in $\underline{C}$, then

Proposition 2.2

$\underline{\text{Cont}}(\underline{A}^{op})$ is right-complete, $\underline{A} \longrightarrow \underline{\text{Cont}}(\underline{A}^{op})$ is a right-continuous full embedding; and every right-continuous $\underline{A} \longrightarrow \underline{E}$, $\underline{E}$ right-complete, may be extended right-continuously to $\underline{\text{Cont}}(\underline{A}^{op})$, uniquely up to natural equivalence.

Proof. All but one statement follow directly from the last proposition. As for the properties of $\underline{A} \longrightarrow \underline{\text{Cont}}(\underline{A}^{op})$ we first observe that if $H_A \longrightarrow T$ is in $\underline{C}$ then since $H_A\colon \underline{A} \longrightarrow \underline{S}^{op}$ is in $\mathbb{F}$ that $(T,H_A) \longrightarrow (H_A,H_A)$ is an isomorphism. Hence there exists $T \longrightarrow H_A$ such that $H_A \longrightarrow T \longrightarrow H_A = 1$ and $T \longrightarrow H_A$ is in $\underline{C}$. Given any map $\langle F \longrightarrow T,\ H_A \longrightarrow T\rangle$ we see that it may be renamed as $\langle F \longrightarrow T \longrightarrow H_A,\ 1\rangle$. In particular the maps in $\underline{\text{Cont}}(\underline{A}^{op})$

from $H_B$ to $H_A$ are represented by maps in $\underline{A}^{op}$ from $H_B$ to $H_A$. That the correspondence is one-to-one follows again from the splitting of $H_A \longrightarrow T$ for $H_A \longrightarrow T$ in $\underline{C}$.

The right-continuity of $\underline{A} \longrightarrow \underline{\text{Cont}}(\underline{A}^{op})$ is obtained as follows: given $T\colon \underline{D} \longrightarrow \underline{A}$, $\underline{D}$ small $\varinjlim T = A$ we wish to show that $\lim H_{TD} \longrightarrow H_A$ is an isomorphism in $\underline{\text{Cont}}(\underline{A}^{op})$. It clearly suffices to show that as a map in $\underline{\text{Small}}(\underline{A}^{op})$ it is in $\underline{C}$. The very definition of $\mathbb{T}$ insures that it is. ∎

For many purposes $\underline{\text{Cont}}(\underline{A}^{op})$ is too big. It need not be locally small. We wish to find a smaller completion.

## III. CONCORDANT FUNCTORS

Given small $\underline{D}_1 \subset \underline{D}_2$ and functor $T\colon \underline{D}_2 \longrightarrow \underline{A}$ we say that $\langle \underline{D}_1, \underline{D}_2, T\rangle$ is a SATURATED PAIR OF DIAGRAMS if for every $A \in \underline{A}$ the induced function

$$\varprojlim_{\underline{D}_2}(TD,A) \longrightarrow \varprojlim_{\underline{D}_1}(TD,A)$$

is an isomorphism. By directly translating, then, the pair is saturated if for every family $\{TD \longrightarrow A\}_{D \in \underline{D}_1}$

such that

$$\begin{array}{ccc} TD & \searrow & \\ \downarrow & & A \\ TD' & \nearrow & \end{array}$$

all $D \longrightarrow D' \in \underline{D}_1$. There exists a unique enlargement of the family $\{TD \longrightarrow A\}_{D \in \underline{D}_2}$ such that

$$\begin{array}{ccc} TD & \searrow & \\ \downarrow & & A \\ TD' & \nearrow & \end{array}$$

all $D \longrightarrow D' \in \underline{D}_2$.

If either $\varinjlim_{\underline{D}_1} TD$ or $\varinjlim_{\underline{D}_2} TD$ exist in $\underline{A}$ then the pair is saturated if and only if they both exist and are equal.

A functor $F\colon \underline{A} \longrightarrow \underline{B}$ is RIGHT CONCORDANT if it carries saturated pairs into saturated pairs:

$\langle \underline{D}_1, \underline{D}_2, T\rangle$ saturated $\Longrightarrow$ $\langle \underline{D}_1, \underline{D}_2, FT\rangle$ saturated.

Proposition 3.1

Right-concordant functors are right continuous.

If $\underline{A}$ and $\underline{B}$ are complete then $F\colon \underline{A} \longrightarrow \underline{B}$ is right-concordant if and only if it is right-continuous. ■

Proposition 3.2

Representable functors $\underline{A}^{op} \longrightarrow \underline{S}$ are left-concordant.

Products of left-concordant functors are left-concordant.

Equalizers of maps between left-concordant functors are left-concordant.

If $\underline{A}' \subset$ Lucid $(\underline{A}^{op})$ contains the representables and *if all functors in* $\underline{A}'$ *are left-concordant, then the Yoneda embedding* $\underline{A} \longrightarrow \underline{A}'$ *is right-concordant.*

*Proof.* For the last statement we need only note that $\underline{A} \longrightarrow \underline{A}'$ is right-concordant if and only if $\underline{A}^{op} \longrightarrow (\underline{A}')^{op} \xrightarrow{(-,T)} \underline{S}$ is left-concordant for every $T \in \underline{A}'$. That it is such is the definition of the left-concordance of T. ∎

*Lemma 3.3*

*For* $\underline{A}$ *pre-bicomplete,* $(F \longrightarrow P') \in$ *Lucid* $(\underline{A}^{op})$ *such that the Isbell conjugate* $F'^* \longrightarrow F^*$ *is an isomorphism and for any left-concordant* $T\colon \underline{A}^{op} \longrightarrow \underline{S}$ *(lucid or not), it is the case that* $(\bar{F},T) \longrightarrow (F,T)$ *is an isomorphism.*

*Proof.* We first display F as a $\varinjlim$ of representables. Since it is petty we can find a set $\{H_{A_i} \longrightarrow F'\}_I$ such that $\sum H_{A_i} \longrightarrow F'$ is onto. For each $\langle i,j\rangle \in I \times I$ we let

$$\begin{array}{ccc} P_{\langle i,j\rangle} & \longrightarrow & H_{A_i} \\ \downarrow & & \downarrow \\ H_{A_j} & \longrightarrow & F' \end{array}$$

be a pullback. Because $F'$ is lucid and $\underline{A}$ has pre-products we know that $P_{\langle i,j\rangle}$ is petty (1.7). Let $\{H_{A_k} \longrightarrow P_{\langle i,j\rangle}\}_{K_{i,j}}$ be such that $\sum_K H_{A_k} \longrightarrow P$ is onto. Define $\underline{D}_1$ to be the category whose objects are $I \cup \bigcup K_{i,j}$ and with maps $k \longrightarrow i$, $k \longrightarrow j$ for every $k \in K_{i,j}$. We define $\underline{D}_1 \longrightarrow \underline{A}$ by sending $i \in I$ to $A_i$, $k \in K_{i,j}$ to $A_k$. Given $K \longrightarrow i \in \underline{D}_1$ we compose $H_{A_k} \longrightarrow P_{i,j} \longrightarrow H_{A_i}$ and determine a map $A_k \longrightarrow A_i$. Similarly for $k \longrightarrow j$. $\underrightarrow{\lim}\,(\underline{D}_1 \longrightarrow \underline{A}) = F'$.

Similarly we may find $\underline{D}_3 \longrightarrow \underline{A}$ such that $\underrightarrow{\lim}\,(\underline{D}_3 \longrightarrow \underline{A})$.

For every $i \in \underline{D}_1$ we may choose $c(i) \in \underline{D}_3$ and

$$\begin{array}{ccc} H_{A_i} & \longrightarrow & F' \\ {\scriptstyle H_{M(i)}}\downarrow & & \downarrow \\ H_{A_{c(i)}} & \longrightarrow & F \end{array} \quad .$$

We define $\underline{D}_2$ to be the category whose objects are the disjoint union of those from $\underline{D}_1$ and $\underline{D}_3$, whose maps are those from $\underline{D}_1$ and $\underline{D}_3$ together with new maps $i \longrightarrow c(i)$, one for each $i \in \underline{D}_1$, and finally the necessary compositions $k \longrightarrow i \longrightarrow c(i)$, $k \longrightarrow j \longrightarrow c(j)$ for $k \in K_{i,j}$. We define $\underline{D}_2 \longrightarrow \underline{A}$ by extending the union of the two previous embeddings $\underline{D}_1 \longrightarrow \underline{A}$, $\underline{D}_3 \longrightarrow \underline{A}$ to $\underline{D}_2 \longrightarrow \underline{A}$ with $i \longrightarrow c(i)$ going to $r(i)$. With no assumptions on the map $F' \longrightarrow F$ it is easily the case that $\underrightarrow{\lim}\,(\underline{D}_2 \longrightarrow \underline{A}) = F$.

$\underline{D}_1$, $\underline{D}_2$ is a saturated pair because $F'^* \longrightarrow F^*$ is an isomorphism. ∎

We consider again the category $\underline{\text{Small}}(\underline{A}^{op})$ and define $\mathbb{F}$ as the family of all right-continuous functors from $\underline{\text{Small}}(\underline{A}^{op})$ which are extensions of right-concordant functors from $\underline{A}$, $\underline{C}$ as the maps carried into isomorphisms by every functor in $\mathbb{F}$ and $\underline{\text{Conc}}(\underline{A}^{op})$ the category of fractions obtained by inverting the maps in $\underline{C}$. By the proof of the last proposition, $\underline{C}$ consists of those maps which become isomorphisms when conjugated.

Proposition 3.4

$\underline{\text{Conc}}(\underline{A}^{op})$ is right-complete, $\underline{A} \longrightarrow \underline{\text{Conc}}(\underline{A}^{op})$ is a right-concordant full embedding and every right-concordant $\underline{A} \longrightarrow \underline{E}$, $\underline{E}$ right-complete, may be extended right-continuously to $\underline{\text{Conc}}(\underline{A}^{op})$; uniquely up to natural equivalence.

Proof. As for 2.2. ▮

$\underline{\text{Conc}}(\underline{A}^{op})$ need not be locally-small. But.

Theorem 3.5

If $\underline{A}$ is pre-bicomplete, then $\underline{\text{Conc}}(\underline{A}^{op})$ is locally-small.

Proof. Given any conjugacy-isomorph $F \longrightarrow F'$ we obtain an induced map $F' \longrightarrow F^{**}$ such that $F \longrightarrow F' \longrightarrow F^{**}$ is the canonical $F \longrightarrow F^{**}$. $F' \longrightarrow F^{**} = F' \longrightarrow F'^{**} \longrightarrow F^{**}$ where the second map is the inverse of the isomorphism

$F^{**} \longrightarrow F'^{**}$.

Lemma 3.51

$\langle G \longrightarrow F_1, F \longrightarrow F_1 \rangle \equiv \langle G \longrightarrow F_2, F \longrightarrow F_2 \rangle$ *if and only if* $G \longrightarrow F_1 \longrightarrow F^{**} = G \longrightarrow F_2 \longrightarrow F^{**}$.

*Proof.* We suppose first that $G \longrightarrow F_1 \longrightarrow F^{**} = G \longrightarrow F_2 \longrightarrow F^{**}$. Consider the pushout $\begin{array}{ccc} F & \longrightarrow & F_1 \\ \downarrow & & \downarrow \\ F_2 & \longrightarrow & F_3 \end{array}$ and note that the induced map $F_3 \longrightarrow F^{**}$ is such that $F_n \longrightarrow F_3 \longrightarrow F^{**} = F_n \longrightarrow F^{**}$, $n = 1,2$. Let $F_4$ be the image of $F_3 \longrightarrow F^{**}$. $F_3 \longrightarrow F_4$ is easily seen to be a conjugacy-isomorph. Hence $G \longrightarrow F_1 \longrightarrow F_4 = G \longrightarrow F_2 \longrightarrow F_4$ and $\langle G \longrightarrow F_1, F \longrightarrow F_1 \rangle \equiv \langle G \longrightarrow F_2, F \longrightarrow F_2 \rangle$. The converse is left as an exercise.

The lemma easily proves the theorem. Indeed, it informs us that $(G,F)$ as defined in *Conc*$(\underline{A}^{op})$ is a subset of $(G,F^{**})$ as defined in *Lucid* $(\underline{A}^{op})$, namely those maps $G \longrightarrow F^{**}$ such that there exists

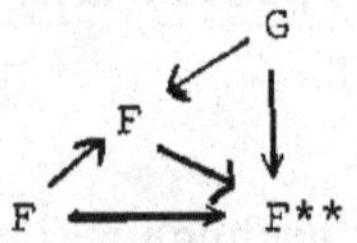


where $F \longrightarrow F_1$ is a conjugacy-isomorph and $F_1 \longrightarrow F^{**}$ is monomorphic. ∎

**Proposition 3.6**

*If* $\underline{A}$ *is small, then* $\text{Conc}(\underline{A}^{op})$ *is the full left-closure in* $(\underline{A}^{op},\underline{S})$ *of representables, and is reflective.*

*Proof.* Given $F \in (\underline{A}^{op},\underline{S})$ let $\bar{F}$ be its reflection in the left-closure of the representables. $F \longrightarrow \bar{F}$ is a conjugacy-isomorphism and for any conjugacy-isomorph $F \longrightarrow F_1$ there exists unique $\begin{array}{ccc} F & \longrightarrow & F_1 \\ & \searrow & \downarrow \\ & & \bar{F} \end{array}$. Hence $(G,F)$ is defined in $\text{Conc}(\underline{A}^{op})$ is canonically the same as $(G,\bar{F})$ as defined in $(\underline{A}^{op},\underline{S})$. $\blacksquare$

**Corollary 3.7**

*If* $\underline{A}$ *is small* $\text{Conc}(\underline{A}^{op})$ *is the category of concordant functors.* $\blacksquare$

We consider now a very special case.

Suppose that $\underline{A}$ is a partially ordered set. In this case $\text{Conc}(\underline{A}^{op}) \simeq \text{Conc}(\underline{A})$ and could, therefore, be called the SYMMETRIC COMPLETION. To verify the symmetry, we note first that a product of representables $\Pi H_i$ has the following characteristic property:

$$(\Pi H_i)(j) = \begin{cases} 1 & \text{if } j \leq i \text{ all } i \in I \\ 0 & \text{otherwise} \end{cases},$$

where 1 is the set with one element, 0 the empty set. There is at most one map between two products of representables, hence $\underline{\text{Conc}}(\underline{A}^{op})$ is precisely the set of products of representables. $\Pi_I H_i \simeq \Pi_{I'} H_i$ if and only if $\{j|j \leq i \text{ all } i \in I\} = \{j|j \leq i \text{ all } i \in I'\}$. Given $I$ we may define $\bar{I} = \{i|i \geq j \text{ for all } j \leq i \text{ all } i \in I\}$, and then $\Pi_I H_i \simeq \Pi_{I'} H_i$ if and only if $\bar{I} = \bar{I}'$. We'll say that a pair of subsets $\langle J,I \rangle$ is a DEDEKIND PAIR if

1) $j \in J,\ i \in I \Longrightarrow j \leq i$

2) $j \leq i$ all $i \in I \Longrightarrow j \in J$

3) $i \geq j$ all $j \in J \Longrightarrow i \in I$.

Given $I$ we may define $J = \{j|j \leq i \text{ all } i \in I\}$ and $\langle J,\bar{I} \rangle$ is a Dedekind pair. We may now note that the isomorphism classes of $\underline{\text{Conc}}(\underline{A}^{op})$ are in natural correspondence with the set of Dedekind pairs, and the existence of maps between objects in $\underline{\text{Conc}}(\underline{A}^{op})$ is reflected by the natural ordering of Dedekind pairs: $\langle J,I \rangle \leq \langle J',I' \rangle$ if and only if $J \subset J'$ if and only if $I' \subset I$. (It helps to note that $J \cap I$ need not be empty, but if non-empty it contains at most one point.) We have obtained a symmetric description of $\underline{\text{Conc}}(\underline{A}^{op})$, namely as the ordered set of Dedekind pairs.

Given a complete $\underline{B}$ and bi-concordant $\underline{A} \longrightarrow \underline{B}$ there are two canonical extensions to $\underline{\text{Conc}}(\underline{A}^{op})$ one left-continuous, the other right-continuous. The two coincide (as a bi-continuous) if and only if $f\colon \underline{A} \longrightarrow \underline{B}$ is UNIFORMALY CONTINUOUS, i.e., if for any Dedekind $\langle J,I \rangle$ it is the case that $\sup_J f(j) = \inf_I f(i)$.

## IV. WHEN DOES PETTY IMPLY LUCID?

Curiosity: 4.1

All petty functors from $\underline{S}$ to $\underline{S}$ or from $\underline{S}^{op}$ to $\underline{S}$ are lucid.

Proof. It suffices to show that for any petty $T$ and pair of transformations $x,y: H \longrightarrow T$ that the equalizer of $x$ and $y$ is petty, where $H$ is representable. A fortiore, it suffices to show that subfunctors or representables are petty. Let $T$ be a subfunctor of $(-,A)$, define $S(T)$ to be $\{\mathrm{Im}(f) \mid f \in TB,\ B \in \underline{S}\}$. Then $T$ is generated by $\{A' \hookrightarrow A\}_{A' \in S}$. Let $T$ be a subfunctor of $(A,-)$, define $Q(T)$ to be the set of equivalence relations on $A$ induced by elements of $T$. Then $T$ is generated by $\{A \longrightarrow A/\equiv\}_{\equiv \in Q}$. ∎

Pathology: 4.2

(Nunke's) The forgetful functor from abelian groups to sets has a quotient (hence very petty) that is not lucid.

(Hedrlin's) For $T$ the terminal functor from semi-groups to set, $T + T$ has a quotient (hence ridiculously petty) that is not lucid.

Proof. Nunke finds a strictly ascending chain of subfunctors of the identity functor. Briefly, they are described as follows: let $K$ be a non-limit cardinal, let $S_K \subset \Pi_K Z$

be the subgroup of K-tuples with support smaller than K, let $A_K = \Pi_K Z/_{S_k}$, let $T_K \subset I$ be the subfunctor generated by $\{x \in A_L \mid L \leq K\}$, let $T = \cup T_K$. T is not petty. I/T then is the example.

Hedrlin can fully embed the big discrete category into semigroups. In particular, there exists a proper class C of semigroups such that for any $A,B \in C$ $(A,B) = \emptyset$. Define $T' \subset T$ as follows:

$$T'X = \begin{cases} * & \text{if } \exists A \in C,\ A \longrightarrow X \\ \emptyset & \text{if } \forall A \in C,\ (A,X) = \emptyset \end{cases} .$$

T′ is not petty. Let

$$\begin{array}{ccc} T' & \longrightarrow & T \\ \downarrow & & \downarrow \\ T & \longrightarrow & P \end{array}$$

be a pushout. P is not lucid. We'll say that a category $\underline{A}$ is PIL if all petty functors from $\underline{A}$ to $\underline{S}$ are lucid. Certainly, if all subfunctors of representables are petty then PIL. The converse is true. Indeed:

<u>Proposition 4.3</u>

<u>$\underline{A}$ is PIL if and only if all subfunctors of petty functors are petty.</u>

<u>Proof</u>. Suppose T′ is a subfunctor of T, T′ not petty, T petty. We construct a non-lucid petty functor. Let

$$\begin{array}{ccc} T' & \longrightarrow & T \\ \downarrow & & \downarrow \\ T & \longrightarrow & P \end{array}$$

be a pushout. $P$ is petty since it is a quotient of $T + T$. But $T'$ is the equalizer of the two maps from $T$ to $P$.

Proposition 4.4

*If* $\underline{A}$ *is* PIL *then any category of the form* $(A,\underline{A})$ *(i.e., objects* $A \longrightarrow A'$, *maps* $A \rightrightarrows {A' \atop \downarrow \atop A''}$ *) is* PIL *and any full subcategory of a* PIL *is a* PIL.

*Proof*. As we have noted, it suffices to verify that subfunctors of representables are petty. This in turn is equivalent, via the usual argument made familiar by Hilbert, to an ascending chain condition, namely that no strictly ascending chains exist indexed by *all* the ordinals. Now if $\underline{A}$ is PIL and $A \longrightarrow B \in (A,\underline{A})$ then the functor represented by $A \longrightarrow B$ may be embedded in the functor $\underline{A}(B,F(-))$ where $F\colon (A,\underline{A}) \longrightarrow \underline{A}$ is the forgetful functor. If $\underline{A}'$ is a full subcategory of $\underline{A}$, $A' \in \underline{A}'$ then the lattice of subfunctors of $\underline{A}'(A',-)$ is a retract of the lattice of subfunctors of $\underline{A}(A,-)$. The inclusion is given by $S \longmapsto$ the subfunctor of $\underline{A}(A,-)$ generated by $\{x \in SA\}_{A \in A'}$, the retraction is induced by restrictions. ∎

Proposition 4.5

*Finite Cartesian products and disjoint union of* PIL's *are* PIL. ∎

Proposition 4.6

*If* $\underline{A}$ *is co-well powered and such that every map factors as an epimorphism followed by a split-monomorphism then* $\underline{A}$ *is* PIL'.

*Proof.* Just as for $\underline{A}$ = Sets. ∎

The writer knows only one PIL not accounted for by the last few propositions. To be more precise, he knows one category $\underline{A}$ not to be obtained as a disjoint union of full subcategories of something of the form $(B,\underline{B})$ where $\underline{B}$ is a Cartesian product of categories satisfying the hypothesis of the last proposition. But if $\underline{B}$ is replaced with $\underline{B} \times \underline{A}^K$, K discrete, then every PIL the writer knows may be so obtained.

The exceptional $\underline{A}$ is the category of sets with a distinguished endomorphism. Because any algebraic theory with one unary operator and a set of constants, yields a full embedding of its algebras into $(F_C, \underline{A})$ (the equations of the theory can make it a proper embedding) we know that such theories yield PIL catogories. But allow just two unary operators and PIL is lost. Indeed if M is either the monoid on two generators a,b with

$$a^3 = a^2b = ab^2 = b^3, \quad a^4 = a^5$$

or the commutative monoid with three generators a,b,c with

$$a^2 = ab = ac = bc = b^2 = c^2, \quad a^3 = a^4$$

then the category of semigroups may be fully embedded in $(A,\mathrm{Sets}^M)$ and $\mathrm{Sets}^M$ is not PIL. Hence the fact that for M the free monoid on one generator it is the case that $\mathrm{Set}^M$ is PIL should not be expected to be easy. We need a few facts about sets with endomorphisms.

## V. THE CATEGORY OF SETS WITH ENDOMORPHISM

Given a set A with an endomorphism $s\colon A \longrightarrow A$, we define a rank function $r\colon A \longrightarrow \mathcal{O}^*$, where $\mathcal{O}^*$ is the "extended ordinals", that is, the ordinals with a maximal element "$\infty$" adjoined. To define r we first define the following transfinite sequence of subsets of A.

$$A_0 = A$$

$$A_{\alpha+1} = sA_\alpha = \{(x) \mid x \in sA_\alpha\}$$

for limit ordinal $\gamma$, $A\gamma = \bigcap_{\alpha<\gamma} A_\alpha$ .

A uniform definition may be given by

$$A\gamma = \bigcap_{\alpha<\gamma} sA_\alpha \;.$$

We understand $\qquad A_\infty = \bigcap_{\alpha\in\theta} A_\alpha$ .

Define $r(x) = \sup \{\alpha \mid x \in A_\alpha\}$.

We may note that $r(x) = \alpha \Longrightarrow x \in A_\alpha$ and $x \notin A_{\alpha+1}$,

$r(x) = 0 \iff x$ has no ancestors.

$r(x) = 1 \iff x$ has a father but no grandfather.

$r(x) = 2 \iff x$ has a grandfather but no great grandfather.

Note that since $A$ is a set there must exist an ordinal $\alpha$ such that $A_\alpha = A_{\alpha+1}$ and hence $A_\alpha = A_\beta$ all $\beta > \alpha$ and $A_\alpha = A_\infty$. (Keep in mind that $A_\infty$ could be empty.) In particular $sA_\infty = A_\infty$. We note then that

$$r(x_0) = \infty \implies \exists x_1, x_2, \ldots \text{ such that } s(x_{n+1}) = x_n \text{ all } n.$$

If $r(x) = \alpha < \infty$ then $x \in A_\alpha$, $s(x) \in A_{\alpha+1}$ and $r(s(x)) > r(x)$. We obtain then the converse of the last implication (because there are no descending chains of ordinals):

$$\exists x_0, x_1, \ldots \text{ such that } s(x_{n+1}) = s(x) \implies (x_0) = \infty \quad .$$

$A_\infty$ is therefore precisely the set of elements with infinite ancestral lines (perhaps periodic, even constant). Note that $r(x) < \omega$ means that there is a bound on the length of ancestral lines. $\omega \leq r(x) < \infty$ means the ancestral lines are unbounded but all finite. $r(x) = \omega$ means unbounded ancestral lines but that each ancester has bounded ancestral lines.

<u>Lemma 5.1</u>

<u>For any</u> $x$ <u>and</u> $\alpha < r(x)$ <u>there exists</u> $y$ <u>such that</u> $s(y) = x$, $\alpha \leq r(y)$.

<u>Proof</u>. $x \in A_{r(x)} = \bigcap_{\beta < r(x)} sA_\beta \subset sA_\alpha$. ▮

Proposition 5.2

*If A and B are sets with distinguished endomorphisms (both called s), f: A ⟶ B is a map such that $f(s(x)) = s\, f(x)$ all x, then $r_B(f(x)) \geq r_A(x)$ all x.*

Proof. If $r_A(x) = \infty$ we can choose an infinite ancestral line, apply f and obtain an infinite ancestral line for f(x), yielding $r_B(f(x)) = \infty$ .

Note that if $r_A(x) = 0$ it is clear that $r_B(f(x)) \geq r_A(x)$. For the general case we argue inductively. Suppose that $\beta$ is such that we know $r_A(y) < \beta \Longrightarrow r_B(f(y)) \geq r_A(y)$. Given $r_A(x) = \beta$, for any $\alpha < \beta$ we choose y such that $s(y) = x$, $r_A(y) \geq \alpha$. It follows that $f(s(y)) = sf(y) = f(x)$, $r_B(f(y)) \geq \alpha$. Hence $r_B(f(x)) = r_B(sf(y)) = r_B(f(x)) \geq \alpha$, thus $r_B(f(x)) \geq \alpha$ all $\alpha < r_A(x)$ and $r_B(f(x)) \geq r_A(x)$. ∎

Given a set A with distinguished endomorphism s: A ⟶ A, we may define an equivalence relation by $x \equiv y$ if there exist n, m, $s^n x = s^m y$. We call a subset of A, PURE if it is closed with respect to $\equiv$. The purification of a subset is the intersection of all pure subsets containing it. A may be partitioned into its minimal pure subsets. In the category of sets with distinguished endomorphisms disjoint union is the categorical sum. Hence the $\equiv$ classes provide the maximal decomposition of A. If A is the only pure subset containing a set A′ we say A is an ESSENTIAL EXTENSION of of $A'$. Note then that given $A' \subset A$, we may define $B$ as the purification of $A'$, $C$ as the compliment of $B$ and obtain $A = B + C$ where $B$ is an essential extension of $A'$.

Lemma 5.3

In the category of sets with distinguished endomorphisms let $A$ be an essential extension of $A'$. A map $f: A' \longrightarrow B$ may be extended to all of $A$ if and only if $r_A(x) \leq r_B(f(x))$ all $x \in A'$.

Proof: $\Longrightarrow$ last lemma. We use Zorn's lemma on the set of partial extensions which maintain the rank inequality. It suffices to show that any such map may be extended just a little (while preserving the rank inequality). Accordingly we show that if $A' \neq A$ we may choose $x \in A - A'$ such that $sx \in A'$ and define a map $\bar{f}: A' \cup \{x\} \longrightarrow B$ such that $\bar{f}|A' = f$, $s\bar{f}(x) = \bar{f}(sx)$ $r_A(x) \leq r_B(\bar{f}(x))$. If $r_A(x) = \infty$ then $r_A(sx) = \infty$ and $r_B(f(sx)) = \infty$. We may pick $y \in B$ such that $sy = x$, $r_B(y) = \infty$ and define $\bar{f}(x) = y$. If $r_A(x) < \infty$ we may by 5.1 find $y \in B$ such that $sy = f(sx)$ $r_B(y) \geq r_A(x)$ because $r_A(x) < r_B(f(sx))$. Define $\bar{f}(x) = y$. ∎

Theorem 5.4

For the category of sets with a distinguished endomorphism any subfunctor of a representable functor is petty.

Proof. We need to show that for any object $A$ and

transfinite sequence of maps $\{f_\alpha : A \longrightarrow B_\alpha\}$, ranging through the ordinal numbers, that there exist $\alpha < \beta$ and a commutative triangle

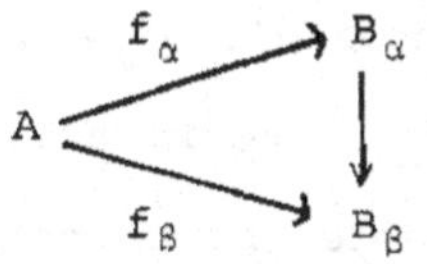

.

(This is equivalent to the conclusion of the theorem as follows: given the transfinite sequence we consider the subfunctors $T \subset (A,-)$ defined by $T(B) = \{A \xrightarrow{f} B \mid \exists \alpha, g,\ gf_\alpha = f\}$. If $T$ is petty let $\{h_i : A \longrightarrow C_i\}_{i \in I}$ be a generating set. For each $i \in I$ let $\alpha(i)$ be such that $\exists g: B_{\alpha(i)} \longrightarrow C_i$ such that $gf_{\alpha(i)} = h_i$. Let $\beta$ be an ordinal larger than $\{\alpha(i)\}$. There exists $i$ such that $A \xrightarrow{f_\beta} B_\beta = A \xrightarrow{h_i} C_i \xrightarrow{g} B_\beta$ and hence a triangle

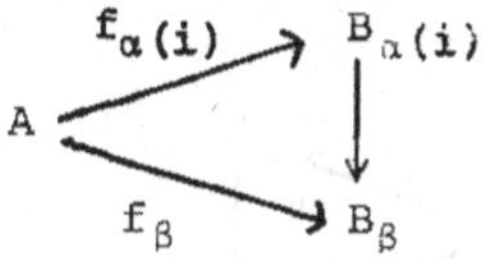

.

Conversely suppose $T$ is a subfunctor of $(A,-)$, and suppose that $T$ is not petty. Let $\{T_\alpha\}$ be a strictly transfinite sequence of subfunctors of $T$. For each ordinal $\alpha$ choose $f_\alpha \in T_{\alpha+1} - T_\alpha$, and obtain a sequence that violates the condition.) We shall in fact show more. Given $\{f_\alpha : A \longrightarrow B_\alpha\}$ we shall find a cofinal subsequence of ordinals $\mathcal{O}' \subset \mathcal{O}$ such that for every $\alpha, \beta \in \mathcal{O}'$, $\alpha < \beta$ there exists a triangle

$$\begin{array}{ccc} & \overset{f_\alpha}{\nearrow} & B_\alpha \\ A & & \downarrow \\ & \underset{f_\beta}{\searrow} & B_\beta \end{array} .$$

Given the transfinite sequence let $C_\alpha \subset B_\alpha$ be the purifications of the image of $f_\alpha$. (We recall that $C_\alpha = \{x \mid \exists n \; s^n x \in \mathrm{Im}(f_\alpha)\}$, where $s: B_\alpha \longrightarrow B_\alpha$ is the distinguished endomorphism of $B_\alpha$). $D_\alpha = B_\alpha - C_\alpha$. $D_\alpha$ is a subobject. $B_\alpha = C_\alpha + D_\alpha$ where $+$ is the categorical sum. The next two lemmas jointly imply the result:

Lemma 5.41

Given a transfinite sequence $\{f_\alpha: A \longrightarrow C_\alpha\}$ where each $C_\alpha$ is an essential extension of $\mathrm{Im}(f_\alpha)$ there exists a cofinal subsequence $\mathcal{O}' \subset \mathcal{O}$ such that for $\alpha, \beta \in \mathcal{O}'$, $\alpha < \beta$ there is a triangle

$$\begin{array}{ccc} & \overset{f_\alpha}{\nearrow} & C_\alpha \\ A & & \downarrow \\ & \underset{f_\beta}{\searrow} & C_\beta \end{array} .$$

Lemma 5.42

Given a transfinite sequence $\{D_\alpha\}$ there exists a cofinal subsequence $\mathcal{O}' \subset \mathcal{O}$ such that for $\alpha, \beta \in \mathcal{O}'$, $\alpha < \beta$ there exists $D_\alpha \longrightarrow D_\beta$.

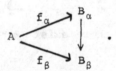

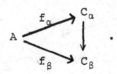

The lemmas imply the main result rather easily. We first find a cofinal subsequence such that there exist maps between the purifications of the images and then a cofinal subsequences of that in which maps exist between the complements of the purifications.

Proof of Lemma 5.41. Each $f_\alpha$ induces a congruence on A: $x \underset{\alpha}{\equiv} y$ if $f_\alpha(y) = f_\alpha(y)$. There are only a set of consequences on A, and hence we can find a cofinal subsequence such that all the congruences are the same. We notationally assume, therefore, that such is already the case. Each $f_\alpha$ factors as $A \longrightarrow A/\equiv \longrightarrow C_\alpha$ where $A/\equiv \longrightarrow C$ is one-to-one and it clearly suffices to solve the problem for $\{A/\equiv \longrightarrow C_\alpha\}$. We may notationally assume, therefore, that the given $\{f_\alpha: A \longrightarrow C_\alpha\}$ is a sequence of one-to-one maps, and in fact are inclusions. Thus we are given a sequence of essential extensions of A, $\{A \subset C_\alpha\}$.

For each $\alpha$ we obtain a function $r_\alpha: |A| \longrightarrow \mathcal{O}^*$. Where $r_\alpha(x) = r_{C_\alpha}(x)$, - the rank function defined earlier. Lemma 5.3 says that it suffices to find a cofinal subsequence $\mathcal{O}' \subset \mathcal{O}$ such that for $\alpha, \beta \in \mathcal{O}'$, $\alpha < \beta$ it is the case that $r_\alpha \leq r_\beta$ (i.e., $r_\alpha(x) \leq r_\beta(x)$ all $x \in |A|$). We switch notation: let I be a set, let $\{r_\alpha: I \longrightarrow \mathcal{O}^*\}$ be a transfinite sequence of function from I to the extended ordinals. We wish to show that there is a cofinal subsequence in which $r_\alpha \leq r_\beta$.

As before, each $\alpha$ produces an equivalent relation on I, we may pass to a cofinal subsequence in which all the equivalence relations are the same and may specialize to the case that the functions are one-to-one. We notationally assume, therefore, that such is already the case.

Each $r_\alpha$ defines a well-ordering on I. ($x_\alpha < y$ if $r_\alpha(x) < r_\beta(y)$). Again there are only a set of orderings on I and we may pass to a cofinal subsequence such that all are the same. We assume therefore that I is well ordered by a relation < and that each $r_\alpha$: I $\longrightarrow$ $\mathcal{O}^*$ is an order-preserving embedding into the extended ordinals.

Either there is a cofinal subsequence in which $\infty \in \mathrm{Im}(r_\alpha)$ or a cofinal subsequence in which $\mathrm{Im}(r_\alpha) \subset \mathcal{O}$ (or both). We may specialize to the latter case as follows: If there is a cofinal subsequence in which $\infty \in \mathrm{Im}(r_\alpha)$ it follows that I has a maximal dement m, that $r_\alpha(m) = \infty$ all $\alpha$ in the subsequence. Let $I' = I-\{m\}$, $r'_\alpha = r_\alpha|I'$ and it suffices to prove the result for $\{r'_\alpha: I' \longrightarrow \mathcal{O}\}$. In any case, therefore, it suffices to assume that the sequence is already such that $\mathrm{Im}(r_\alpha) \subset \mathcal{O}$.

Either $\cup_\alpha \mathrm{Im}(r_\alpha)$ is a set or not. In the first case it is clear that only a set of functions appear in the sequence and that there is a cofinal subsequence in which all the functions are the same and we would be done. In the second case we define $k \in I$ to be the first (with respect to the well-ordering on I) element such that $\{r_\alpha(k)\}$ is not a set. Let $I' = \{i|i < k\}$. Then $\cup_\alpha r_\alpha(I')$ is a set. For each function $f: I' \longrightarrow \cup_\alpha r_\alpha(I')$ define $V_f = \{r_\alpha(k) | r_\alpha | I' = f\}$. Because $\cup_f V_f = \cup_\alpha r_\alpha(k)$ there must exist $f: I' \longrightarrow \mathcal{O}$ such that $V_f$ is not a set. Let $\mathcal{O}' = \{\alpha | r_\alpha | I' = f\}$. $\mathcal{O}'$ is cofinal. We may notationally assume, therefore that the given sequence $\{r_\alpha\}$ is such that for all $\alpha, \beta$ $r_\alpha | I' = r_\beta | I'$ and that $\{r_\alpha(k) | \alpha \in \mathcal{O}\}$ is not a set. Now define a function $\delta: \mathcal{O} \longrightarrow \mathcal{O}$ as follows: $\delta(0) = 0$; $\delta(\alpha) = \min\{\beta | r_\beta(k) \geq \cup_{\delta(\sigma)} r(I)$ $\sigma < \alpha\}$. It is easy to check that for $\alpha < \beta$ $r_{\delta(\alpha)} \leq r_{\delta(\beta)}$ and we are done.

<u>Proof of Lemma 5.42.</u> Given $\{D_\alpha\}$ define $r(D_\alpha) = \sup\{r(x) | x \in D_\alpha\}$. Either there is a cofinal subsequence $\mathcal{O}'$ such that $r(D_\alpha) = \infty$ all $\alpha \in \mathcal{O}'$ or a cofinal $\mathcal{O}'$ such that $r(D_\alpha) \neq \infty$ all $\alpha \in \mathcal{O}'$ (or both). We treat the two cases separately.

In the first case we may assume that the given sequence is such that $r(D_\alpha) = \infty$ all $\alpha$. For each $\alpha$, define $P_\alpha$ as the set of integers $\{n | \exists x \in D_\alpha \ s^n(x) = x\}$. Only a set of possibilities exist for $P_\alpha$ and there is a cofinal subsequence such that they are all the same. We pass to such a subsequence and prove that it satisfies the conclusions of the lemma.

Assume then that $\{D_\alpha\}$ is given, $r(D_\alpha) = \infty$ all $\alpha$, and for any integer $n$, and pair of ordinals $\alpha, \beta$ it is the case that $\exists x \in D_\alpha \ s^n(x) = x \iff \exists x \in D_\beta \ s^n(x) = x$ we wish to show that for any pair of ordinals there exists $f: D_\alpha \longrightarrow D_\beta$. Clearly the problem is no longer a problem about transfinite sequences but about pairs. Given $D_\alpha$ and $D_\beta$, we partition $D_\alpha$ into its minimal pure subobjects. $D_\alpha = \cup_{i \in I} E_i$, each $E_i$ an indecomposable object. Disjoint union is the categorical sum and it suffices to show that there exists a map $E_i \longrightarrow D_\beta$ each i. If $r(E_i) < \infty$ we pick an arbitrary point $x \in E_i$ and a point $y \in D_\beta$ such that $r(y) = \infty$. We define $f(x) = y$, $f(s^n(x)) = s^n(y)$. If we define $E_i' = \{s^n(x) | n = 0,1,\ldots\}$ we note that $E_i$ is an essential extension of $E_i'$, that for $s^n(x) \in E_i'$ $r(s^n(x)) < r_{D_\beta}(f(s^n(x)))$ (because $r(s^n(x)) < \infty$ and $(r_{D_\beta}) s^n(y) = \infty$). Hence lemma yields a map $E_i \longrightarrow D_\beta$. If $r(E_i) = \infty$ either there exists $x \in E_i$ and $n > 0$ such that $s^n(x) = x$ or not. In the latter case we repeat the argument above, taking any $x$ in $E_i$. Otherwise we let $n$ be the smallest positive integer such that there exists $x \in E_i$, $s^n(x) = x$, and choose $y \in D_\beta$ such that $s^n(y) = y$ (finally using the assumption $P_\alpha = P_\beta$). Define $E_i' = \{x, s(x), \ldots s^{n-1}(x)\}$ and $f: E_i' \longrightarrow D_\beta$ by $f(s^i(x)) = s^i(y)$. Lemma 5.3 now provides a map $E_i \longrightarrow D_\beta$.

For the remaining case we assume that $\{D_\alpha\}$ is given, $r(D_\alpha) < \infty$ all $\alpha$. Either $\{r(D_\alpha) | \alpha \in \mathcal{O}\}$ is a set or not. In the latter case we may pick a cofinal subsequence such that for $\alpha < \beta$ $r(D_\alpha) < r(D_\beta)$. We show that this provides the solution as follows: Partition $D_\alpha$ into its minimal pure subobjects. $D_\alpha = \cup E_i$. It suffices to show that for each $i$ there is a map $E_i \longrightarrow D_\beta$. Pick any $x \in E_i$. Choose $y \in D_\beta$ such that $r_{D_\beta}(y) > r(D_\alpha)$. Let $E_i' = \{s^n(x) \mid n = 0,1,\ldots\}$ and $f\colon E_i' \longrightarrow D_\beta$ be the function $f(s^n(x)) = s^n(y)$. Because $r_{D_\alpha}(s^n(x)) < r_{D_\beta}(f(s^n(x))$ lemma 5.3 yields a map $f\colon E_i \longrightarrow D_\beta$. In the former case, there is only a set of ordinals that appear $\{r_{D_\alpha}(x) \mid x \in D_\alpha\}$ is a set. For each $\alpha$, and $x \in D_\alpha$ we define the sequence $t_{\alpha,x}\colon \omega \longrightarrow 0$ by $t_{\alpha,x}(n) = r_{D_\alpha}(s^n(x))$. For each $\alpha$ we obtain a set of $\mathfrak{S}_\alpha$ of sequences $\{t_{\alpha,x} \mid x \in D_\alpha\}$. The condition on the values of $r_{D_\alpha}(x)$ imply that there are only a set of possibilities for $\mathfrak{S}_\alpha$ and hence there is a cofinal subsequence such that they are all the same. We show that this provides a solution as follows: Given $D_\alpha, D_\beta, \mathfrak{S}_\alpha = \mathfrak{S}_\beta$ we partition $D_\alpha$ into its minimal pure subobjects. $D_\alpha = \cup E_i$. It suffices to show that there exists a map $E_i \longrightarrow D_\beta$ each $i$. Pick $x \in E_i$ the sequence $t_{\alpha,x}\colon \omega \longrightarrow 0$ appears in $\mathfrak{S}_\beta$, hence there exists $y \in D_\beta$ such that $t_{\alpha,x} = t_{\beta,y}$. Define $E_i' = \{s^n(x) \mid n = 0,1,\ldots\}$, $f\colon E_i' \longrightarrow D_\beta$ by $f(s^n(x)) = s^n(y)$. Then $r_{E_i}(s^n(x)) = r_{D_\beta}(f(s^n(y)))$ because $r_{E_i}x(s^n(x)) = r_{D_\alpha}(s^n(x)) = t_{\alpha,x}(n) = t_{\beta,y}(n)$ $= D_\beta(s^n(y)) = D_\beta(f(s^n(x)))$. And the lemma 5.3 yields a map $E_i \longrightarrow D_\beta$. ∎

The techniques of the proof above can be used to

prove the following propositions in which are used the definitions:

$r_D(x)$ the rank as defined for lemma 5.3

$r(D) = \sup\{r_D(x) \mid x \in D\}$

$t_{D,x}$ the sequence $\omega \longrightarrow \mathcal{O}^*$ defined by $t_{D,x}(n) = D(s^n(x))$.

$\$_D$ the set of sequences $\{t_{D,x} \mid x \in D\}$

$\$_{D_1} \leq \$_{D_2}$ iff for all $t \in \$_{D_1}$ there exists $t' \in \$_{D_2}$, $t \leq t'$.

$P_D = \{n \mid \exists x \in D,\ s^n(x) = x\}$.

<u>Proposition 5.5</u>

<u>Given objects,</u> $D_1, D_2$ <u>in the category of sets with distinguished endomorphism there exists a map</u> $f\colon D_1 \longrightarrow D_2$ <u>if and only if</u>

$$r(D_1) < r(D_2)$$

<u>or</u> $\qquad r(D_1) = r(D_2) = \infty$ and $P(D_1) \subset P(D_2)$

<u>or</u> $\qquad r(D_1) = r(D_2) < \infty$ and $\$(D_1) \leq \$(D_2)$ . ∎

It may also be pointed out that every ascending sequence $t\colon \omega \longrightarrow \mathcal{O}^*$ appears, as follows:

For any limit ordinal $\alpha$, let $A$ be the set of ascending sequences from $\omega$ to the ordinals less than $\alpha$. Define $s\colon A \longrightarrow A$ by $(st)(n) = t(n + 1)$. It may be easily verified that $A_\beta$, as used for the definition of $r$, is the subset of sequences whose initial values are greater than or equal to

$\beta$. Hence $r_A(t) = t(0)$ and $r_A(s^n(t)) = t(n)$. An immediate corollary is that the terminal functor on the opposite category has a non-petty subfunctor, namely that which is empty only on objects of rank $\infty$.

Finally we note that the set $\mathbb{S}_D$ is larger than it need be. Define sequences $t, t' \colon \omega \longrightarrow \mathcal{O}$ to be stably equivalent, $t \sim t'$ if there exist integers $n, m$ such that $t(i + n) = t'(i + m)$ all $i$. Any two sequences arising from an indecomposable object are stably equivalent. The set $[\mathbb{S}]$ of stable equivalence types may be used instead of the set of all sequences. It is amusing that the pre-ordered family of indecomposable objects of rank $\omega$ (the ordering given by the existence of maps) is equivalent to the "orders of infinity" of real variable analysis.

# THE CATEGORICAL COMPREHENSION SCHEME

by

John W. Gray

## INTRODUCTION

This paper is based on an attempt to find an analogue in category theory to the comprehension scheme of set theory which says, essentially, that given a property, there is a set consisting exactly of the elements having that property. Lawvere has translated this into a statement about adjoint functors

$$(\text{Sets}, X) \rightleftarrows 2^X$$

which are determined by substitution. If, instead of 2-valued functions on a set, one considers set-valued functors on a category, then he showed that there is a similar pair of adjoint functors

$$(\text{Cat}, \mathcal{X}) \rightleftarrows \text{Sets}^{\mathcal{X}^{op}}$$

where the functor from right to left assigns to $F: \mathcal{X}^{op} \longrightarrow \text{Sets}$ the corresponding fibred category over $\mathcal{X}$ with discrete fibres. It is natural to ask if this extends to $\text{Cat}^{\mathcal{X}^{op}}$ with values being arbitrary (split, normal) fibrations over $\mathcal{X}$. In §1, we review and reformulate fibred categories, in §2, we show, using properties of comma categories that there is such a pair of adjoint functors

$$(\text{Cat}, \mathcal{X}) \rightleftarrows \text{Cat}^{\mathcal{X}},$$

and in §3, we discuss what it would mean for this to be an instance of the comprehension scheme. The answer is that it is not, and the reason is that in this context the comprehension scheme is equivalent to asserting that

$$\varinjlim (\mathcal{B} \xrightarrow{1_{\mathcal{B}}} \mathrm{Cat}) = \mathcal{B} \in \mathrm{Cat} \qquad *$$

where $1_{\mathcal{B}}$ is the constant functor with value $\mathbb{1} \in \mathrm{Cat}$. This is false, unless $\mathcal{B}$ is discrete.

The difficulty appears to lie in applying the notions of functor categories, comma categories, adjointness and limits in a delicate and interlocking way to Cat, which is intrinsically a 2-category. A discussion of the definition and some of the properties of 2-categories will be found in §4. It seems reasonable that if the above notions were suitably altered to take account of the 2-category structure, then one might hope to recapture a form of the comprehension scheme. Thus, in §5 on pro 2-functors and bifibrations, we describe the basic construction in terms of which one can introduce the proper notion of 2-comma categories and hence of super 2-functor categories (§6). This determines the notion of 2-adjointness (§7). (It should not be supposed that this coincides with adjointness enriched in Cat as in Linton [AC] or Eilenberg and Kelly [CC].) In particular, one needs an amusing version of the Yoneda lemma. 2-adjointness, of course, determines the notion of 2-colimits in Cat, which are computed in §8. Finally, in §9, we return to the comprehension scheme and find that it does work in considerably greater generality than indicated above, the basic tool being the calculation of 2-Kan extensions of Cat-valued functors.

In an appendix, we discuss the implications of these results for a categorical foundations of mathematics. It is interesting and perhaps significant that the form of adjointness for 2-categories appears to be forced by requiring that the analogue of equation * should hold, and that this in turn is forced by asking for the comprehension scheme. If one believes that the comprehension scheme is a basic ingredient of mathematical thought, then the entire theory presented here is already rigidly determined. In fact, this is how I felt in writing down the theory. This was, of course, helped by the fact that I had already found most of the constructions during a year spent at the Forschungsinstitut für Mathematik of the ETH in Zürich, while on an NSF Fellowship. Actually, I first described the "basic construction" of §5 at Oberwolfach in 1964, but it was only after hearing Lawvere's discussion of the comprehension scheme during this conference that I realized (almost instantly) how these constructions fitted together. Such is the power of the comprehension scheme. I should also remark that in [BC], Bénabou promises that functor categories and adjointness are different for bicategories. From remarks of Tierney, it seems likely that Bénabou has results analogous to some of those presented here; in particular, he apparently is aware of the Yoneda-like lemma in §8.

# PART I THE PROBLEM

## §1 CATEGORICAL FIBRATIONS

For a review of the Grothendieck theory of categorical fibrations ([SGA]) and other aspects of the theory, see [FCC]. We reformulate the notions in the form we shall use them here. Let $P\colon \mathcal{E} \longrightarrow \mathcal{B}$ be a functor and for each object $B \in |\mathcal{B}|$, let $\mathcal{E}_B = P^{-1}(B)$ with the inclusion functor $J_B\colon \mathcal{E}_B \longrightarrow \mathcal{E}$.

Definition

$P\colon \mathcal{E} \longrightarrow \mathcal{B}$ is an i-fibration, $i = 0,1$ if, for each $f\colon A \longrightarrow B$ in $\mathcal{B}$, there exist

i) functors

| $(i = 0)$ | $(i = 1)$ |
|---|---|
| $\mathcal{E}_A \xrightarrow{f_*} \mathcal{E}_B$ | $\mathcal{E}_B \xrightarrow{f^*} \mathcal{E}_A$ |

ii) natural transformations

| | |
|---|---|
| $\theta_f\colon J_A \longrightarrow J_B \circ f_*$ | $\theta_f\colon J_B \circ f^* \longrightarrow J_A$ |

with $P(\theta_f) = f$, such that given any $m\colon D \longrightarrow E$ in $\mathcal{E}$ with $P(m) = f$, then there is a unique factorization

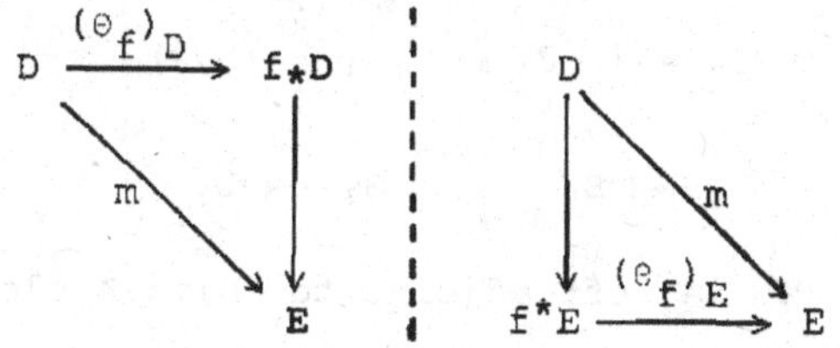


An i-cleavage for $P$ is a choice of the functors and natural

transformations. It is called split-normal if

$$(i_B)_* = E_B \quad\Big|\quad (i_B)^* = E_B$$

$$(gf)_* = g_* f_* \quad\Big|\quad (gf)^* = f^* g^*$$

To account for the terminology, let $\mathbb{1}$ denote the category with a single identity morphism and $\mathbb{2}$ the category illustrated by

$$\overset{0}{\bullet} \longrightarrow \overset{1}{\bullet}$$

with $\partial_i\colon \mathbb{1} \longrightarrow \mathbb{2}$ the functor given by $\partial_i(\mathbb{1}) = i$; $i = 0,1$. As usual, given two functors $F_i\colon A_i \longrightarrow B$, $i = 0,1$, the comma category $(F_0,F_1)$ is defined to be the inverse limit of the diagram

$$A_0 \xrightarrow{\ F_0\ } B \xleftarrow{\ B^{\partial_0}\ } B^{\mathbb{2}} \xrightarrow{\ B^{\partial_1}\ } B \xleftarrow{\ F_1\ } A_1 \ .$$

(See [CCFM], [FCC] and §2.) In particular, $(P,B)$ and $(B,P)$ are defined, and there are induced functors

$$S_0 = \{E^{\partial_0}, P^{\mathbb{2}}\}\colon E^{\mathbb{2}} \longrightarrow (P,B) \quad\Big|\quad S_1 = \{P^{\mathbb{2}}, E^{\partial_1}\}\colon E^{\mathbb{2}} \longrightarrow (B,P) \ .$$

It was shown in [FCC] that $P\colon E \longrightarrow B$ is an i-fibration if and only if there is a functor

$$L_0\colon (P,B) \longrightarrow E^{\mathbb{2}} \quad\Big|\quad L_1\colon (B,P) \longrightarrow E^{\mathbb{2}}$$

such that

$$S_0 L_0 = (P,B) \quad\Big|\quad S_1 L_1 = (B,P)$$

$$L_0 \longrightarrow\!\!\!\!\!| \ S_0 \quad\Big|\quad S_1 \longrightarrow\!\!\!\!\!| \ L_1$$

($M \longrightarrow\!\!\!\!\!| \ N$ means $M$ is left adjoint to $N$). A cleavage is equivalent to a choice of $L$, and split normality can be described in terms of equations satisfied by L. In earlier terminologies

1-fibration = fibration [SGA] = fibration [FCC]

0-fibration = cofibration [SGA] = opfibration [FCC] .

For split normal i-fibrations, it makes sense to talk about cleavage preserving functors over $\mathcal{B}$; i.e., commutative triangles

$$\begin{array}{ccc} \mathcal{E} & \xrightarrow{F} & \widetilde{\mathcal{E}} \\ & {}_{P}\searrow \quad \swarrow_{\widetilde{P}} & \\ & \mathcal{B} & \end{array}$$

such that

$$\begin{array}{ccc} (P,\mathcal{B}) & \xrightarrow{L_0} & \mathcal{E}^2 \\ {\scriptstyle (F,\mathcal{B},\mathcal{B}^2,\mathcal{B},\mathcal{B})}\downarrow & & \downarrow{\scriptstyle F^2} \\ (\widetilde{P},\mathcal{B}) & \xrightarrow{\widetilde{L}_0} & \widetilde{\mathcal{E}}^2 \end{array} \qquad \Bigg| \qquad \begin{array}{ccc} (\mathcal{B},P) & \xrightarrow{L_1} & \mathcal{E}^2 \\ \downarrow & & \downarrow \\ (\mathcal{B},\widetilde{P}) & \xrightarrow{\widetilde{L}_1} & \widetilde{\mathcal{E}}^2 \end{array}$$

commutes. (See §2 for the notation.) The contravariant functor which assigns to $\mathcal{B}$ the category $\mathrm{Split}_i(\mathcal{B})$ of split normal i-fibrations over $\mathcal{B}$ (with small fibres) - made into a functor by pullbacks - is "2-representable" by Cat (the category of small categories).

<u>Proposition</u>

There are adjoint equivalences

$$\mathrm{Split}_0(\mathcal{B}) \rightleftarrows \mathrm{Cat}^{\mathcal{B}} \qquad\qquad \mathrm{Split}_1(\mathcal{B}) \rightleftarrows \mathrm{Cat}^{\mathcal{B}^{op}} \; .$$

<u>Proof</u>. Given a split normal 0-fibration $P\colon \mathcal{E} \longrightarrow \mathcal{B}$,

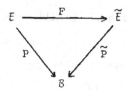

the corresponding functor is

$$\mathcal{B} \longrightarrow \text{Cat}: B \longrightarrow \mathcal{E}_B: f \longrightarrow f_* \ .$$

It is immediate that a cleavage preserving functor determines a natural transformation between the corresponding cat-valued functors.

Conversely, the functor in the opposite direction is given by pulling back a universal 0-fibration over Cat (this was pointed out by Lawvere in [ETH]) which we provisionally denote by $\widetilde{\text{Cat}}$. (Later it will be called $|[\mathbb{1}\ , \text{Cat}]|$.) Here $\widetilde{\text{Cat}}$ is the category whose objects are

$$\{(\mathcal{A},A)\ |A \in |\mathcal{A}| \ \text{ and } \ \mathcal{A} \in |\text{Cat}|\}$$

and whose morphisms are given by

$$\text{Hom}((\mathcal{A},A),(\mathcal{B},B)) = \{(H,h)|H: \mathcal{A} \longrightarrow \mathcal{B} \ \text{ and } \ h: H(A) \longrightarrow B\} \ .$$

Composition is $(K,k)(H,k) = (KH,kK(h))$. If

$$P: \widetilde{\text{Cat}} \longrightarrow \text{Cat}: (\mathcal{A},A) \longrightarrow \mathcal{A}: (H,h) \longrightarrow H \ ,$$

then a lifting functor L is easily described showing that P is a split normal 0-fibration.

If $F: \mathcal{B} \longrightarrow \text{Cat}$, then the corresponding split normal 0-fibration over $\mathcal{B}$ is the pullback

$$\begin{array}{ccc} \mathcal{E}_F & \longrightarrow & \widetilde{\text{Cat}} \\ {\scriptstyle P_F}\downarrow & & \downarrow \\ \mathcal{B} & \xrightarrow{\ F\ } & \text{Cat} \end{array}$$

while if $G: \mathcal{B}^{op} \longrightarrow \text{Cat}$, then the corresponding split normal 1-fibration $\tilde{P}_G: \tilde{\mathcal{E}}_G \longrightarrow \mathcal{B}$ comes from the pullback

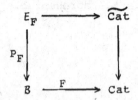

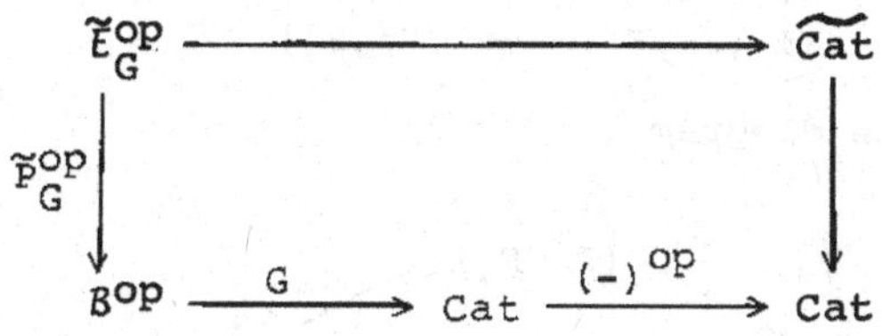


## §2 COMMA CATEGORIES

In [CCC], a number of operations on comma categories and their properties are catalogued. We mention here those that we need.

i) Commutative diagrams

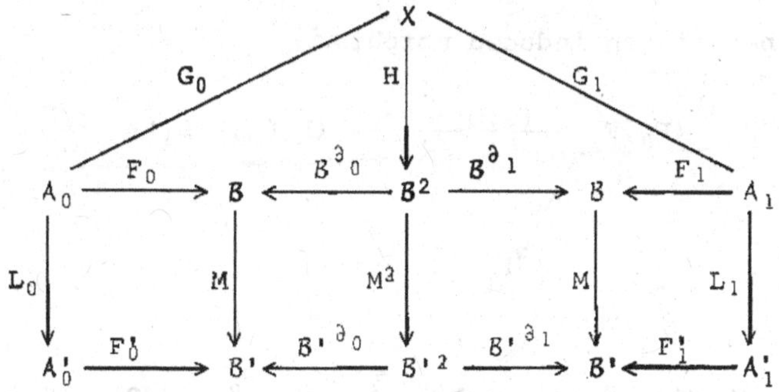


induce functors

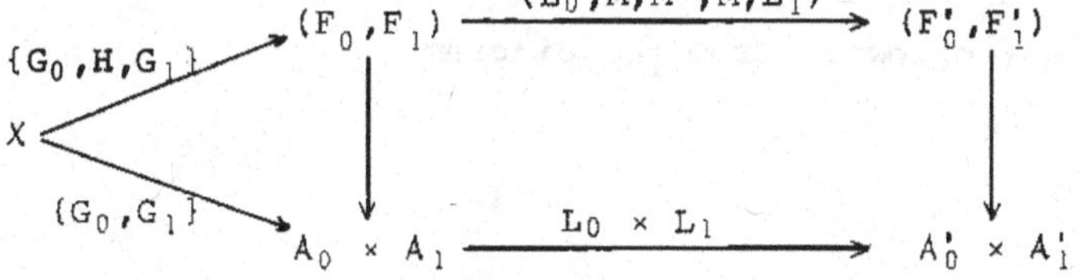


over $A_0 \times A_1$ and $L_0 \times L_1$ respectively. In general, given an arbitrary morphism

$$\alpha: (F_0, F_1) \longrightarrow (F_0', F_1')$$

over $L_0 \times L_1$ , the diagram

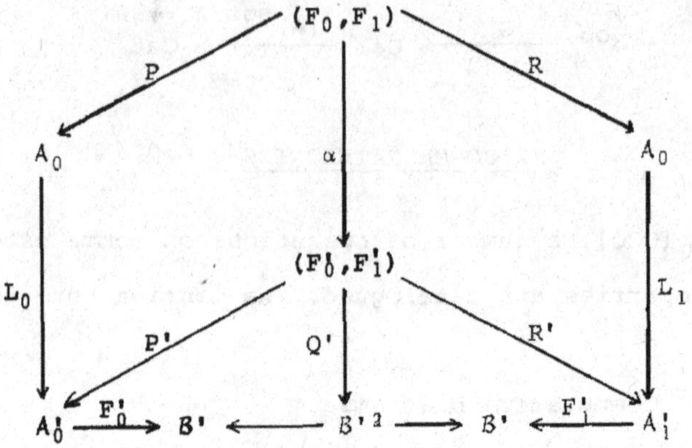

shows that there is an induced morphism

$$(F_0, F_1) \xrightarrow{\{P, Q'\alpha, R\}} (F_0'L_0, F_1'L_1)$$

$$A_0 \times A_1$$

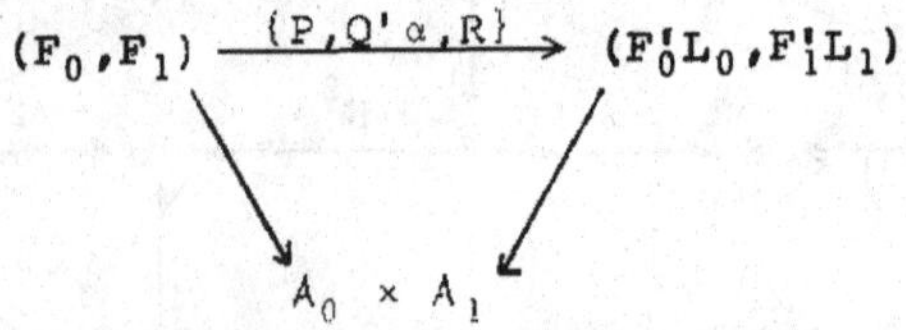

over $A_0 \times A_1$ .

ii) Given $F_i: A_i \longrightarrow B$, $i = 0,1,2$, there is a functor

$$(F_0, F_1) \underset{A_1}{\times} (F_1, F_2) \xrightarrow{0} (F_0, F_2)$$

defined as follows: From the diagram

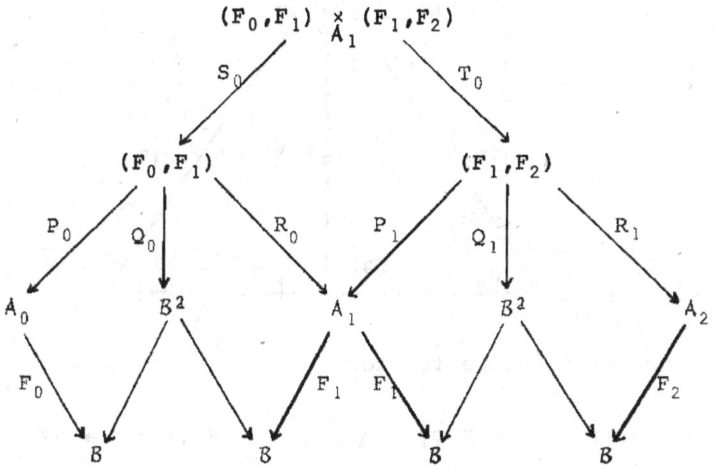


we deduce the existence of a functor $W$ making the diagram

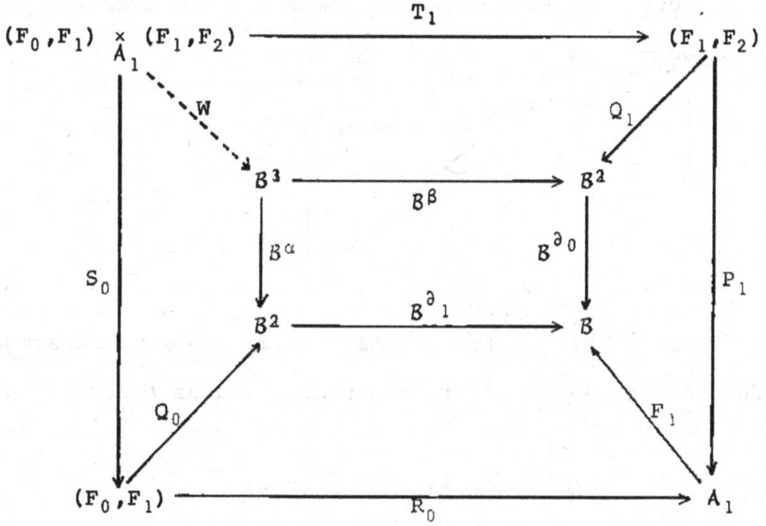


commutative. It is easily checked that this $W$ also gives a commutative diagram

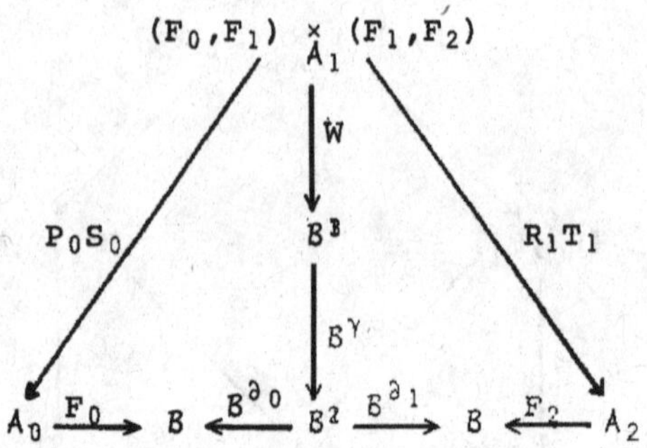


and hence determines a functor

$$"\circ" = \{P_0S_0, B^{\gamma}W, R_1T_1\}: (F_0,F_1) \underset{A_1}{\times} (F_1,F_2) \longrightarrow (F_0,F_2)$$

which is "associative" in an obvious sense.

<u>Note</u>. $\alpha$, $\beta$, and $\gamma$ designate the indicated morphisms in the category **3** .

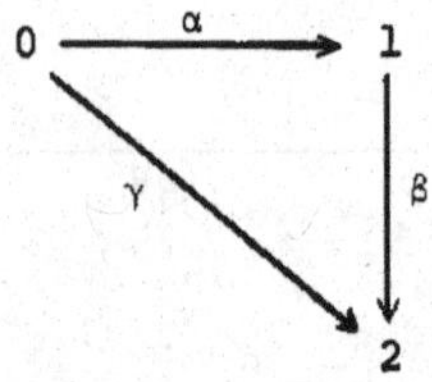


iii) A natural transformation $m: F \longrightarrow G$, regarded as a functor $m: A \longrightarrow B^2$ is the same thing as a functor of the form

$$\tilde{m} = \{A, m, A\}: A \longrightarrow (F,G) \ ;$$

i.e.,

$$\begin{array}{ccc} A & \xrightarrow{\tilde{m}} & (F,G) \\ & {\scriptstyle\Delta}\searrow \quad \swarrow{\scriptstyle\{P,R\}} & \\ & A \times A & \end{array}$$

commutes. Thus the set of natural transformations from F to G is the pullback

$$\begin{array}{ccc} \mathrm{Nat}(F,G) & \longrightarrow & (F,G)^{A} \\ \downarrow & & \downarrow \\ 1 & \xrightarrow{\ulcorner\Delta\urcorner} & (A \times A)^{A} \end{array}$$

iv) If $F_i\colon A_i \longrightarrow B$ and $G_i\colon A_i \longrightarrow B$, $i = 0,1$ are functors and if $m_0\colon G_0 \longrightarrow F_0$ and $m_1\colon F_1 \longrightarrow G_1$ are natural transformations, then there is an induced functor

$$\overline{(m_0,m_1)}\colon (F_0,F_1) \longrightarrow (G_0,G_1)$$

over $A_0 \times A_1$ given by the composition

$$\begin{array}{ccccc} (F_0,F_1) & \xrightarrow{\sim} & A_0 \underset{A_0}{\times} (F_0,F_1) \underset{A_1}{\times} A_1 & \xrightarrow{\tilde{m}_0 \times (F_0,F_1) \times \tilde{m}_1} & (G_0,F_0) \underset{A_0}{\times} (F_0,F_1) \underset{A_1}{\times} (F_1,G_1) \\ & \searrow{\scriptstyle\overline{(m_0,m_1)}} & & & \downarrow{\scriptstyle "\circ"} \\ & & & & (G_0,G_1) \end{array}$$

In [CCC] there is (hopefully) a basis for the many relations satisfied by these operations.

In [FCC], it is shown that if $M\colon A \longrightarrow X$ is any functor, then the functor $P_M\colon (M,X) \longrightarrow X$ given by

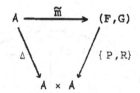

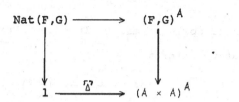

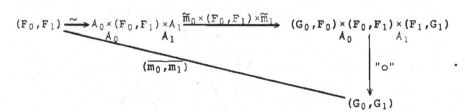

$$\begin{array}{ccccc} (M,X) & \longrightarrow & X^2 & \xrightarrow{X^{\partial_1}} & X \\ \downarrow & & \downarrow {\scriptstyle X^{\partial_0}} & & \\ A & \xrightarrow{M} & X & & \end{array} \qquad P_M : (M,X) \longrightarrow X$$

has a canonical split normal 0-cleavage. Furthermore, there is a commutative diagram

$$\begin{array}{ccc} A & \xrightarrow{Q_M} & (M,X) \\ & {\scriptstyle M} \searrow \quad \swarrow {\scriptstyle P_M} & \\ & X & \end{array}$$

.

Here $Q_M$ is left adjoint to the projection $(M,X) \longrightarrow A$. This is universal in the sense that given any commutative diagram

$$\begin{array}{ccc} A & \xrightarrow{Q} & E \\ & {\scriptstyle M} \searrow \quad \swarrow {\scriptstyle P} & \\ & X & \end{array}$$

where $P$ has a given split normal 0-cleavage, then there is a unique cleavage preserving functor $H: (M,X) \longrightarrow E$ with $HQ_M = Q$ and $PH = P_M$.

From the diagram

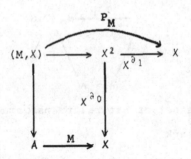

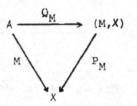

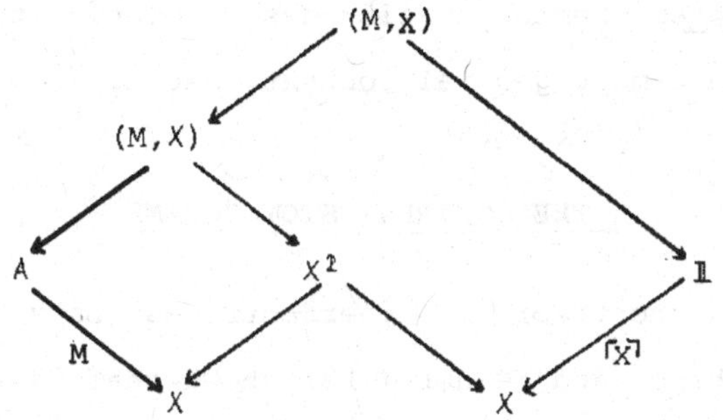


it follows that the fibre of $P_M$ over $X \in |X|$ is $(M,\ulcorner X \urcorner)$ and that $P_M$ corresponds, via the equivalence of §1 to the functor

$$X \longrightarrow \text{Cat}: X \longrightarrow (M,\ulcorner X \urcorner): f \longrightarrow (\overline{M,f}) .$$

From the above operations and their properties, it is easily checked that this gives a functor

$$(\text{Cat},X) \xrightarrow{\Phi} \text{Cat}^X$$

where $\Phi(A \xrightarrow{M} X) = (M,-) \in \text{Cat}^X$. Conversely, the operation that assigns to a functor $G: X \longrightarrow \text{Cat}$ the category $E_G \longrightarrow X$ over $X$ (as in §1), forgetting that it is a 0-fibration, gives a functor

$$\text{Cat}^X \xrightarrow{\check{\Phi}} (\text{Cat},X) .$$

The following result is immediate from the preceding universal mapping property.

<u>Proposition</u>

$\Phi$ is left adjoint to $\check{\Phi}$

$$(\text{Cat},X) \underset{\check{\Phi}}{\overset{\Phi}{\rightleftarrows}} \text{Cat}^X .$$

Note. For a detailed description of the adjunction maps in a more general context, see §9.

## §3 THE COMPREHENSION SCHEME

In the theory of hyperdoctrines, Lawvere (this volume, perhaps, and [CVHOL]) has shown that one sometimes has a pair of adjoint functors like the above pair which fits into a much richer framework, called the comprehension scheme. It works, for instance, in the case of sets for a pair of adjoint functors

$$(\mathrm{Sets}, X) \rightleftarrows 2^X$$

and in the case of small categories for

$$(\mathrm{Cat}, \mathcal{X}) \rightleftarrows \mathrm{Sets}^{\mathcal{X}^{op}} .$$

In our context, the comprehension scheme would require that the functor $\Phi$ above be calculated in the following manner. A functor $F: \mathcal{A} \longrightarrow \mathcal{X}$ induces a functor

$$\mathrm{Cat}^F : \mathrm{Cat}^{\mathcal{X}} \longrightarrow \mathrm{Cat}^{\mathcal{A}}$$

which is just composition with $F$. This functor has a left adjoint which (following Lawvere) we denote by

$$\Sigma F: \mathrm{Cat}^{\mathcal{A}} \longrightarrow \mathrm{Cat}^{\mathcal{X}} .$$

For any $M: \mathcal{A} \longrightarrow \mathrm{Cat}$, $\Sigma F(M): \mathcal{X} \longrightarrow \mathrm{Cat}$ is called the (left) Kan extension of $M$ along $F$ (see [AF]). It is easily described, being the functor whose value at $X \in |\mathcal{X}|$ is given by

$$[\Sigma F(M)](X) = \varinjlim ((F,X) \longrightarrow \mathcal{A} \xrightarrow{M} \mathrm{Cat}) .$$

It satisfies, for any $N: X \longrightarrow \mathrm{Cat}$,

$$\mathrm{Nat}(\Sigma F(M),N) \approx \mathrm{Nat}(M,NF) .$$

Now given a composition

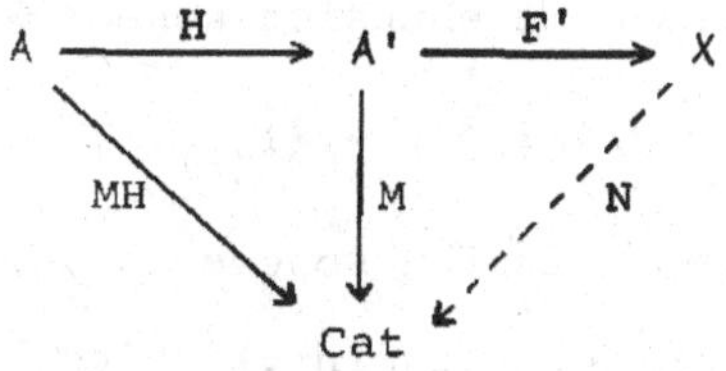


with $F = F'H$, then composition with $H$ induces a function

$$\begin{array}{ccc} \mathrm{Nat}(M,NF') & \longrightarrow & \mathrm{Nat}(MH,NF'H) \\ \wr & & \wr \\ \mathrm{Nat}(\Sigma F'(M),N) & \dashrightarrow & \mathrm{Nat}(\Sigma F(MH),N) \end{array}$$

and hence, taking $N = \Sigma F'(M)$, one deduces the existence of a natural transformation

$$\Sigma F(MH) \longrightarrow \Sigma F'(M)$$

let $1_A: A \longrightarrow \mathrm{Cat}$ denote the constant functor with value the category $\mathbb{1} \in |\mathrm{Cat}|$. Then, for any object $F: A \longrightarrow X$ of $(\mathrm{Cat},X)$,

$$\Sigma F(1_A): X \longrightarrow \mathrm{Cat}$$

is an object of $\mathrm{Cat}^X$, and for any morphism

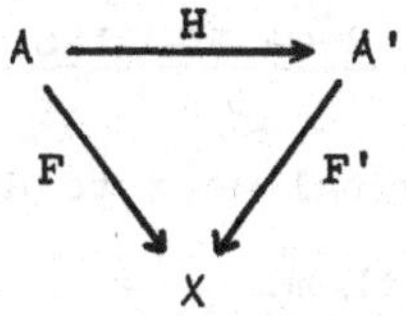


of $(\mathrm{Cat},X)$, one has (since $1_{A'}H = 1_A$) a morphism

$\Sigma F(1_A) \longrightarrow \Sigma F'(1_{A'})$. It is easily checked that this gives a functor

$$\Sigma(-)(1_{(-)}) : (\mathrm{Cat}, X) \longrightarrow \mathrm{Cat}^X .$$

The analogue of the comprehension scheme would say that

$$\Phi \stackrel{?}{=} \Sigma(-)(1_{(-)}) .$$

(Actually, the comprehension scheme is the requirement that $\Sigma(-)(1_{(-)})$ have a right adjoint.) Since we know what $\Phi$ is, we arrive at the proposed equation

$$\underrightarrow{\lim}\,((F,X) \longrightarrow A \xrightarrow{1_A} \mathrm{Cat}) \stackrel{?}{=} (F,X) .$$

Or, equivalently, for any small category $B$,

$$\lim\,(B \xrightarrow{1_B} \mathrm{Cat}) \stackrel{?}{=} B \in |\mathrm{Cat}| .$$

If $B$ is discrete, this holds, which accounts for the second instance of the comprehension scheme mentioned at the beginning of this section. If $B$ is not discrete, this equation is <u>false</u>, and therefore there is no categorical comprehension scheme in this sense.

# PART II. THE SOLUTION

## §4 2-CATEGORIES

There are several ways to describe 2-categories. We mention a number of them.

4.1

i) The elementary theory of abstract 2-categories.

In the spirit and notation of Lawvere's elementary theory of abstract categories, there is an intrinsic description of 2-categories. Let

$$\Delta_0,\ \Delta_1 \text{ and } \Gamma$$

$$\tilde{\Delta}_0\ \ \tilde{\Delta}_1 \text{ and } \tilde{\Gamma}$$

be two independent sets of operators for domain, codomain and composition, called the strong and the weak category structures, respectively. Each triple is to satisfy the axioms for the elementary theory of abstract categories ([CCFM], p. 2) and in addition there are four axioms,

a) $\Delta_i(\tilde{\Delta}_j(x)) = \tilde{\Delta}_j(\Delta_i(x)) \qquad i,j = 0,1$

b) $\Gamma(x,y;u) \Longrightarrow \Gamma(\tilde{\Delta}_i(x),\tilde{\Delta}_i(y);\tilde{\Delta}_i(u)), \qquad i = 0,1$

$\tilde{\Gamma}(x,y;u) \Longrightarrow \tilde{\Gamma}(\Delta_i(x),\Delta_i(y);\Delta_i(u)), \qquad i = 0,1$

c) $\Gamma(x,y;u)$ and $\Gamma(x',y';u')$ and $\tilde{\Gamma}(x,x';v)$ and

$\tilde{\Gamma}(y,y';v')$ and $\Gamma(v,v';f)$ and $\tilde{\Gamma}(u,u';g) \Longrightarrow f = g$

d) $x = \Delta_0 x \Longrightarrow x = \tilde{\Delta}_0 x$ .

The first three axioms are symmetric in the strong and weak structures and lead to the theory of double categories as in Ehresmann, [CS]. The third axiom says that in the picture

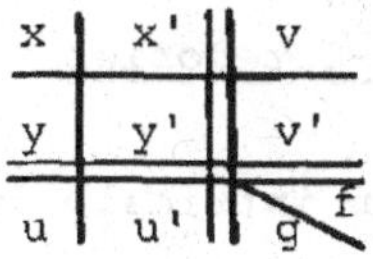


$\Longrightarrow f = g$

where strong composition is vertical and weak horizontal, the resulting values are the same. The fourth asymmetrical axiom, which distinguishes 2-categories from double categories (Ehresmann [CS]) says that strong objects are weak objects. In this paper we generally denote strong composition by juxtaposition and weak composition by dot (•).

ii) The basic theory of abstract 2-categories.

We forgo any extended discussion of this theory here, mostly because of ignorance. (But, see the appendix.) The general idea is that it is a cartesian closed (meta) 2-category built from $\underset{\sim}{\mathbf{2}}_2$ in the same fashion that the basic theory of abstract categories is built from $\mathbf{2}_1$. Here, $\mathbf{2}_0 = 1$, $\mathbf{2}_1 = \mathbf{2}$ and $\underset{\sim}{\mathbf{2}}_2$ is the 2-category illustrated by

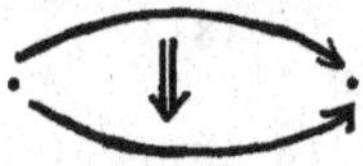

Note that corresponding to a 2-category $\underset{\sim}{A}$, there are three naturally associated categories, the strong category, the weak category, and the "underlying" category $|\underset{\sim}{A}|$ which is the strong structure restricted to the weak objects. This is the universal "locally discrete" part of $\underset{\sim}{A}$; i.e., its weak structure is discrete. Part of the point of this paper is that there is more "basic" structure than is indicated by this analogy.

## 4.2 Categorical Description of 2-Categories

Suppose we are given an intrinsic description of

categories. For the moment, we take this to mean the basic theory of abstract categories of Lawvere [CCFM], together with the axiom that there exists an object Cat which is itself a model of the basic theory and which is reasonably complete. We denote the subcategory of discrete objects of Cat by Sets, and the "set of objects" functor by

$$| \ |: \mathrm{Cat} \longrightarrow \mathrm{Sets} \ .$$

Note that Sets lacks some of the properties one might want for the category of sets, but that is not the problem that concerns us here. If we restrict attention to categories $A$ with small hom sets; i.e., with Hom functors

$$\mathrm{Hom}_A\colon A^{op} \times A \longrightarrow \mathrm{Sets}$$

then, since Cat is a cartesian closed category over Sets, we may speak of Cat-categories in the sense of Eilenberg-Kelly, [CC]; that is, categories $A$ together with a factorization of the Hom functor through Cat

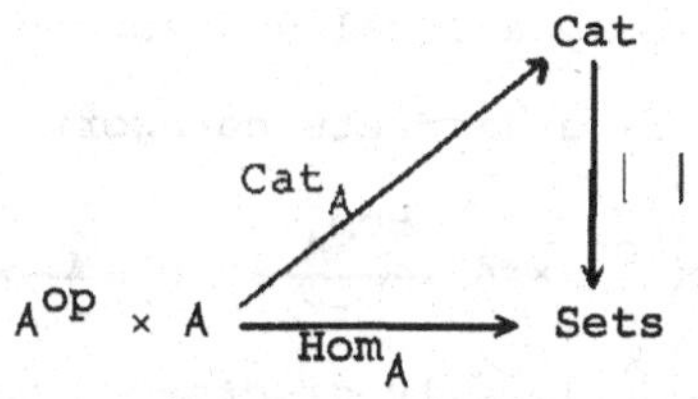


and a "composition rule" for any three objects (i.e., a functor)

$$\mathrm{Cat}_A(A,B) \times \mathrm{Cat}_A(B,C) \xrightarrow{\ \ell\ } \mathrm{Cat}_A(A,C)$$

which is natural in all three variables, is strictly associative, has strict units, and reduces to ordinary composition on objects. A 2-*category* is precisely such a Cat-category. In this description, the objects of $\mathrm{Cat}_{\mathcal{A}}(A,B)$ are the morphisms of $\mathcal{A}$ and are frequently called the 1-cells of $\mathcal{A}$, while the morphisms of $\mathrm{Cat}_{\mathcal{A}}(A,B)$ are called the 2-cells of $\mathcal{A}$ (Bénabou [BC]). In terms of the intrinsic description, the objects of $\mathcal{A}$ correspond to the strong objects, the 1-cells to the weak objects, and the 2-cells to the basic entities.

We shall denote 2-categories by underlined script letters $\underset{\sim}{\mathcal{A}}$, $\underset{\sim}{\mathcal{B}}$, etc., and their underlying ordinary categories (i.e., forget $\mathrm{Cat}_{\mathcal{A}}(-,-)$) by $|\underset{\sim}{\mathcal{A}}|$, etc. $\underset{\sim}{\mathrm{Cat}}$ will denote Cat with its canonical structure as a 2-category; thus $|\underset{\sim}{\mathrm{Cat}}| = \mathrm{Cat}$. A 2-functor between 2-categories $\underset{\sim}{\mathcal{A}}$ and $\underset{\sim}{\mathcal{B}}$ is a Cat-functor; i.e., an ordinary functor $F\colon |\underset{\sim}{\mathcal{A}}| \longrightarrow |\underset{\sim}{\mathcal{B}}|$ together with functors

$$F_{A,B}\colon \mathrm{Cat}_{\mathcal{A}}(A,B) \longrightarrow \mathrm{Cat}_{\mathcal{B}}(FA,FB)$$

which commute with composition and reduce to the given values of $F$ on objects of $\mathrm{Cat}_{\mathcal{A}}(A,B)$.

In this context, a locally discrete 2-category is one for which $\mathrm{Cat}_{\mathcal{A}}(A,B)$ is a discrete category; hence

$$\mathrm{Cat}_{\mathcal{A}}(-,-) = (\mathcal{A}^{\mathrm{op}} \times \mathcal{A} \xrightarrow{\mathrm{Hom}_{\mathcal{A}}} \mathrm{Sets} \hookrightarrow \mathrm{Cat}) \ .$$

Since 2-functors between locally discrete 2-categories are just ordinary functors, we may and often shall identify categories with locally discrete 2-categories.

There are a number of obvious constructions on 2-categories. Finite limits and colimits clearly exist, and there are two kinds of opposite 2-categories:

i) Weak opposite ${}^{op}\underset{\sim}{A}$; $Cat_{{}^{op}\underset{\sim}{A}}(A,B) = [Cat_{\underset{\sim}{A}}(A,B)]^{op}$

ii) Strong opposite $\underset{\sim}{A}^{op}$; $Cat_{\underset{\sim}{A}^{op}}(A,B) = Cat_{\underset{\sim}{A}}(B,A)$

and, of course, their combination ${}^{op}\underset{\sim}{A}^{op}$ . In general, for any functor $K: Cat \longrightarrow Cat$ which is product preserving, ${}^{K}\underset{\sim}{A}$ will denote the 2-category with the same objects as $\underset{\sim}{A}$ and in which

$$Cat_{{}^{K}\underset{\sim}{A}}(A,B) = K[Cat_{\underset{\sim}{A}}(A,B)]$$

except that we make the obvious simplifications $X_{\underset{\sim}{A}} = (-)^{X}\underset{\sim}{A}$ and, as above ${}^{op}\underset{\sim}{A} = (-)^{op}\underset{\sim}{A}$ .

Given two 2-categories, $\underset{\sim}{A}$ and $\underset{\sim}{B}$, there is a "functor 2-category" $\underset{\sim}{B}^{\underset{\sim}{A}}$ as follows:

i) objects are 2-functors $F: \underset{\sim}{A} \longrightarrow \underset{\sim}{B}$ .

ii) given two such 2-functors, $F$ and $G$, then

$$Cat_{\underset{\sim}{B}^{\underset{\sim}{A}}}(F,G)$$

is the inverse limit of the diagram of categories made up of interconnected pieces of the form

$$\begin{array}{ccccc} & & & & Cat(FB,GB) \\ & & & & \Big\downarrow \ulcorner 0 \urcorner \\ & & & & Cat(FA,GB)^{Cat(FA,FB)} \\ & & & & \Big\downarrow (-)^{F_{A,B}} \\ Cat(FA,GA) & \xrightarrow{\ulcorner 0 \urcorner} & Cat(FA,GB)^{Cat(GA,GB)} & \xrightarrow[(-)^{G_{A,B}}]{} & Cat(FA,GB)^{Cat(A,B)} \end{array}$$

for every A and B in $\underset{\sim}{A}$ . It is easily checked that the 1-cells of $\underset{\sim}{B}^{\underset{\sim}{A}}$ are the Cat-natural transformations of [CC]. This functor 2-category is adjoint to direct products in the sense that there is an evaluation 2-functor

$$\underset{\sim}{A} \times \underset{\sim}{B}^{\underset{\sim}{A}} \xrightarrow{ev} \underset{\sim}{B}$$

satisfying the usual universal mapping property. This adjointness is enriched in Cat; actually in 2-Cat; i.e.

$$\underset{\sim}{B}^{(\underset{\sim}{A} \times \underset{\sim}{X})} \approx (\underset{\sim}{B}^{\underset{\sim}{A}})^{\underset{\sim}{X}} .$$

Here 2-Cat refers to some (possibly nonexistent) 2-category of (small) 2-categories. It itself is then a 3-category; i.e., is enriched in 2-Cat. Note that if $\underset{\sim}{B}$ is locally discrete then so is $\underset{\sim}{B}^{\underset{\sim}{A}}$ and if both are locally discrete then $\underset{\sim}{B}^{\underset{\sim}{A}}$ coincides with the usual functor category. Finally, if $\underset{\sim}{A}$ is locally discrete, then

$$|\underset{\sim}{B}^{\underset{\sim}{A}}| = |\underset{\sim}{B}|^{|\underset{\sim}{A}|} .$$

## 4.3 Set-theoretical Description of 2-Categories

There are at least three ways to accomplish this.

i) Start with a set theoretic description of categories in which Cat is the category of small categories, or U-categories for some universe U . Then follow the description in 4.2.

ii) Start with the elementary theory of abstract 2-categories and define a 2-category to be a model of this theory; i.e., a set (or class) equipped with suitable operations satisfying the axioms. A 2-functor is then a function preserving the operations.

iii) See the description of hypercategories in Eilenberg-Kelly [CC], p. 425.

## §5 PRO 2-FUNCTORS AND BIFIBRATIONS

Definition

A pair of functors

$$\mathcal{A} \xleftarrow{P} \mathcal{E} \xrightarrow{Q} \mathcal{B}$$

is called a (1,0)-bifibration if

i) $P$ is a 1-fibration and $Q$ is a 0-fibration.

ii) Let $\mathcal{E}^A = P^{-1}(A)$ and $\mathcal{E}_B = Q^{-1}(B)$. Then $P|\mathcal{E}_B$ is a 1-fibration and $Q|\mathcal{E}^A$ is a 0-fibration for all $A \in \mathcal{A}$ and $B \in \mathcal{B}$.

iii) The inclusion functors $\mathcal{E}^A \longrightarrow \mathcal{E}$ and $\mathcal{E}_B \longrightarrow \mathcal{E}$ are maps of fibrations in the sense of [FCC].

A cleavage for a bifibration is a choice of all the functors $f_*$ or $f^*$ whose existence is postulated in i) and ii). A cleavage is called split-normal if

iv) It is split-normal for each fibration, $P$, $Q$, $P|\mathcal{E}_B$ and $Q|\mathcal{E}^A$.

v) The inclusion functors of iii) are cleavage preserving,

vi) For any $f: A' \longrightarrow A$ in $\mathcal{A}$ and $g: B \longrightarrow B'$ in $\mathcal{B}$, $f^*: \mathcal{E}^A \longrightarrow \mathcal{E}^{A'}$ and $g_*: \mathcal{E}_B \longrightarrow \mathcal{E}_{B'}$ are cleavage preserving.

Note that as in the case of fibrations, these conditions can be expressed by equations involving the lifting functors.

A <u>cleavage preserving morphism</u> between split-normal (1,0)-bifibrations $\{P,Q\}$ and $\{P',Q'\}$ is a commutative diagram

$$\begin{array}{ccc} \mathcal{E} & \xrightarrow{\quad T \quad} & \mathcal{E}' \\ & {\scriptstyle \{P,Q\}} \searrow \quad \swarrow {\scriptstyle \{P',Q'\}} & \\ & \mathcal{A} \times \mathcal{B} & \end{array}$$

such that $T$ is cleavage preserving for both fibration structures. The category of split-normal (1,0)-bifibrations over $\mathcal{A}$ and $\mathcal{B}$ will be denoted by $\text{Split}_{(1,0)}(\mathcal{A},\mathcal{B})$.

$\text{Split}_{(i,j)}(\mathcal{A},\mathcal{B})$ is defined analogously, but only the (0,1) case is interesting since

$$\text{Split}_{(i,i)}(\mathcal{A},\mathcal{B}) = \text{Split}_i(\mathcal{A} \times \mathcal{B}) .$$

<u>Proposition</u>

There is an adjoint equivalence

$$\text{Split}_{(1,0)}(\mathcal{A},\mathcal{B}) \underset{\Psi'}{\overset{\Psi}{\rightleftarrows}} \text{Cat}^{\mathcal{A}^{op} \times \mathcal{B}} .$$

<u>Proof</u>. Given a split-normal (1,0)-bifibration $\{P,Q\}$, then $\Psi(P,Q): \mathcal{A}^{op} \times \mathcal{B} \longrightarrow \text{Cat}$ is the functor whose

value on $(A,B)$ is $E^A_B = E^A \cap E_B$ and on $(f: A' \longrightarrow A,\ g: B \longrightarrow B')$ is

$$f^*g_* = g_*f^*: E^A_B \longrightarrow E^{A'}_{B'}$$

$\Psi$ extends to a functor in an obvious way.

The functor $\Psi'$ is a specialization of the basic construction on which the whole theory depends, to which we now turn.

Definition

Let $\underset{\sim}{A}$ and $\underset{\sim}{B}$ be 2-categories. A pro-2-functor from $\underset{\sim}{A}$ to $\underset{\sim}{B}$ is a 2-functor

$$F: \underset{\sim}{A}^{op} \times \underset{\sim}{B} \longrightarrow \underset{\sim}{Cat} \quad .$$

We shall describe a construction which assigns to $F$ a 2-category $\underset{\sim}{E}_F$ over $\underset{\sim}{A} \times \underset{\sim}{B}$ whose underlying category $|\underset{\sim}{E}_F|$ is split-normal (1,0)-bifibred over $|\underset{\sim}{A}|$ and $|\underset{\sim}{B}|$ . This construction actually establishes an equivalence between $\underset{\sim}{Cat}^{\underset{\sim}{A}^{op} \times \underset{\sim}{B}}$ and a suitable category of 2-bifibrations. We forgo the definition of this notion and the proof that $\Psi'(F) = |\underset{\sim}{E}_F|$. However, the construction of $\underset{\sim}{E}_F$ is of crutial importance here and we give it in two forms, a category-theoretic form and a set-theoretic form.

Observe first that $F: \underset{\sim}{A}^{op} \times \underset{\sim}{B} \longrightarrow \underset{\sim}{Cat}$ can be regarded as an object function together with functors

$$Cat_{\underset{\sim}{A}}(A',A) \times Cat_{\underset{\sim}{B}}(B,B') \longrightarrow F(A',B')^{F(A,B)} \quad .$$

In particular, specializing to the identity morphisms of A or B and using the cartesian closed structure of Cat, we deduce functors

$$F(A,B) \times \mathrm{Cat}_{\underset{\sim}{B}}(B,B') \xrightarrow{o_B} F(A,B')$$

$$\mathrm{Cat}_{\underset{\sim}{A}}(A',A) \times F(A,B) \xrightarrow{o_A} F(A',B)$$

which we regard as operations of $\underset{\sim}{A}$ and $\underset{\sim}{B}$ on F . We also forgo a formal treatment of pro-2-functors as modules over $\underset{\sim}{A}$ and $\underset{\sim}{B}$, and merely point out (to no one's surprise) that diagrams like

$$\begin{array}{ccc} F(A,B) \times \mathrm{Cat}_{\underset{\sim}{B}}(B,B') \times \mathrm{Cat}_{\underset{\sim}{B}}(B',B'') & \longrightarrow & F(A,B) \times \mathrm{Cat}_{\underset{\sim}{B}}(B,B'') \\ \downarrow & & \downarrow \\ F(A,B') \times \mathrm{Cat}_{\underset{\sim}{B}}(B',B'') & \longrightarrow & F(A,B'') \end{array}$$

commute.

<u>The Basic Construction</u>. Let $F\colon \underset{\sim}{A}^{op} \times \underset{\sim}{B} \longrightarrow \underset{\sim}{\mathrm{Cat}}$ . Then $\underset{\sim}{E}_B$ is the 2-category whose objects are triples (A,X,B) where $A \in \underset{\sim}{A}$, $B \in \underset{\sim}{B}$ and $X \in F(A,B)$. Given two objects, (A,X,B) and (A',X',B'), one gets two functors

$$\mathbb{1} \times \mathrm{Cat}_{B}(B,B') \xrightarrow{\ulcorner X \urcorner \times \mathrm{id}} F(A,B) \times \mathrm{Cat}_{B}(B,B') \xrightarrow{o_B} F(A,B')$$

$$\mathrm{Cat}_{A}(A,A') \times \mathbb{1} \xrightarrow{\mathrm{id} \times \ulcorner X' \urcorner} \mathrm{Cat}(A,A') \times F(A',B') \xrightarrow{o_A} F(A,B')$$

with codomain F(A,B'), which therefore have a comma category. We set

$$\mathrm{Cat}_{\underset{\sim}{E}_F}((A,X,B),(A',X',B')) = (o_B(\ulcorner X \urcorner \times \mathrm{id}),\ o_A(\mathrm{id} \times \ulcorner X' \urcorner)).$$

Composition is illustrated by the adjoining large diagram.

If $F$ and $G$ are two pro-2-functors and if $\varphi\colon F \longrightarrow G$ is a Cat-natural transformation; i.e., a morphism in $\underset{\sim}{Cat}^{\underset{\sim}{A}^{op} \times \underset{\sim}{B}}$, then for each $(A,B)$

$$\varphi_{A,B}\colon F(A,B) \longrightarrow G(A,B)$$

is a functor and the diagrams

$$\begin{array}{ccc} F(A,B) \times Cat(B,B') & \longrightarrow & F(A,B') \\ \downarrow & & \downarrow \\ G(A,B) \times Cat(B,B') & \longrightarrow & G(A,B') \end{array} \qquad \begin{array}{ccc} Cat(A',A) \times F(A,B) & \longrightarrow & F(A',B) \\ \downarrow & & \downarrow \\ Cat(A',A) \times G(A,B) & \longrightarrow & G(A',B) \end{array}$$

commute. Hence $\varphi$ determines a 2-functor

$$\Phi\colon \underset{\sim}{E}_F \longrightarrow \underset{\sim}{E}_G$$

where $\Phi(A,X,B) = (A,\varphi_{A,B}(X),B)$ and

$$\Phi_{(A,X,B),\ (A',X',B')}\colon Cat_{\underset{\sim}{E}_F}((A,X,B),\ (A',X',B')) \longrightarrow$$

$$\longrightarrow Cat_{\underset{\sim}{E}_G}((A,\varphi(X),B),\ (A',\varphi(X'),B'))$$

is the induced functor between comma categories as in §2, i), whose existence follows from the preceeding commutative diagrams. Commutativity with composition is an easy diagram chase.

Besides this category-theoretic description, we seem unable (because of incompetence, presumably) to dispense with a set theoretic description later on. Thus, the objects of $Cat_{\underset{\sim}{E}_F}((A,X,B),\ (A',X',B'))$ - that is, the 1-cells of $\underset{\sim}{E}_F$ - are triples $(f,\varphi,g)$ where $f\colon A \longrightarrow A'$ in $A$ ,

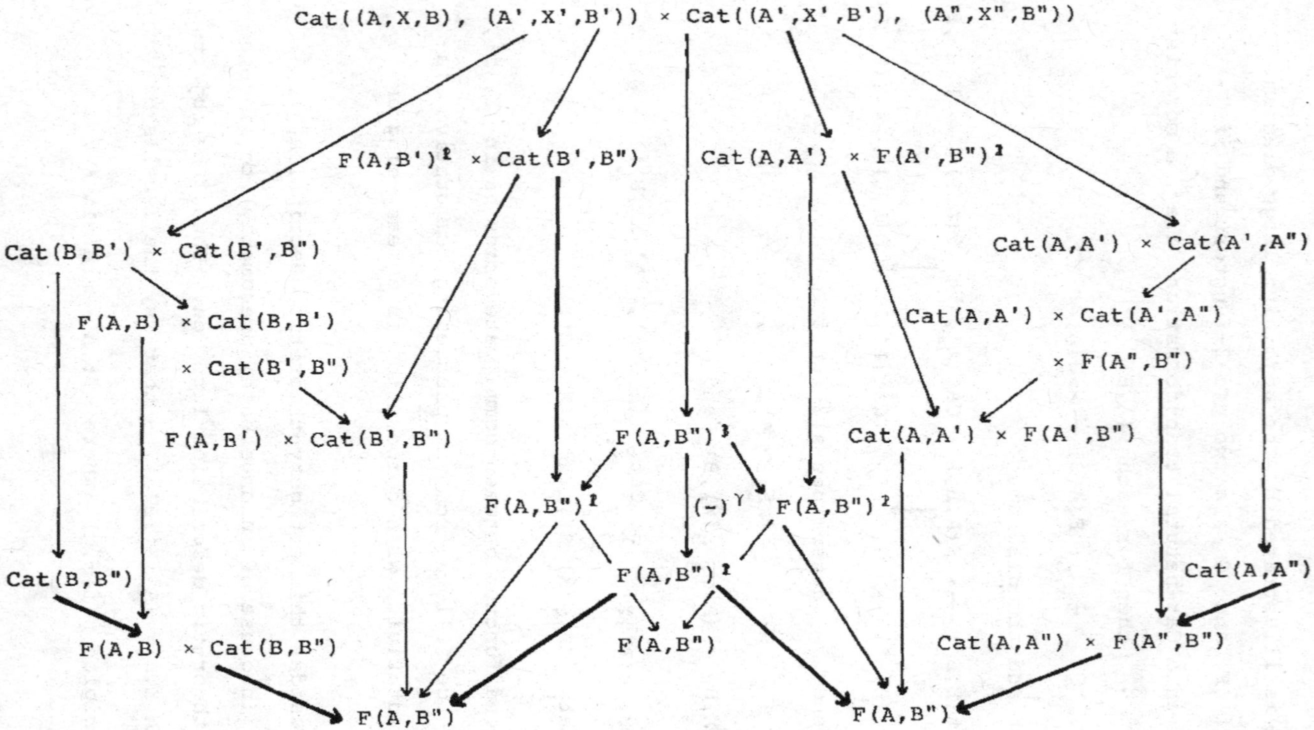


Yields $Cat((A,X,B),(A',X',B')) \times Cat((A',X',B'),(A",X",B")) \xrightarrow{0} Cat((A,X,B),(A",X",B"))$

$g\colon B \longrightarrow B'$ in $\mathcal{B}$ and $\varphi\colon F(A,g)(X) \longrightarrow F(f,B')(X')$ in $F(A,B')$. The morphisms of $\mathrm{Cat}_{\underset{\sim}{E}_F}((A,X,B),\ (A',X',B'))$ - that is the 2-cells of $\underset{\sim}{E}_F$ from $(f,\varphi,g)$ to $(f',\varphi',g')$ - are pairs $(\sigma,\tau)$ where $\sigma\colon f \longrightarrow f'$ in $\mathrm{Cat}_{A}(A,A')$, $\tau\colon g \longrightarrow g'$ in $\mathrm{Cat}_{\mathcal{B}}(B,B')$ and the diagram

$$\begin{array}{ccc} F(A,g)(X) & \xrightarrow{\varphi} & F(f,B')(X') \\ {\scriptstyle F(A,\tau)(X)}\downarrow & & \downarrow{\scriptstyle F(\sigma,B')(X)} \\ F(A,g')(X) & \xrightarrow{\varphi'} & F(f',B')(X') \end{array}$$

commutes. The formula for composition of morphisms is

$$(h,\lambda,k)(f,\varphi,g) = (hf, \lambda f \cdot k\varphi, kg)$$

and of 2-cells is

$$(\sigma',\tau')(\sigma,\tau) = (\sigma'\sigma, \tau'\tau) \ .$$

<u>Proposition</u>

If $H\colon \underset{\sim}{A}' \longrightarrow \underset{\sim}{A}$, $K\colon \underset{\sim}{\mathcal{B}}' \longrightarrow \underset{\sim}{\mathcal{B}}$ and $F\colon A^{op} \times \mathcal{B} \longrightarrow \underset{\sim}{\mathrm{Cat}}$, then

$$\begin{array}{ccc} \underset{\sim}{E}_F \circ (H^{op} \times K) & \longrightarrow & \underset{\sim}{E}_F \\ \downarrow & & \downarrow \\ \underset{\sim}{A}' \times \underset{\sim}{\mathcal{B}}' & \longrightarrow & \underset{\sim}{A} \times \underset{\sim}{\mathcal{B}} \end{array}$$

is a pullback.

<u>Examples and Remarks</u>.

i) As observed before, $\Psi'\colon \mathrm{Cat}^{A^{op} \times \mathcal{B}} \longrightarrow \mathrm{Split}_{(1,0)}(A,\mathcal{B})$ assigns to

$F: A^{op} \times B \longrightarrow Cat$ the category $|\underset{\sim}{E}_F|$ with its canonical projections onto $A$ and $B$. Note that if $\underset{\sim}{A}$ and $\underset{\sim}{B}$ are locally discrete and $F$ is set-valued, then $\underset{\sim}{E}_F$ is locally discrete; but otherwise, not.

ii) If $A$ is a category and

$$Hom_A: A^{op} \times A \longrightarrow Sets \subset Cat\ ,$$

then $\underset{\sim}{E}_{Hom_A} = A^2$ .

iii) Let $\underset{\sim}{A}$ be a 2-category with

$$Cat_{\underset{\sim}{A}}: \underset{\sim}{A}^{op} \times \underset{\sim}{A} \longrightarrow \underset{\sim}{Cat}\ .$$

We define $\underset{\sim}{Fun}_A = \underset{\sim}{E}_{Cat_A}$ with

$$(\delta_i)_A: \underset{\sim}{Fun}_A \longrightarrow \underset{\sim}{A},\ i = 0,1$$

the canonical projections. We shall also write

$$Fun_A = |\underset{\sim}{Fun}_A|\ .$$

Note that if $F: \underset{\sim}{A} \longrightarrow \underset{\sim}{B}$ is a 2-functor then there is a commutative diagram

$$\begin{array}{ccc} \underset{\sim}{Fun}_A & \xrightarrow{\underset{\sim}{Fun}_F} & \underset{\sim}{Fun}_B \\ \downarrow & & \downarrow \\ \underset{\sim}{A} \times \underset{\sim}{A} & \longrightarrow & \underset{\sim}{B} \times \underset{\sim}{B} \end{array}$$

so that $\underset{\sim}{Fun}_{(-)}$ can be regarded as an endofunctor on 2-Cat. As such, it participates in triples analogous to those for $(-)^2$ on Cat.

iv) In particular, for

$$Cat^{op} \times Cat \xrightarrow{(-)^{(-)}} Cat$$

the corresponding 2-category is denoted by $\underset{\sim}{\text{Fun}}$ and the projections by $\delta_i: \underset{\sim}{\text{Fun}} \longrightarrow \underset{\sim}{\text{Cat}}$, $i = 0,1$. The objects of $\underset{\sim}{\text{Fun}}$ are functors $(A \overset{X}{\longrightarrow} B)$, the morphisms (1-cells) of $\underset{\sim}{\text{Fun}}$ are diagrams

$$
\begin{array}{ccc}
A & \overset{F}{\longrightarrow} & A' \\
X \downarrow & \quad \varphi & \downarrow X' \\
B & \underset{G}{\longrightarrow} & B'
\end{array}
$$

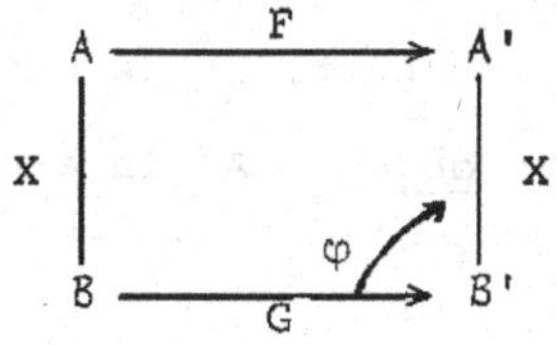

where $\varphi: GX \longrightarrow X'F$ is a natural transformation, composition being

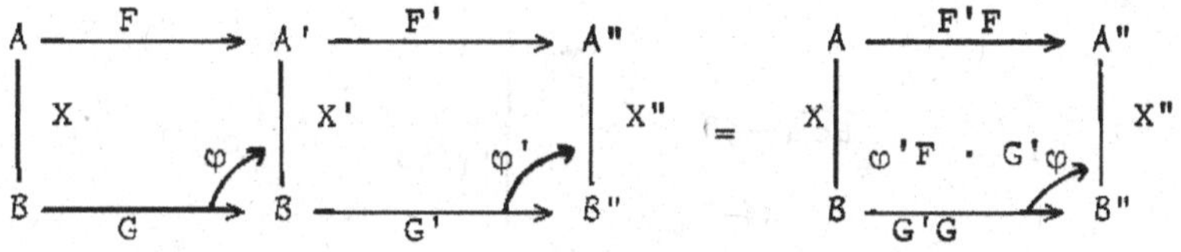

$$
\begin{array}{ccccc}
A & \overset{F}{\longrightarrow} & A' & \overset{F'}{\longrightarrow} & A'' \\
X \downarrow & \varphi & X' \downarrow & \varphi' & \downarrow X'' \\
B & \underset{G}{\longrightarrow} & B' & \underset{G'}{\longrightarrow} & B''
\end{array}
\;=\;
\begin{array}{ccc}
A & \overset{F'F}{\longrightarrow} & A'' \\
X \downarrow & \varphi'F \cdot G'\varphi & \downarrow X'' \\
B & \underset{G'G}{\longrightarrow} & B''
\end{array}
$$

and the 2-cells of $\underset{\sim}{\text{Fun}}$ are diagrams

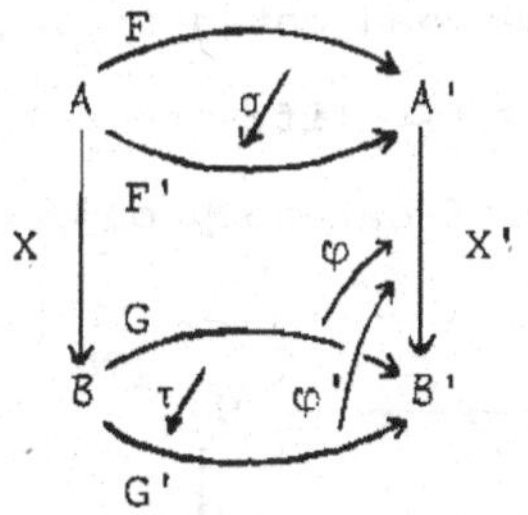

where $\sigma: F \longrightarrow F'$ and $\tau: G \longrightarrow G'$ are natural transformations such that the diagram (of natural transformations)

$$
\begin{array}{ccc}
GX & \overset{\varphi}{\longrightarrow} & X'F \\
\tau X \downarrow & & \downarrow X'\sigma \\
G'X & \underset{\varphi'}{\longrightarrow} & X'F'
\end{array}
$$

commutes.

## §6 2-COMMA CATEGORIES AND SUPER FUNCTOR CATEGORIES

In order to place the constructions we are about to make in their proper context, it is necessary to explain the notion of a category object $A$ in a category $\mathcal{A}$ .

Definition

A category object in $\mathcal{A}$ is an object $A \in \mathcal{A}$ together with a factorization

$$\begin{array}{ccc} & & \mathrm{Cat} \\ & \overset{\widetilde{\mathrm{Hom}}(-,A)}{\nearrow} & \Big|\ |(-)^{2}| \\ \mathcal{A}^{op} & \xrightarrow[\mathrm{Hom}(-,A)]{} & \mathrm{Sets} \end{array} .$$

A functorial morphism between category objects is a morphism $f$ in $\mathcal{A}$ such that $\mathrm{Hom}(-,f)$ lifts to a natural transformation $\widetilde{\mathrm{Hom}}(-,f)$. The category of category objects in $\mathcal{A}$ is then the pullback

$$\begin{array}{ccc} \mathcal{A}_C & \longrightarrow & \mathrm{Cat}^{\mathcal{A}^{op}} \\ \downarrow & & \downarrow |(-)^{2}|^{\mathcal{A}^{op}} \\ \mathcal{A} & \xrightarrow[\text{Yoneda}]{} & \mathrm{Sets} \end{array} .$$

We may assume, as in [CCFM], an isomorphism

$$\mathrm{Cat} \approx \mathrm{Sets}^{(\{\mathbf{4}\}^{op})}$$

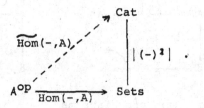

the right hand side denoting the category of limit preserving functors from $\{\mathbf{4}\}^{op}$ to Sets. It follows (in fact, for any such "limit theory") that if $\mathcal{A}$ has finite limits, then

$$\mathcal{A}_C \approx \mathcal{A}^{(\{\mathbf{4}\}^{op})} .$$

In this representation $|(-)^{\mathbf{2}}|$ becomes the operation which assigns to $F\colon \{\mathbf{4}\}^{op} \longrightarrow$ Sets its value at $\mathbf{2}$, $F(\mathbf{2})$. The structure of $\mathcal{A}_C$ is then easily deduced. A category object is determined by an object $A \in \mathcal{A}$ with the following structure:

i) Two morphisms $\bar{\partial}_i\colon A \longrightarrow A$, $i = 0,1$ such that $\bar{\partial}_i\bar{\partial}_j = \bar{\partial}_j$ .

ii) Let $I \xrightarrow{\tau} A \underset{A}{\overset{\bar{\partial}_i}{\rightrightarrows}} A$ be an equalizer. (It is the same for $i = 0,1$ ). Then $\bar{\partial}_i = \tau\partial_i$ for $\partial_i\colon A \longrightarrow I$ and $\partial_i\tau = I$ . Let

$$\begin{array}{ccc} A' & \xrightarrow{\beta} & A \\ {\scriptstyle\alpha}\downarrow & & \downarrow{\scriptstyle\partial_0} \\ A & \xrightarrow[\partial_1]{} & I \end{array} \qquad \text{and} \qquad \begin{array}{ccc} A'' & \xrightarrow{\nu} & A' \\ {\scriptstyle\mu}\downarrow & & \downarrow{\scriptstyle\alpha} \\ A' & \xrightarrow[\beta]{} & A \end{array}$$

be pullbacks.

iii) There is a morphism $\gamma\colon A' \longrightarrow A$ with $\partial_0\gamma = \partial_0\alpha$ and $\partial_1\gamma = \partial_1\beta$ .

iv) $\gamma\{\bar{\partial}_0, A\} = A$ and $\gamma\{A, \bar{\partial}_1\} = A$ .

v) $\gamma\{\alpha\mu, \gamma\nu\} = \gamma\{\gamma\mu, \beta\nu\}\colon A'' \longrightarrow A$

Note: $\{-,-\}$ denotes induced morphisms into pullbacks (or out of pushouts).

Cocategory objects are defined dually. The axioms about $\mathbf{2}$ in [CCFM] just say that $\mathbf{2}$ is a cocategory object in the universe. Hence $A^{\mathbf{2}}$ is a category object in Cat for any $A$; and, in fact, $(-)^{\mathbf{2}}$ is a category object in $\text{Cat}^{\text{Cat}}$; i.e., as an endofunctor on Cat, its values on functors being "functorial." This structure is all that is needed to describe a "comma category" construction with the usual properties. This structure can be described starting with any category object, but it is of course especially rich for category objects in non-trivial functor categories. For details, see [CCC].

<u>Proposition</u>

$\underset{\sim}{\text{Fun}}_{(-)}: 2\text{-}\underset{\sim}{\text{Cat}} \longrightarrow 2\text{-}\underset{\sim}{\text{Cat}}$ is a category object in $(2\text{-}\underset{\sim}{\text{Cat}})^{(2\text{-}\underset{\sim}{\text{Cat}})}$.

<u>Proof</u>.

i) Let $\tau: \text{Id} \longrightarrow \underset{\sim}{\text{Fun}}_{(-)}$ be the Cat-natural transformation such that $\tau_{\underset{\sim}{A}}: \underset{\sim}{A} \longrightarrow \underset{\sim}{\text{Fun}}_A$ is the 2-functor given by $\tau_{\underset{\sim}{A}}(A) = (A,\text{id}_A,A)$, $\tau_{\underset{\sim}{A}}(f) = (f,\text{id}_f,f)$ and $\tau_A(\sigma) = (\sigma,\tilde{\sigma})$. (A categorical description of $\tau_{\underset{\sim}{A}}$ is easily given.) Clearly $\tilde{\delta}_i\tau = \text{id}$, so, if we set $\tilde{\delta}_i = \tau\delta_i$, then $\tilde{\delta}_i\tilde{\delta}_j = \tilde{\delta}_j$ and

$$\text{Id} \xrightarrow{\tau} \underset{\sim}{\text{Fun}}_{(-)} \underset{\text{id}}{\overset{\tilde{\delta}_i}{\rightrightarrows}} \underset{\sim}{\text{Fun}}_{(-)}$$

is an equalizer for $i = 0,1$.

ii) Let

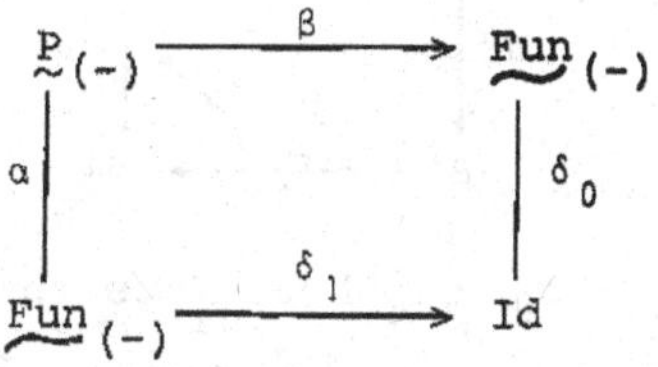


be a pullback (of endofunctors on 2-$\underset{\sim}{Cat}$). Then there is a Cat-natural transformation

$$\underset{\sim}{P}_{(-)} \longrightarrow \underset{\sim}{Fun}_{(-)}$$

such that $\delta_0\gamma = \delta_0\alpha$ and $\delta_1\gamma = \delta_1\beta$. This can be described by a diagram similar to the composition diagram for $\underset{\sim}{E}_F$. In set-theoretic terms, objects of $\underset{\sim}{Fun}_A$ are of the form $(A,f,A')$ where $f\colon A \to A'$ is a 1-cell in $A$, so objects of $\underset{\sim}{P}_A$ are pairs of the form $\{(A,f,A'), (A',g,A'')\}$. We set

$$\gamma_A\{(A,f,A'), (A'g,A'')\} = (A,gf,A'') .$$

A 1-cell of $\underset{\sim}{P}_A$ is a diagram

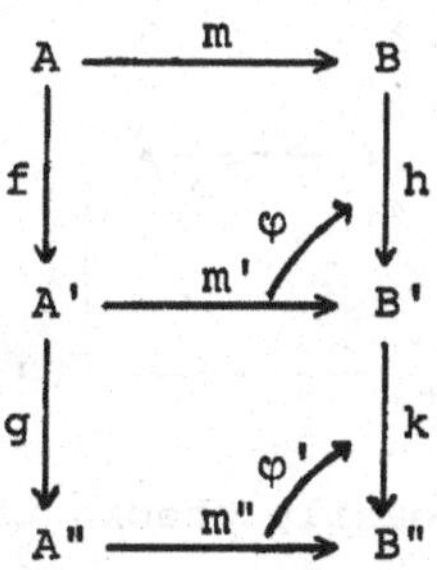


where $\varphi$ and $\varphi'$ are 2-cells in $A$. We define $\gamma_A$ on this 1-cell to be the one cell

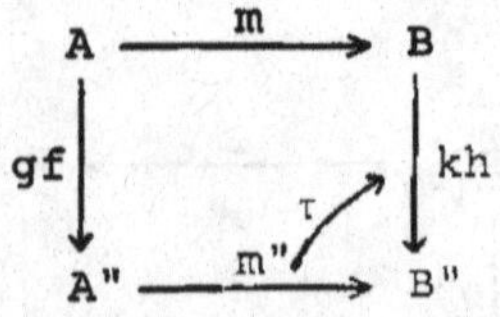


of $\mathrm{Fun}_A$ , where $\tau = k\varphi \cdot \varphi' f$ . Note that this operation gives $|\mathrm{Fun}_A|$ a different 2-category structure; a fact which is equivalent to $\gamma$ being a functor satisfying the stated properties. Finally, a 2-cell of $P_A$ looks like

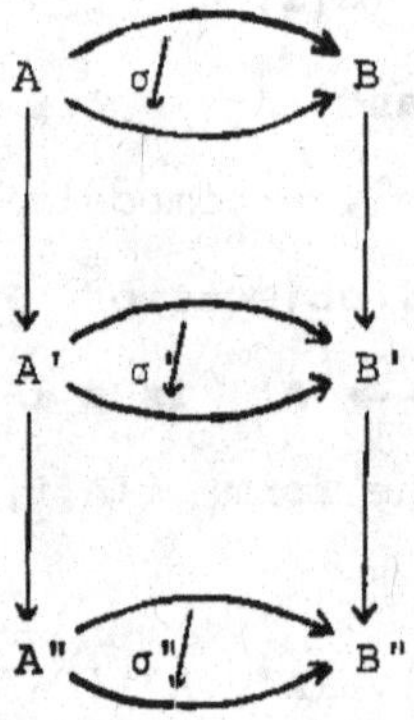


and $\gamma_A$ on this is $(\sigma, \sigma")$. $\gamma_A$ is then a 2-functor for each $\underset{\sim}{A}$ and natural in $\underset{\sim}{A}$ .

iii) If

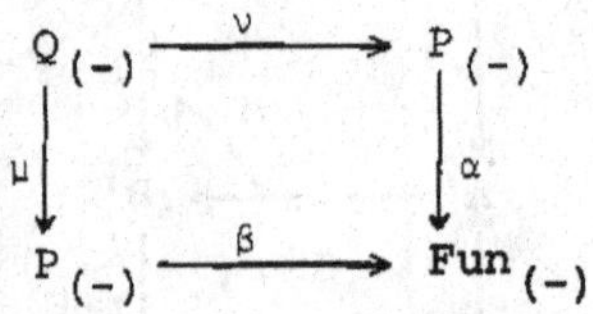


is a pullback, then one easily checks that

$$\gamma\{\alpha\mu, \gamma\nu\} = \gamma\{\gamma\mu, \beta\nu\} .$$

Definition

Let $F_i: \underset{\sim}{A}_i \longrightarrow \underset{\sim}{B}$, $i = 0,1$ , be 2-functors. Then $[F_0,F_1]$ is the inverse limit of the diagram

$$\underset{\sim}{A}_0 \xrightarrow{F_0} \underset{\sim}{B} \xleftarrow{\delta_0} \underset{\sim}{\mathrm{Fun}}_B \xrightarrow{\delta_1} \underset{\sim}{B} \xleftarrow{F_1} \underset{\sim}{A}_1 \, .$$

$[F_0,F_1]$ is called a 2-comma category. (The general hierarchy is $(F_0,F_1) = F_0 \underset{B}{\times} F_1$ for sets, $(F_0,F_1)_1 = (F_0,F_1)$, $(F_0,F_1)_2 = [F_0,F_1]$, etc.) As in §2 we have various operations on 2-comma categories.

i) Induced functors

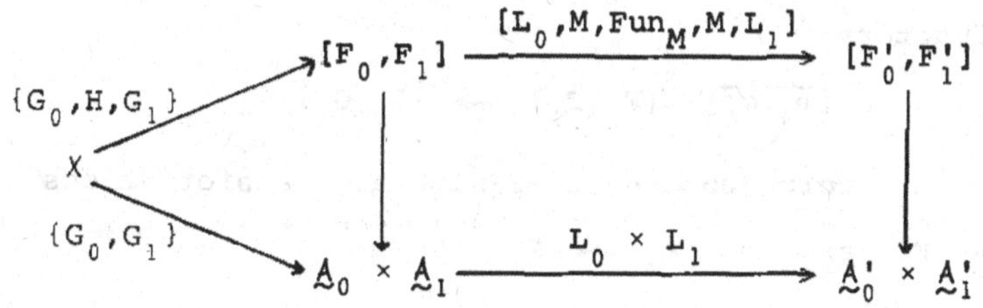


over $\underset{\sim}{A}_0 \times \underset{\sim}{A}_1$ and $L_0 \times L_1$; and $\alpha: [F_0,F_1] \longrightarrow [F'_0,F'_1]$ over $L_0 \times L_1$ induces

$$\bar{\alpha} = \{P,Q'\alpha,R\}: [F_0,F_1] \longrightarrow [F'_0L_0,F'_1L_1]$$

exactly as in §2, except that $B^2$ is replaced by $\mathrm{Fun}_B$ .

ii) There is an associative composition

$$[F_0,F_1] \underset{\underset{\sim}{A}_1}{\times} [F_1,F_2] \longrightarrow [F_0,F_2]$$

derived exactly as in §2 with $B^3$ replaced by $\underset{\sim}{P}_B$ .

iii) The correspondence in §2, iii) becomes a definition.

Definition

If $F,G\colon \underset{\sim}{A} \longrightarrow \underset{\sim}{B}$ are 2-functors then a 2-natural transformation is a 2-functor

$$\begin{array}{ccc} \underset{\sim}{A} & \xrightarrow{m} & [F,G] \\ & {\scriptstyle\Delta}\searrow \quad \swarrow & \\ & \underset{\sim}{A} \times \underset{\sim}{A} & \end{array}$$

over $\underset{\sim}{A} \times \underset{\sim}{A}$ . Note that a 2-natural transformation is not a Cat-natural transformation, in general.

iv) Exactly as in §2, using composition one derives induced functors

$$[\overline{m_0,m_1}]\colon [F_0,F_1] \longrightarrow [G_0,G_1]$$

over $A_0 \times A_1$ corresponding to 2-natural transformations $m_0\colon G_0 \longrightarrow F_0$ and $m_1\colon F_1 \longrightarrow G_1$ .

Definition

Let $\underset{\sim}{A}$ and $\underset{\sim}{B}$ be 2-categories. Then the super functor category $\underset{\sim}{\mathrm{Fun}}(\underset{\sim}{A},\underset{\sim}{B})$ is the 2-category whose objects are 2-functors from $\underset{\sim}{A}$ to $\underset{\sim}{B}$ and such that

$$\begin{array}{ccc} \mathrm{Cat}_{\underset{\sim}{\mathrm{Fun}}(\underset{\sim}{A},\underset{\sim}{B})}(F,G) & \longrightarrow & [F,G]^{\underset{\sim}{A}} \\ \downarrow & & \downarrow \\ \mathbb{1} & \xrightarrow{\ulcorner \Delta \urcorner} & (\underset{\sim}{A} \times \underset{\sim}{A})^{\underset{\sim}{A}} \end{array}$$

is a pullback. Ostensibly, this pullback is a 2-category,

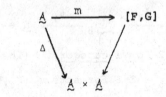

but it is easily seen to be locally discrete. Note that $[F,G]^{\underset{\sim}{A}}$ means the 2-functor category described in §4.2. Composition is the induced morphism in the diagram

$$\begin{array}{ccc}
\mathrm{Cat}(F_0,F_1) \times \mathrm{Cat}(F_1,F_2) & \longrightarrow & ([F_0,F_1] \underset{\underset{\sim}{A}}{\times} [F_1,F_2])^{\underset{\sim}{A}} \\
\searrow \quad \downarrow \scriptstyle{(\text{dashed})} & & \downarrow \\
\mathrm{Cat}(F_0,F_1) & \longrightarrow & [F_0,F_2]^{\underset{\sim}{A}} \\
\downarrow & & \downarrow \\
\mathbb{1} & \xrightarrow{\ulcorner \Delta \urcorner} & (\underset{\sim}{A} \times \underset{\sim}{A})^{\underset{\sim}{A}}
\end{array}$$

Using this composition in the usual fashion, one deduces the following result.

<u>Proposition</u>. $\underset{\sim}{\mathrm{Fun}}(-,-): (2\text{-}\underset{\sim}{\mathrm{Cat}})^{op} \times (2\text{-}\underset{\sim}{\mathrm{Cat}}) \longrightarrow 2\text{-}\underset{\sim}{\mathrm{Cat}}$ is a 2-functor.

For future calculations, we need to know in horrendous set-theoretical detail exactly what $\underset{\sim}{\mathrm{Fun}}(\underset{\sim}{A},\underset{\sim}{B})$ looks like. From the above description one <u>deduces</u> that if $F$ and $G$ are 2-functors from $\underset{\sim}{A}$ to $\underset{\sim}{B}$ then a 2-natural transformation $\varphi: F \longrightarrow G$ assigns

i) to an object $A \in \underset{\sim}{A}$, a morphism

$$\varphi_A: F(A) \longrightarrow G(A) \text{ in } \underset{\sim}{B}$$

ii) to a morphism (1-cell) $f: A \longrightarrow B$ of $\underset{\sim}{A}$, a diagram

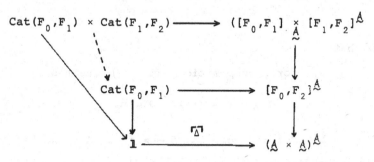

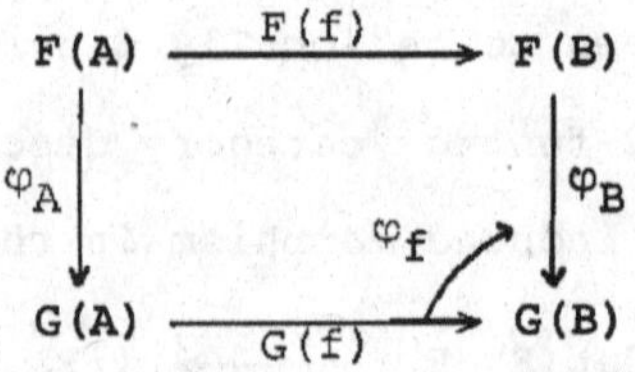


such that

a) for a composition $gf$ in $\underset{\sim}{A}$ one has

$$\varphi_{gf} = \varphi_g F(f) \cdot G(g)\varphi_f$$

b) for a 2-cell $\sigma\colon f \longrightarrow f'$ in $\underset{\sim}{A}$ , one has a commutative diagram

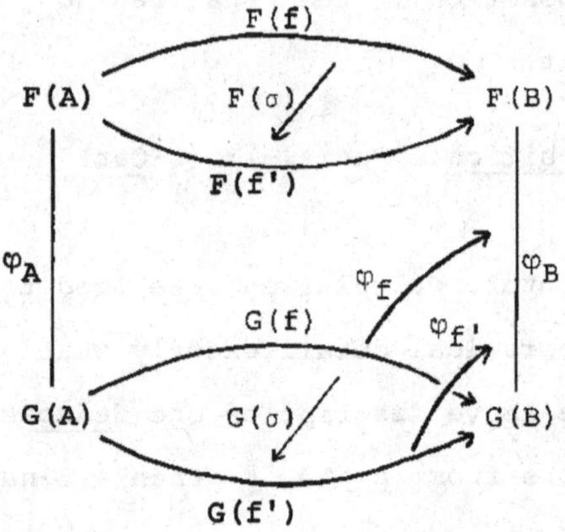


i.e., in $\mathrm{Cat}_{\underset{\sim}{B}}(F(A), G(B))$, the diagram

$$\begin{array}{ccc} G(f)\varphi_A & \xrightarrow{\ \varphi_f\ } & \varphi_B F(f) \\ {\scriptstyle G(\sigma)\varphi_A}\downarrow & & \downarrow{\scriptstyle \varphi_B F(\sigma)} \\ G(f')\varphi_A & \xrightarrow{\ \varphi_{f'}\ } & \varphi_B F(f') \end{array}$$

commutes.

If $\varphi$ and $\psi$ are 2-natural transformations from $F$ to $G$ (i.e., 1-cells of $\underset{\sim}{\mathrm{Fun}}(\underset{\sim}{A},\underset{\sim}{B})$), then a 2-cell $t: \varphi \longrightarrow \psi$ in $\underset{\sim}{\mathrm{Fun}}(\underset{\sim}{A},\underset{\sim}{B})$ assigns to each $A \in \underset{\sim}{A}$ a 2-cell $t_A: \varphi_A \longrightarrow \psi_A$ in $\underset{\sim}{B}$ such that for any 2-cell $\sigma: f \longrightarrow f'$ in $\underset{\sim}{A}$, the diagram

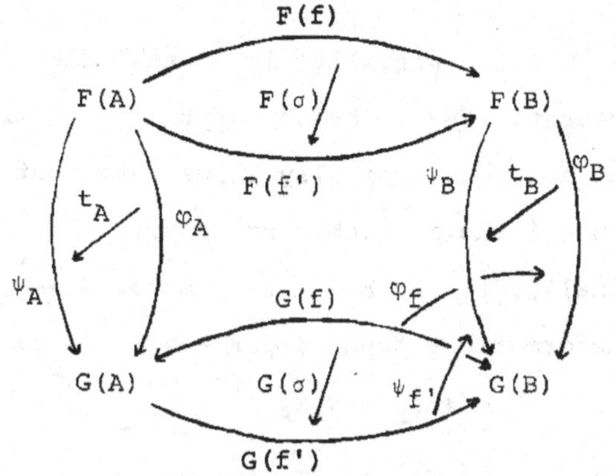


commutes. This commutativity can be expressed either as commutativity of the cube

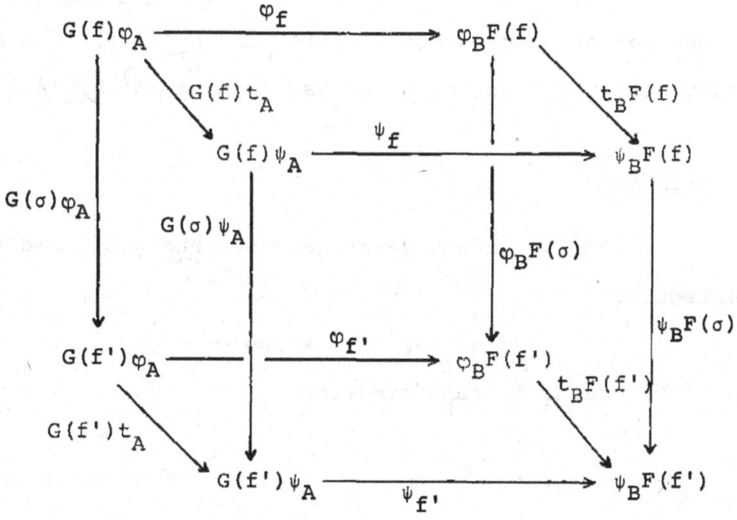

or of the square

$$\begin{array}{ccc} G(f)\varphi_A & \xrightarrow{\varphi_f} & \varphi_B F(f) \\ {\scriptstyle G(\sigma)t_A}\downarrow & & \downarrow{\scriptstyle t_B F(\sigma)} \\ G(f')\psi_A & \xrightarrow{\psi_{f'}} & \psi_B F(f') \end{array}$$

in $\mathrm{Cat}_{\underset{\sim}{B}}(F(A),G(B))$. Actually, it is sufficient to require only the commutativity of the top of the cube for all $f$, but this does not so graphically illustrate that we have, in fact, used up all the available structure.

Finally, if $\varphi: F \longrightarrow G$ and $\psi: G \longrightarrow H$ are 2-natural transformations then $\psi\varphi: F \longrightarrow H$ is given by

$$(\psi\varphi)_A = \psi_A\varphi_A$$

$$(\psi\varphi)_f = \psi_B\varphi_f \cdot \psi_f\varphi_A \ .$$

The composition of 2-cells is simply $(st)_A = s_A t_A$.

The prefix 2- will refer to $\underset{\sim}{\mathrm{Fun}}(\underset{\sim}{A},\underset{\sim}{B})$; thus a 2-subfunctor means a monomorphism in $\underset{\sim}{\mathrm{Fun}}(\underset{\sim}{A},\underset{\sim}{B})$ and a 2-natural equivalence means an isomorphism in $\underset{\sim}{\mathrm{Fun}}(\underset{\sim}{A},\underset{\sim}{B})$.

Proposition

$\underset{\sim}{B}^{\underset{\sim}{A}}$ is a sub 2-category of $\underset{\sim}{\mathrm{Fun}}(\underset{\sim}{A},\underset{\sim}{B})$ and the imbedding

$$(-)^{(-)} \longrightarrow \mathrm{Fun}(-,-)$$

is a Cat-natural transformation.

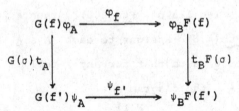

Exercise. Compare $\underset{\sim}{A}^{\mathbf{2}_2}$, $\underset{\sim}{\text{Fun}}(\mathbf{2}_2,\underset{\sim}{A})$ and $\underset{\sim}{\text{Fun}}_{\underset{\sim}{A}}$. In particular, show $\underset{\sim}{\text{Fun}}_{\underset{\sim}{A}} \approx {}^{op}\underset{\sim}{\text{Fun}}(\mathbf{2},{}^{op}\underset{\sim}{A})$.

Examples.

1) $[H,K] \longrightarrow A_0 \times A_1$ is isomorphic as a split (1,0)-bifibred category and as a 2-category to $\mathcal{E}\text{Cat}_{\mathcal{B}}(H(-),K(-)) \longrightarrow A_0 \times A_1$.

2) $|[\mathbb{1},\underset{\sim}{\text{Cat}}]| \approx \widetilde{\text{Cat}}$ (see §1).

3) For any $F\colon X \longrightarrow \text{Cat}$, $E_F = [\mathbb{1},F]$. In particular, if $F = X \longrightarrow \mathbb{1} \xrightarrow{\mathbb{A}} \text{Cat}$, then $[\mathbb{1},F] = \mathbb{A} \times X$.

4) $\underset{\sim}{\text{Fun}} = [\underset{\sim}{\text{Cat}},\underset{\sim}{\text{Cat}}]$ and $\underset{\sim}{\text{Fun}}_A = [\underset{\sim}{A},\underset{\sim}{A}]$.

5) $\underset{\sim}{\text{Fun}}(\mathbb{1},\underset{\sim}{B}) \approx \underset{\sim}{B}$ and for any $\underset{\sim}{A}$, the constant functor $\tau_A\colon \underset{\sim}{A} \longrightarrow \mathbb{1}$ induces $\underset{\sim}{\text{Fun}}(\tau_A,B)\colon \underset{\sim}{B} \longrightarrow \underset{\sim}{\text{Fun}}(\underset{\sim}{A},\underset{\sim}{B})$.

6) If $\underset{\sim}{A}$ and $\underset{\sim}{B}$ are locally discrete, then $\underset{\sim}{\text{Fun}}(\underset{\sim}{A},\underset{\sim}{B}) = \underset{\sim}{B}^{\underset{\sim}{A}}$; i.e., for functors between locally discrete categories

$$[F,G] = (F,G) .$$

7) The 2-categories $[F,G]$ are useful for certain notions in category theory. Thus, let $1\colon \mathbb{1} \longrightarrow \text{Cat}$ be the terminal object of Cat and let $\mathbb{B}\colon \mathbb{1} \longrightarrow \text{Cat}$ be any other object. Then $\mathcal{B} = [1,\mathbb{B}]$ is the category in the universe that looks like the object $\mathbb{B}$ of Cat. In the notation of [CCFM], p. 17, $\mathcal{B}$ $(\in)$ Cat and $\mathbb{B} = \mathcal{B}_{\text{Cat}}$. There is a canonical imbedding

$$\mathcal{B} = [1,\mathbb{B}] \xrightarrow{[1,\text{Cat},\text{Fun},\text{Cat},\mathbb{1}]} [\text{Cat},\mathbb{B}]$$

and $\mathcal{B}$ is cocomplete if and only if this functor has a left adjoint

$$\varinjlim : [\mathrm{Cat},\mathbb{B}] \longrightarrow \mathcal{B} .$$

This shows, for instance, that if $F: A \longrightarrow [\mathrm{Cat},\mathbb{B}]$ is a functor such that $\varinjlim_A F = F_\infty$ exists in $[\mathrm{Cat},\mathbb{B}]$, then denoting the values of $F$ by $F(A): \mathbb{D}_A \longrightarrow \mathbb{B}$ and $F_\infty: \mathbb{D}_\infty \longrightarrow \mathbb{B}$ we get that in $\mathcal{B}$,

$$\varinjlim_A (\varinjlim_{\mathbb{D}_A} F(A)) = \varinjlim_{\mathbb{D}_\infty} F_\infty .$$

The proof follows from the diagram

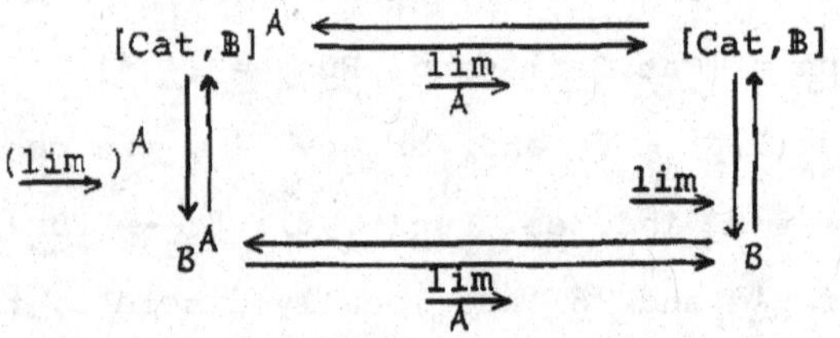


Note that $[\mathrm{Cat},\mathbb{B}]$ was introduced by Grothendieck in [SGA], §11,b where it is denoted by $\mathrm{Cat}\|\mathbb{B}$ . It was utilized by Giraud [MD] and Hofmann [CCEF], who also discusses the functor

$$\varinjlim : [\mathrm{Cat},\mathbb{B}] \longrightarrow \mathcal{B} .$$

As still another use, note that if $\mathbb{B}$ and $\mathbb{C}$ are two objects of $\mathrm{Cat}$ , then

$$[\mathbb{B},\mathbb{C}] = \mathcal{C}^{\mathcal{B}}$$

(assuming that $\mathrm{Cat}$ is full in the universe).

## §7 2-ADJOINTNESS

Given two functorial morphisms between a pair of category objects one can describe adjointness in terms of the corresponding comma category constructions and derive the usual properties. (See [CCC].) In the present case, it works out as follows.

<u>Definition</u>

Let $F: \underset{\sim}{B} \longrightarrow \underset{\sim}{A}$ and $U: \underset{\sim}{A} \longrightarrow \underset{\sim}{B}$ be 2-functors. A 2-<u>adjunction</u> is an isomorphism

$$\begin{array}{ccc} [F,\underset{\sim}{A}] & \xrightarrow{\ \Phi\ } & [\underset{\sim}{B},U] \\ & \searrow \quad \swarrow & \\ & \underset{\sim}{B} \times \underset{\sim}{A} & \end{array}$$

of 2-categories over $\underset{\sim}{B} \times \underset{\sim}{A}$ .

<u>Proposition</u>

If $\Phi$ is also a morphism of split (1,0)-bifibrations, then $\Phi$ corresponds to an ordinary Cat-enriched adjunction between $F$ and $U$ .

<u>Proof</u>. Since $[F,A] \cong \underset{\sim}{E}_{Cat_A}(F(-),-)$ and $[\underset{\sim}{B},U] \cong \underset{\sim}{E}_{Cat_A}(-,U(-))$ we have

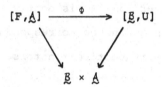

$$\begin{array}{ccc} \underset{\sim}{E}_{Cat_A(F(-),-)} & \xrightarrow{\Phi} & \underset{\sim}{E}_{Cat_B(-,U(-))} \\ & \searrow \quad \swarrow & \\ & \underset{\sim}{B} \times \underset{\sim}{A} & \end{array}$$

Under the hypotheses, $\Phi$ then corresponds to a Cat-natural isomorphism

$$Cat_A(F(-),-) \xrightarrow{\sim} Cat_B(-,U(-)) \quad .$$

<u>Remark</u>. In the case of ordinary categories, $(F,A) = E_{Hom(F(-),-)}$ is a split (1,0)-bifibration with discrete i-fibres, $i = 0,1$. A functor between two such discrete bifibrations over $B \times A$ is easily seen to be a morphism of bifibrations and hence corresponds to a natural transformation of the respective functors. Thus an ordinary adjunction

$$\begin{array}{ccc} (F,A) & \xrightarrow{\sim} & (B,U) \\ & \searrow \quad \swarrow & \\ & B \times A & \end{array}$$

is <u>always</u> (assuming $A$ and $B$ are locally small) the same as a natural equivalence

$$Hom_A(F(-),-) \xrightarrow{\sim} Hom_B(-,U(-)) \quad .$$

In the case of 2-categories these conditions are not equivalent, as we shall see by example later.

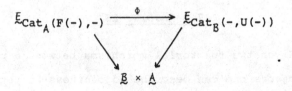

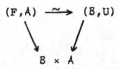

## §8 2-COLIMITS IN CAT

The notion of limits depends on the notions of "functor categories" and of "adjointness", being the adjoint on one side or the other of the constant imbedding

$$B \longrightarrow B^A \quad .$$

In our case, we equally well have the constant imbedding (see §6, example 5),

$$\Delta_{\underset{\sim}{B}}^{\underset{\sim}{A}} = \underset{\sim}{\mathrm{Fun}}(\tau_{\underset{\sim}{A}}, \underset{\sim}{B}) : \underset{\sim}{B} \longrightarrow \underset{\sim}{\mathrm{Fun}}(\underset{\sim}{A}, \underset{\sim}{B})$$

and we define the 2-<u>limit</u> <u>functor</u> (resp., the 2 <u>colimit</u> <u>functor</u>) to be the right (resp., left) 2-adjoint of $\Delta_{\underset{\sim}{B}}^{\underset{\sim}{A}}$, when it exists. We wish to calculate 2-colimits in $\underset{\sim}{\mathrm{Cat}}$ . Thus

$$2\text{-}\underrightarrow{\lim}_{\underset{\sim}{A}} : \underset{\sim}{\mathrm{Fun}}(\underset{\sim}{A}, \underset{\sim}{\mathrm{Cat}}) \longrightarrow \underset{\sim}{\mathrm{Cat}}$$

is the unique (up to a 2-isomorphism) 2-functor such that there is an equivalence

$$\begin{array}{ccc} [2\text{-}\underrightarrow{\lim}_{\underset{\sim}{A}}, \underset{\sim}{\mathrm{Cat}}] & \xrightarrow{\;\simeq\;} & [\underset{\sim}{\mathrm{Fun}}(\underset{\sim}{A}, \underset{\sim}{\mathrm{Cat}}), \Delta_{\underset{\sim}{\mathrm{Cat}}}^{\underset{\sim}{A}}] \\ & \searrow \quad \swarrow & \\ & \underset{\sim}{\mathrm{Fun}}(\underset{\sim}{A}, \underset{\sim}{\mathrm{Cat}}) \times \underset{\sim}{\mathrm{Cat}} & \end{array}$$

<u>Theorem</u>

i) Let $F : A \longrightarrow \mathrm{Cat}$. Then $2\text{-}\underrightarrow{\lim}_A F = [\mathbb{1}, F]$ .

ii) In general, if $F : \underset{\sim}{A} \longrightarrow \mathrm{Cat}$ is a 2-functor then $2\text{-}\underrightarrow{\lim}_{\underset{\sim}{A}} F = |[\mathbb{1}, F]|$ .

Corollary

For $1_A : A \longrightarrow \mathbb{1} \xrightarrow{1} \text{Cat}$ , $2\text{-}\underrightarrow{\lim}_A 1_A \approx A$ .

Note. The theorem says that $2\text{-}\underrightarrow{\lim}_A F$ is the split 0-fibred category over $A$ determined by $F$ . This only really makes sense if $A$ is small and Cat is cat-complete ([CCFM], p. 17). However, the theorem can also be read as asserting the existence of $2\text{-}\underrightarrow{\lim}_A F$ in the universe, providing $[\mathbb{1},F]$ exists. (See the appendix.) We have separated the theorem into two cases and will only prove part i) for a small category $A = [1,\mathbb{A}]$, $\mathbb{A} \in \text{Cat}$. (See §6, Example 7.) This proof has a nice conceptual form in terms of the following lemma, whereas the only proof I know for ii) is an explicit construction of the required equivalence. In the lemma, $(\underset{\sim}{\text{Cat}},\mathbb{A})_0$ denotes the full subcategory of $(\underset{\sim}{\text{Cat}},\mathbb{A})$ determined by split-normal 0-fibrations over $A$ .

Lemma (The Yoneda-like Lemma)

Let $A = [1,\mathbb{A}]$, $\mathbb{A} \in \text{Cat}$ . Then there is a commutative diagram

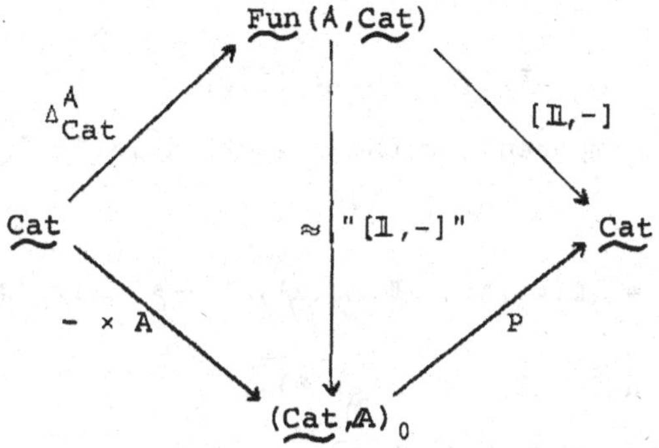


where the vertical functor is an equivalence.

<u>Proof of Lemma</u>. The commutativity of the left hand triangle is Example 3, §6, while that of the right hand triangle is trivial. Note that P denotes the usual projection of a comma category on its first component. In Example 3, §6, it was also pointed out that if F: A ⟶ Cat, then $[\mathbb{1},F] = E_F$ , which is a split 0-fibration over A . Furthermore every split 0-fibration over A arises this way, so we need only worry about morphisms and 2-cells.

If φ: F ⟶ G is a 2-natural transformation in Fun(A,Cat), then

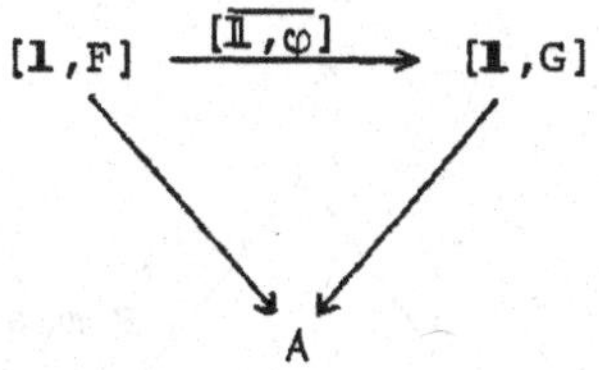


is a morphism over $A = \mathbb{1} \times A$ . Furthermore, if t: φ ⟶ ψ is a 2-cell in Fun(A,Cat), then t determines a natural transformation

$$\bar{t}\colon [\overline{1,\varphi}] \longrightarrow [\overline{1,\psi}]$$

over $\mathcal{A}$ whose component at the object $(\mathbb{1},a,A) \in [\mathbb{1},F]$ is given by

$$\bar{t}_{(\mathbb{1},a,A)} = (\mathbb{1},t_a,A)\colon (\mathbb{1},\varphi_A(a),A) \longrightarrow (\mathbb{1},\psi_A(a),A)$$

and $(\bar{t},\mathcal{A})$ is a 2-cell of $(\underset{\sim}{\mathrm{Cat}},\mathbb{A})_0$.

Conversely, suppose

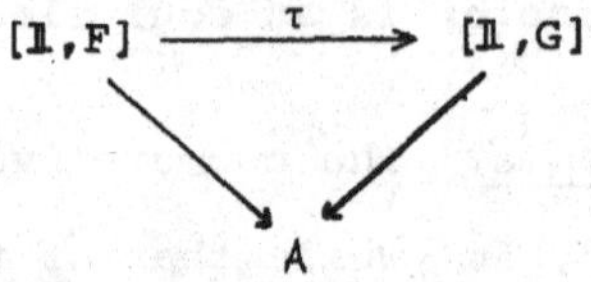


commutes (i.e., is a 1-cell of $(\underset{\sim}{\mathrm{Cat}},\mathbb{A}_0)$. We must show that there is a unique $\varphi\colon F \longrightarrow G$ with $\tau = [\overline{\mathbb{1},\varphi}]$. Define $\varphi$ to be the 2-natural transformation whose component at $A \in \mathcal{A}$ is the functor $\varphi_A\colon F(A) \longrightarrow G(A)$ such that

i) If $a \in F(A)$, then $\varphi_A(a) = \tau(\mathbb{1},a,A)$.

ii) If $f$ is a morphism in $F(A)$, then $\varphi_A(f) = \tau(\mathbb{1},f,A)$; i.e., $\varphi_A = \tau|[\mathbb{1},F]_A = \tau|F(A)$. To define $\varphi_m$ on a morphism $m\colon A \longrightarrow B$ in $\mathcal{A}$, observe that $\tau$ assigns to the morphism

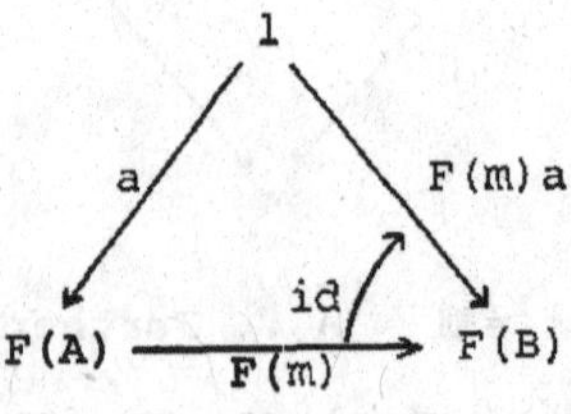

in $[\mathbb{1},F]$, a morphism

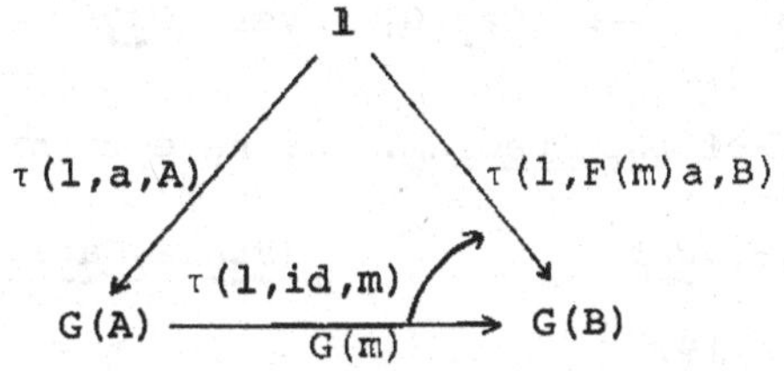


in $[\mathbb{1},G]$. We set $(\varphi_m)_a = \tau(1,id,m)$. Then $\varphi$ is a 2-natural transformation such that $\tau = [\overline{\mathbb{1},\varphi}]$.

Finally, if $s\colon \tau \longrightarrow \tau'$ is a natural transformation over $A$, let $t\colon \varphi \longrightarrow \varphi'$ be the 2-cell between the corresponding 2-natural transformations whose component $t_A\colon \varphi_A \longrightarrow \varphi'_A$ is the natural transformation with components

$$\begin{array}{cccc} (t_A)_a\colon & \varphi_A(a) & \longrightarrow & \varphi'_A(a) \\ \| & \| & & \| \\ s_{(\mathbb{1},a,A)}\colon & \tau(\mathbb{1},a,A) & \longrightarrow & \tau'(\mathbb{1},a,A) \end{array}$$

Then clearly $s = \bar{t}$ .

<u>Note</u>. This lemma says that, while $\mathrm{Cat}^A$ corresponds to split-normal 0-fibrations and cleavage preserving morphisms, $\mathrm{Fun}(A,\underset{\sim}{\mathrm{Cat}})$ corresponds to all functors between such fibrations over $A$ . Thus, it generalizes the proposition of §1. On the other hand, the correspondence between functors $\tau\colon [\mathbb{1},F] \longrightarrow [\mathbb{1},G]$ over $A$ and natural transformations $\varphi\colon F \longrightarrow G$ looks like the Yoneda lemma (§7) which

gives a correspondence between 2-natural transformations and 2-functors $[\underset{\sim}{\mathrm{Cat}},F] \longrightarrow [\underset{\sim}{\mathrm{Cat}},G]$ over $\underset{\sim}{\mathrm{Cat}} \times A$ .

<u>Proof of the Theorem.</u> We have vertical isomorphisms

$$\begin{array}{ccc} [[1,-],\underset{\sim}{\mathrm{Cat}}] & & [\underset{\sim}{\mathrm{Fun}}(A,\underset{\sim}{\mathrm{Cat}}),\Delta^{A}_{\mathrm{Cat}}] \\ \wr & & \wr \\ [P,\underset{\sim}{\mathrm{Cat}}] & \overset{\Phi}{\dashrightarrow} & [(\underset{\sim}{\mathrm{Cat}},\mathbb{A})_0,- \times \mathbb{A}] \end{array}$$

so it is sufficient to establish an equivalence $\Phi$ as indicated, over $(\underset{\sim}{\mathrm{Cat}},\mathbb{A})_0 \times \underset{\sim}{\mathrm{Cat}}$ . An object of $[P,\underset{\sim}{\mathrm{Cat}}]$ can be represented by a diagram

$$\begin{array}{ccc} E & \xrightarrow{M} & X \\ {\scriptstyle Q}\downarrow & & \\ A & & \end{array}$$

where $Q$ is a split normal 0-fibration, while an object of $[(\underset{\sim}{\mathrm{Cat}},\mathbb{A}),- \times \mathbb{A}]$ looks like a diagram

$$\begin{array}{ccc} E & \xrightarrow{N} & X \times A \\ & {\scriptstyle Q}\searrow \quad \swarrow{\scriptstyle pr} & \\ & A & \end{array}$$

On objects, set $\Phi(M,Q) = \{M,Q\}: E \longrightarrow X \times A$ . This clearly gives a bijection between objects. Morphisms on each side look like

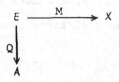

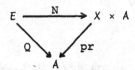

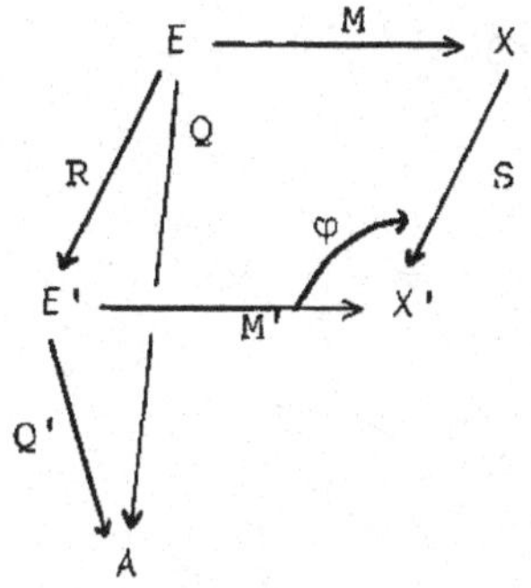


and

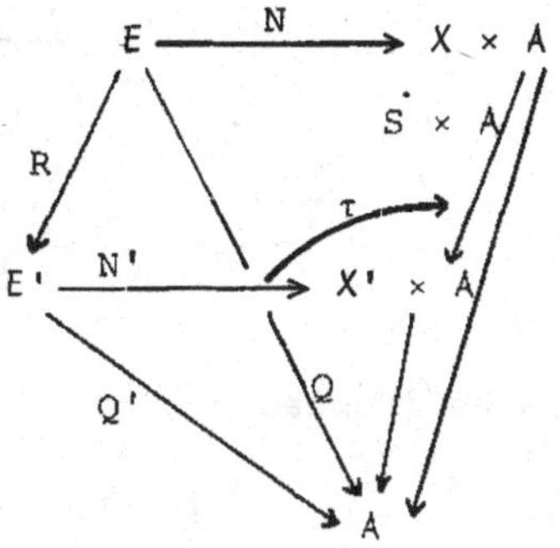


where $pr\tau = id$. Set

$$\Phi(R,\varphi,S) = (R,\{\varphi,Q\},S \times A) \quad .$$

This gives a bijection between 1-cells. Finally, a 2-cell on the left is a pair $(\rho,\sigma)$ where $\rho\colon R \longrightarrow R'$ and $\sigma\colon S \longrightarrow S'$ are natural transformations compatible with $\varphi$ and $\varphi'$. Set $\Phi(\rho,\sigma) = (\rho,\sigma \times A)$. Then $\Phi$ is an equivalence of 2-categories.

Examples.

i) Since $\mathbf{1}$ is terminal in Cat , there is a natural transformation $F \longrightarrow 1_A$ for any functor $F\colon A \longrightarrow$ Cat . The induced functor

$$2\text{-}\underrightarrow{\lim}_A F \longrightarrow 2\text{-}\underrightarrow{\lim}_A 1_A$$

is just the canonical projection

$$[\mathbb{1},F] \longrightarrow A$$

ii) Since the constant imbedding factors via

$$\begin{array}{ccc} \underset{\sim}{B} & \xrightarrow{\Delta} & \underset{\sim}{\mathrm{Fun}}(A,\underset{\sim}{B}) \\ & {\scriptstyle \Delta}\searrow \quad \nearrow{\scriptstyle \subset} & \\ & \underset{\sim}{B}^{A} & \end{array}$$

there is an induced functor

$$\begin{array}{ccc} [\underset{\sim}{B}^{A},\Delta] & \longrightarrow & [\underset{\sim}{\mathrm{Fun}}(A,\underset{\sim}{B}),\Delta] \\ \wr & & \wr \\ [\varinjlim_{A},\underset{\sim}{B}] & \longrightarrow & [2\text{-}\varinjlim_{A},\underset{\sim}{B}] \end{array}$$

which, by the Yoneda lemma, corresponds to a 2-natural transformation

$$2\text{-}\varinjlim_{A} \longrightarrow \varinjlim_{A} .$$

As an example, $A = (\mathrm{Rings})^{op}$ and $F(R)$ is the category of R-modules. Then $2\text{-}\varinjlim F$ is the category of all modules over all rings, while $\varinjlim F$ is the category of abelian groups, since $Z$ is a terminal object of $(\mathrm{Rings})^{op}$.

iii) Sets $\subset$ Cat is not closed under 2-colimits since if $F: A \longrightarrow$ Sets then

$$2\text{-}\varinjlim F = [\mathbb{1},F] = (1,F)$$

and this is discrete only if $A$ is discrete.

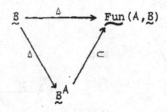

iv) 2-limits in Cat can also be calculated and turn out to be the category of sections of $[\mathbb{1},F]$ over $A$. If, in the definition of 2-natural transformation, all $\varphi_f$ were required to be equivalences, then the corresponding notion of "2-limit" would give "cartesian sections" which long ago was called $\varprojlim$ in [SGA].

## §9 THE 2-COMPREHENSION SCHEME

We must first describe the 2-Kan extension.

**Definition**

Let $F: \underset{\sim}{A} \longrightarrow \underset{\sim}{B}$ be a 2-functor. Then the left 2-adjoint (when it exists) to

$$F^* = \underset{\sim}{\text{Fun}}(F,\underset{\sim}{X}): \underset{\sim}{\text{Fun}}(\underset{\sim}{B},\underset{\sim}{X}) \longrightarrow \underset{\sim}{\text{Fun}}(\underset{\sim}{A},\underset{\sim}{X})$$

is called the (left) 2-Kan extension of $F$. It is denoted by $\Sigma_2 F$.

**Proposition**

Given $H: \underset{\sim}{A} \longrightarrow \underset{\sim}{X}$, then $\Sigma_2 F(H): \underset{\sim}{B} \longrightarrow \underset{\sim}{X}$ is the 2-functor whose value on any $B \in \underset{\sim}{B}$ is given by

$$[\Sigma_2 F(H)](B) = 2\text{-}\varinjlim([F,B] \longrightarrow \underset{\sim}{A} \xrightarrow{H} \underset{\sim}{X})$$

**Corollary**

For $1_A = (\underset{\sim}{A} \longrightarrow \mathbb{1} \xrightarrow{1} \underset{\sim}{\text{Cat}})$, we have

$$\Sigma_2 F(1_A)(-) = [F,-]: \underset{\sim}{B} \longrightarrow \underset{\sim}{\text{Cat}} .$$

**Proofs.** The corollary follows from the proposition by using the corollary to the theorems in §8. The proposition is proved by verifying that the usual construction still makes sense. Thus, define

$$[\Sigma_2 F(H)](-) = 2\text{-}\varinjlim([F,-] \longrightarrow \underset{\sim}{A} \xrightarrow{H} \underset{\sim}{X}) .$$

Then we must produce an equivalence

$$\Phi\colon [\Sigma_2 F, \underset{\sim}{\mathrm{Fun}}(\underset{\sim}{B},\underset{\sim}{X})] \longrightarrow [\underset{\sim}{\mathrm{Fun}}(\underset{\sim}{A},\underset{\sim}{X}), F^*]$$

over $\underset{\sim}{\mathrm{Fun}}(\underset{\sim}{A},\underset{\sim}{X}) \times \underset{\sim}{\mathrm{Fun}}(\underset{\sim}{B},\underset{\sim}{X})$. To do so, we first describe the objects on both sides over a given $H\colon \underset{\sim}{A} \longrightarrow \underset{\sim}{X}$ and $K\colon \underset{\sim}{B} \longrightarrow \underset{\sim}{X}$. An object of $[\underset{\sim}{\mathrm{Fun}}(\underset{\sim}{A},\underset{\sim}{X}), F^*]$ over $H,K$ is a 2-natural transformation $\varphi\colon H \longrightarrow KF$. An object of $[\Sigma_2 F, \underset{\sim}{\mathrm{Fun}}(\underset{\sim}{B},\underset{\sim}{X})]$ over $(H,K)$ is a 2-natural transformation $\lambda\colon \Sigma_2 F(H) \longrightarrow K$. To understand $\lambda$, observe that its component at $B \in \underset{\sim}{B}$ is a morphism

$$\lambda_B\colon [\Sigma_2 F(H)](B) \longrightarrow K(B)$$

in $\underset{\sim}{X}$. But, by definition of $[\Sigma_2 F(H)](B)$, any morphism

$$m\colon [\Sigma_2 F(H)](B) \longrightarrow X$$

in $\underset{\sim}{X}$, corresponds to a 2-natural transformation $m'$ from the 2-functor $[F,B] \longrightarrow \underset{\sim}{A} \xrightarrow{H} \underset{\sim}{X}$ to the constant 2-functor determined by $X \in \underset{\sim}{X}$. $m'$ assigns to an object $(F(A) \longrightarrow B)$ in $[F,B]$ a morphism $H(A) \longrightarrow X$ in $\underset{\sim}{X}$ and to a morphism

$\kappa$:

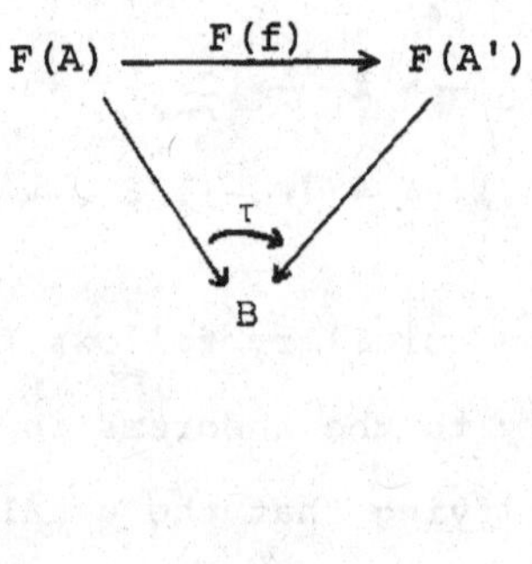

in [F,B] a diagram

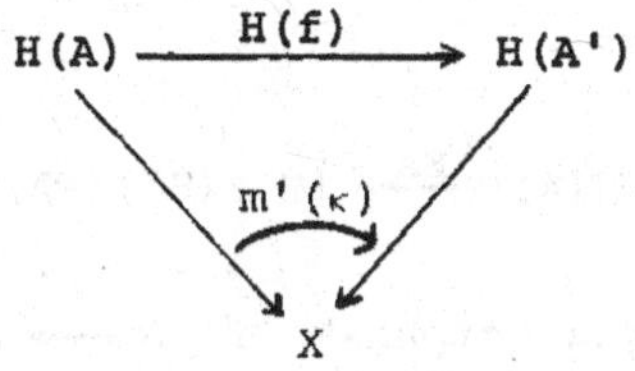


in $\underset{\sim}{X}$ .

Now, given $\varphi: H \longrightarrow KF$ , define

$$\lambda = \Phi'(\varphi): \Sigma_2 F(H) \longrightarrow K$$

to be the 2-natural transformation whose component

$$\lambda_B: [\Sigma_2 F(H)](B) \longrightarrow K(B)$$

corresponds to the 2-natural transformation $\lambda'_B$ whose value on $\kappa$ above is the composed diagram

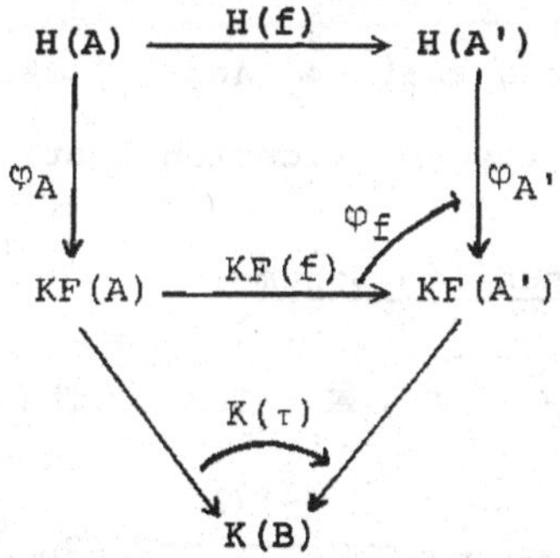


Conversely, given $\lambda: \Sigma_2 F(H) \longrightarrow K$ , define

$$\varphi = \Phi(\lambda): H \longrightarrow KF$$

to be the 2-natural transformation constructed as follows:

i) For each object $A \in \underset{\sim}{A}$ , $F(A): F(A) \longrightarrow F(A)$ is an object of [F,FA], so the adjunction morphism

$$\varepsilon_{FA}\colon H(A) \longrightarrow [\Sigma_2 F(H)](FA)$$

is defined. We set

$$\varphi_A = (H(A) \xrightarrow{\varepsilon_{FA}} [\Sigma_2 F(H)](FA) \xrightarrow{\lambda_{FA}} KFA\ .$$

ii) For a morphism $f\colon A \longrightarrow A'$, $\varphi_f$ is the composed square

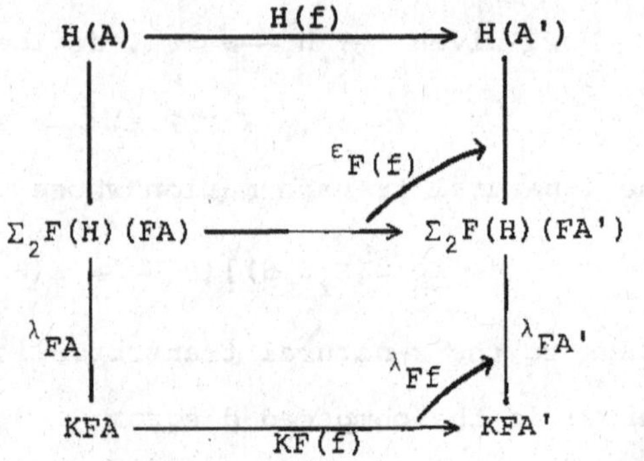


It is easily checked that $\Phi$ and $\Phi'$ are 2-functors. We omit the more lengthy verification that $\Phi' = \Phi^{-1}$ .

<u>Theorem (2-Comprehension Scheme)</u>

Let $X = [\mathbb{1},\mathbb{X}]$, $\mathbb{X} \in \mathrm{Cat}$ . Then the 2-functors

$$[{}^{op}\underset{\sim}{Cat},\mathbb{X}] \underset{[\ ,-]}{\overset{\Sigma_2(-)(1)}{\rightleftarrows}} {}^{op}\underset{\sim}{Fun}(X,\underset{\sim}{Cat})$$

have the following property: There exist 2-functors

$$[\Sigma_2(-)(1),{}^{op}\underset{\sim}{Fun}(X,\underset{\sim}{Cat})] \underset{\Phi'}{\overset{\Phi}{\rightleftarrows}} [[{}^{op}\underset{\sim}{Cat},\mathbb{X}],[\mathbb{1},-]]$$

such that

i) $\Phi$ is (enriched) left adjoint to $\Phi'$

ii) Restricted to the subcategories

$$[\Sigma_2(-)(1), {}^{op}(\mathrm{Cat}^X)] \underset{\Phi'}{\overset{\Phi}{\rightleftarrows}} [[{}^{op}\mathrm{Cat}, \mathbb{X}), [\mathbf{1}, -]]$$

$\Phi' = \Phi^{-1}$ and hence on these subcategories $\Sigma_2(-)(1)$ is left 2-adjoint to $[\mathbf{1},-]$

iii) Restricted further to

$$(\Sigma_2(-)(1), \mathrm{Cat}^X) \underset{\Phi'}{\overset{\Phi}{\rightleftarrows}} ((\mathrm{Cat}, \mathbb{X}), [\mathbf{1}, -])$$

$\Phi$ defines an ordinary adjunction

$$(\mathrm{Cat}, \mathbb{X}) \underset{[\mathbb{1},-]}{\overset{\Sigma_2(-)(1)}{\rightleftarrows}} \mathrm{Cat}^X \; .$$

Lemma

$$\Sigma_2(-)(1_{(-)}) : [{}^{op}\underset{\sim}{\mathrm{Cat}}, \mathbb{X}] \longrightarrow {}^{op}(\underset{\sim}{\mathrm{Cat}}^X) \subset {}^{op}\underset{\sim}{\mathrm{Fun}}(X, \underset{\sim}{\mathrm{Cat}})$$

Proof of Lemma. The 2-Kan extension of $1_A$ along $F$ as a functor in $F$ is very sensitive to variances and only works as indicated. An object of $[{}^{op}\underset{\sim}{\mathrm{Cat}}, \mathbb{X}]$ is a functor $F: A \longrightarrow X$. Since $A$ and $X$ are locally discrete as 2-categories, $\Sigma_2 F(1_A)$ is the functor

$$(F,-) : X \longrightarrow \mathrm{Cat} \; ;$$

i.e., an object ${}^{op}\underset{\sim}{\mathrm{Cat}}^X$. A morphism in $[{}^{op}\underset{\sim}{\mathrm{Cat}}, \mathbb{X}]$ is a diagram

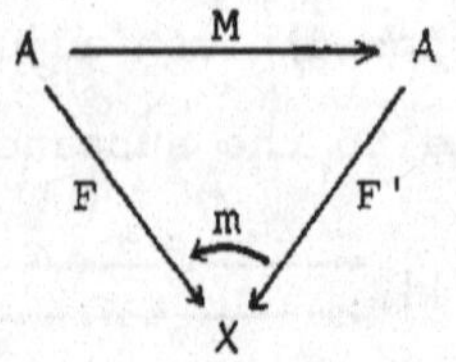


where $m\colon F'M \longrightarrow F$ is a natural transformation. This determines a natural transformation

$$\{M,m\}\colon (F,-) \longrightarrow (F',-)$$

whose component

$$\{M,m\}_X\colon (F,X) \longrightarrow (F',X)$$

is the functor which takes the object $FA \longrightarrow X$ of $(F,X)$ to the object

$$F'MA \xrightarrow{m_A} FA \longrightarrow X$$

of $(F',X)$; i.e., $\{M,m\}_X$ is the composition

$$(F,X) \xrightarrow{\overline{(m,X)}} (F'M,X) \longrightarrow (F'X)\ .$$

It is easily seen that $\{M,m\}$ is natural, and not just 2-natural, so that

$$\Sigma_2(M,m)(1_A) = \{M,m\}$$

is a morphism of ${}^{op}\mathbf{Cat}^X$. Finally, a 2-cell in $[{}^{op}\mathbf{Cat},\mathbf{X}]$ is a diagram

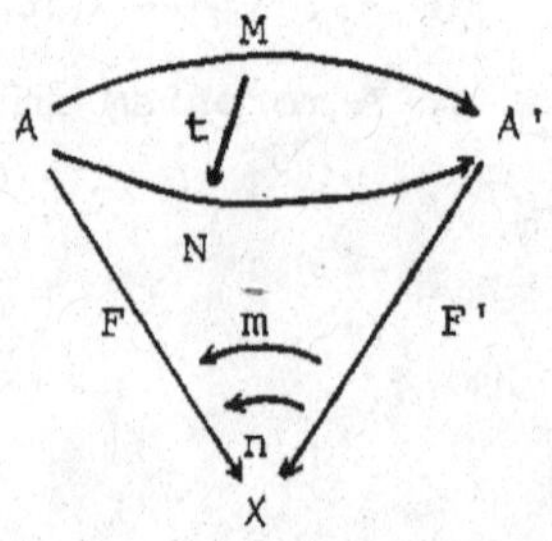

where $t\colon N \longrightarrow M$ is a natural transformation such that $m \cdot F't = n$. This determines a 2-cell in ${}^{op}(\mathrm{Cat}^{X})$ from $\{M,n\}$ to $\{N,n\}$; i.e., a 2-cell in $\mathrm{Cat}^{X}$ from $\{N,n\}$ to $\{M,m\}$, whose component at $X$ is the natural transformation

$$\{N,n\}_X \longrightarrow \{M,m\}_X$$

whose value on $FA \longrightarrow X$ in $(F,X)$ is the morphism

$$\begin{array}{ccc} F'NA & \xrightarrow{\;F't\;} & F'MA \\ & {}_{n_A}\searrow \qquad \swarrow_{m_A} & \\ & FA & \\ & \downarrow & \\ & X & \end{array}$$

in $(F',X)$.

<u>Proof of Theorem</u>. By the Yoneda-like lemma of §8,

$$[\mathbb{1},-]\colon \mathrm{Fun}(X,\mathrm{Cat}) \longrightarrow (\mathrm{Cat},\mathbb{X}) \subset [\mathrm{Cat},\mathbb{X}]$$

and hence, if we wish, we may regard it as a 2-functor

$$[\mathbb{1},-]\colon {}^{op}\mathrm{Fun}(X,\mathrm{Cat}) \longrightarrow [{}^{op}\mathrm{Cat},\mathbb{X}]\,.$$

Using this and the preceeding lemma, it follows that the restrictions of the functors map as indicated.

The functors $\Phi$ and $\Phi'$

$$[\Sigma_2(-)(1),{}^{op}\mathrm{Fun}(X,\mathrm{Cat})] \underset{\Phi'}{\overset{\Phi}{\rightleftarrows}} [[{}^{op}\mathrm{Cat},\mathbb{X}],[\mathbb{1},-]]$$

are described as follows: An object on the left hand side is a 2-natural transformation $\sigma\colon (F,-) \longrightarrow K$, whereas an object on the right hand side is a diagram

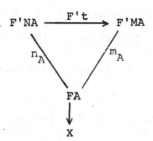

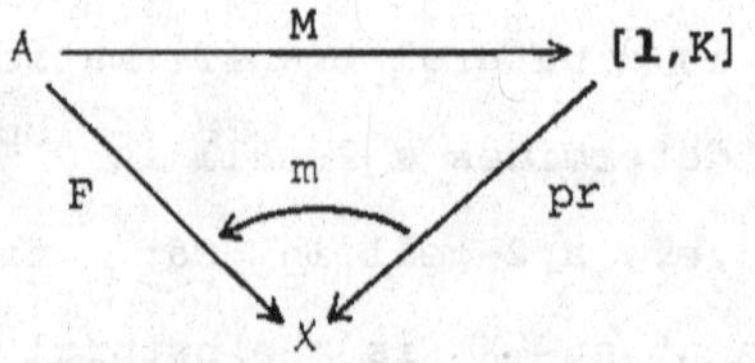


The functor $M$ is described by a pair of components, $M = \{M_1, M_2\}$ where $M_1: \mathcal{A} \longrightarrow \mathcal{X}$ is a functor, and for $A \in \mathcal{A}$, $M_2(A) \in K(M_1(A))$, while for $f: A \longrightarrow A'$ in $\mathcal{A}$,

$$M_2(f): K(M_1(f))(M_2(A)) \longrightarrow M_2(A') .$$

With the description, $m$ is a natural transformation, $m: M_1 \longrightarrow F$.

Given $\sigma: (F,-) \longrightarrow K$, define $\Phi(\sigma)$ to be the object for which

$$M(A) = \{F(A), \sigma_{FA}(id_{FA})\} .$$

If $g: A \longrightarrow A'$, let $\tilde{g}$ be the map

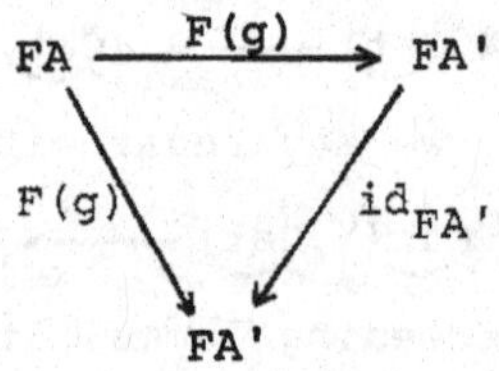


in $(F,A')$ and set

$$M(g) = \{F(g), \sigma_{FA'}(\tilde{g})^0(\sigma_{F(g)})\, id_{F(A)}\} .$$

Finally, $m: M_1 \longrightarrow F$ is the identity.

Conversely, given $M$, $m$, define $\Phi'(M,m)$ to be the 2-natural transformation $\sigma: (F,-) \longrightarrow K$ such that $\sigma_X: (F,X) \longrightarrow K(X)$ is the functor whose value on an object $(A, f: F(A) \longrightarrow X)$ in $(F,X)$ is $K(fm_A)(M_2(A))$ and on a map

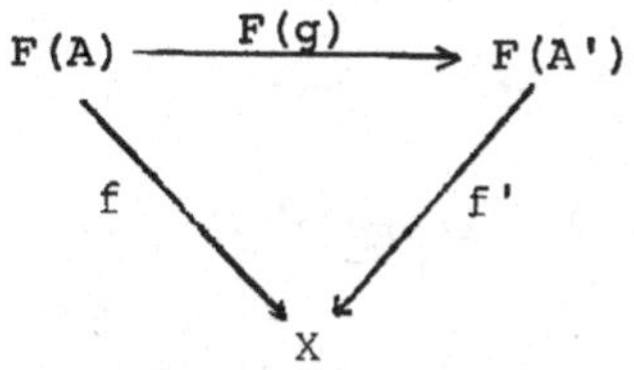


is $K(f'm_{A'})(M_2(g))$.

The action of $\Phi$ and $\Phi'$ on morphisms is considerably more complicated. The descriptions of this and of the enriched natural transformations

$$\Phi \circ \Phi' \longrightarrow \mathrm{id}, \quad \mathrm{id} \longrightarrow \Phi'\Phi$$

as well as the verification that these operations behave as indicated will be published elsewhere.

<u>Note</u>. The distribution of weak dualizations in the general statement of the theorem is a bit mysterious. As with ordinary adjointness, the possibility of dualization gives rise to a number of types of 2-adjointness which do not as yet deserve special names. It is possible that if one concentrated on 1-fibrations as basic rather than 0-fibrations, then this statement might come out more naturally. However, along the way one would find - for this approach - that

$$2\text{-}\varinjlim\,(A \longrightarrow \mathbf{1} \xrightarrow{1} \mathrm{Cat}) = A^{op} \, .$$

It seemed to me that getting $A$ as the answer to this without an artificial dualization was more desirable than getting the neatest comprehension scheme, but this may be overly provincial. In any case, the phenomena exhibited by the proof of the comprehension scheme seem to be genuine and not artifacts of the notation.

## APPENDIX

### Some Remarks on a Categorical Foundations of Mathematics

The possibility, indicated by Lawvere in [CCFM] of giving a foundation for mathematics in categorical terms raises the interesting problem of finding the correct form for such an axiomatization. The answer of course depends on the intended uses. I would like to suggest here that there is in fact a whole hierarchy of theories and that their mutual inter-expressability poses a strong restraint on the form of any one of them.

Specifically, there is the hierarchy of n-categories, where an n-category is a "category" whose "hom-objects" are (n-1)-categories. Here 0-categories are sets, 1-categories are categories, 2-categories are as described in this paper, etc. At the top are $\omega$-categories whose "hom-objects" are simplicial sets (as suggested by Epstein). This sequence should have certain properties.

i) There should be an elementary theory of abstract n-categories. In this paper, we have given this theory for $n = 2$ and the general case is easily derived from that. Nothing new happens until $\omega$ and it is not clear that this is finitely axiomatizable.

ii) There should be a basic theory of abstract

n-categories. The case $n = 1$ is not clear yet and in this paper there is an indication of what is needed for $n = 2$. However, certain things can be said in general.

a) There is a generating n-category $\mathbf{2}_n$ which has two i-cells, $i < n$ and a unique n-cell. (Except, $\mathbf{2}_0 = 1$.) This is finitely describable, but $\mathbf{2}_\omega$ is apparently not.

b) Finite limits and colimits exist and there are "functor n-categories" adjoint to cartesian product. Besides this, there is an increasing sequence of $(n - 1)$ more "hyper functor n-categories" constructed as in the case $n = 2$ presented here. There are almost certainly associated tensor products, but their structure is so complicated that at the moment, for $n = 2$, I cannot even tell if $- \otimes \underset{\sim}{A}$ is adjoint or 2-adjoint to $\underline{\text{Fun}}(\underset{\sim}{A},-)$. I do not know the description of these hyper functor categories in the basic theory.

c) There are a number of functors from n-categories to "locally discrete" n-categories.

d) There is an object which is a model of the basic theory; i.e., an n-category of (small) n-categories. It is an (n+1)-category in, apparently, $n$ different ways.

This is certainly not an exhaustive list of desirable properties. There are some further requirements that are even more crutial. If we regard the basic theory of abstract n-categories as a description of n-categories in terms of themselves, then we also require the following.

iii) m-categories should be expressable in terms of

n-categories for $m < n$. By induction, this reduces to the simply expressed requirement that the axioms of the basic theory of n-categories, restricted to "locally discrete" n-categories should imply the axioms of the basic theory of (n-1)-categories.

iv) n-categories should be expressable in terms of m-categories for $n > m$. Clearly, n-categories can be described in terms of 0-categories = sets, but one would hope that the elegance of the description would increase with decreasing $n - m$. In particular, there should be an analogue of the relation $\mathrm{Cat} \approx \mathrm{Sets}^{(\{4\}^{op})}$ between models of (n-1)-categories and n-categories.

We suggest that a proper foundation of mathematics is an elementary axiomatization of the hierarchy described here.

The actual status of the program outlined above is rather meager compared with its grandiose intentions. Lawvere has given a beautiful discussion of the interconnections between the cases $n = 0$ and $n = 1$, ([CCFM] but it is known that the basic theory of abstract categories as presented there is inadequate for the results claimed.

In this paper we have tried to see what happens if the case $n = 2$ is included, by trying to discuss constructions for 2-categories both in terms of sets and of categories. And, of course, there is interplay going down as exemplified by the examples of §6 and the main result of this paper. However, it seems that there is a glaring inadequacy in the basic theory of categories; namely, we cannot construct the universal 0-fibration (i.e., $\widetilde{\mathrm{Cat}}$ of §1) in the basic theory. What is needed is an axiom that looks like the operation $\cup x$ in set-theory. Let Cat be a model of the basic theory which is full in the universe, and for any $\mathbb{A} \in$ Cat, let $A$ be the corresponding category in the universe that looks like $\mathbb{A}$. ([CCFM], p. 17.)

<u>Axiom</u>

There exists $P: \widetilde{\mathrm{Cat}} \longrightarrow \mathrm{Cat}$ with the following properties:

i) Given $\mathbb{A} \in$ Cat, there is an imbedding $I_A: A \longrightarrow \widetilde{\mathrm{Cat}}$ such that

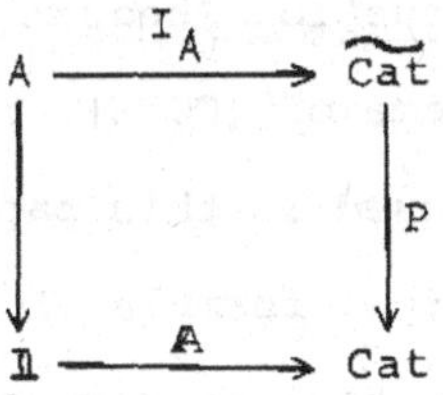


commutes.

ii) Given $F: A \longrightarrow B$, there is $H: A \times \mathbf{2} \longrightarrow \widetilde{\mathrm{Cat}}$ such that

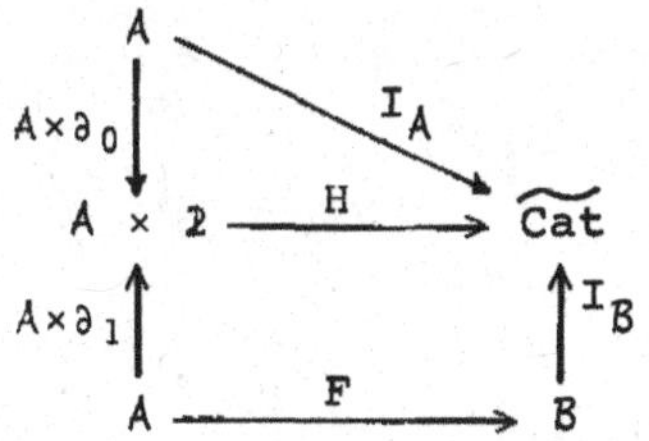


and

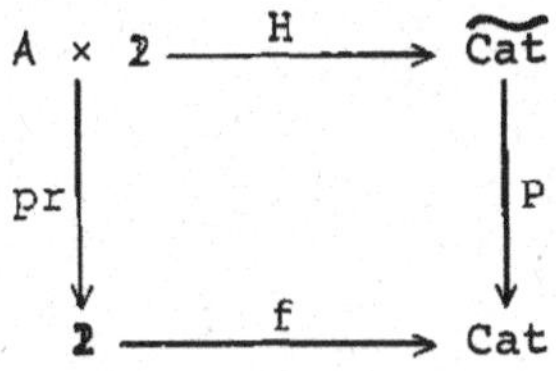

commute.

iii) If $\widetilde{\text{Cat}}' \xrightarrow{P'} \text{Cat}$ also satisfies i) and ii), then there is a unique functor over Cat,

$$\begin{array}{ccc} \widetilde{\text{Cat}} & \longrightarrow & \widetilde{\text{Cat}}' \\ & {\scriptstyle P}\searrow \quad \swarrow{\scriptstyle P'} & \\ & \text{Cat} & \end{array}$$

preserving the structure described in i) and ii).

Hopefully, this is enough to give the basic construction of §5. If not, this could be described by an axiom scheme saying that for each $F: A^{op} \times B \longrightarrow \text{Cat}$ there is a category $E_F \longrightarrow A \times B$ with properties analogous to those of $\widetilde{\text{Cat}}$. We assume that once categories of these sizes are available, then using the Category Construction Theorem and the Predicative Functor-Construction Scheme of [CCFM], it is possible to construct the 2-categories used in this paper within category theory by giving categorical formulas for the objects, for the categories Cat(A,B) and for the composition.

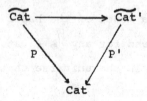

# BIBLIOGRAPHY

[BC] Bénabou, Jean. "Introduction to Bicategories, Reports of the Midwest Category Seminar", *Lecture Notes in Mathematics*, Vol. 47, Springer-Verlag Inc., New York, (1967).

[CS] Ehresmann, C. *Catégories et Structures*, Dunod, Paris, (1965).

[CC] Eilenberg, S. and Kelly, G. M. "Closed Categories", Proceedings of the Conference on Categorical Algebra, La Jolla 1965, Springer-Verlag Inc., New York.

[MD] Giraud, Jean. "Méthode de la Descente", *Bull. Soc. Math. France*, Mémoire 2; VIII + 150 p., (1964).

[GOD] Godement, R. *Topologie Algébrique et Théorie des Faisceaux*, Paris, Hermann, (1958).

[FCC] Gray, J. W. "Fibred and Cofibred Categories", Proceedings of the Conference on Categorical Algebra, La Jolla 1965, Springer-Verlag Inc., New York.

[CCC] Gray, J. W. *The Calculus of Comma Categories*, (to appear).

[SGA] Grothendieck, A. *Catégories Fibrées et Descente*, Séminaire de Géométrie Algébrique, Institute des Hautes Études Scientifiques, Paris, (1961).

[CCEF] Hofmann, K. H. "Categories with Convergences, Exponential Functors and the Cohomology of Compact Abelian Groups", *Math. Zeitschr.*, 104; 106-140, (1968).

[AF] Kan, D. M. "Adjoint Functors", *Trans. Amer. Math. Soc.*, 87; 295-329, (1958).

[CCFM] Lawvere, F. W. "The Category of Categories as a Foundation for Mathematics", Proceedings of the Conference on Categorical Algebra, La Jolla 1965, Springer-Verlag Inc., New York.

[ETH] Lawvere, F. W. Lecture, ETH, Zürich, November 17, 1966, (unpublished).

[CVHOL] Lawvere, F. W. *Category-Valued Higher Order Logic*, (to appear).

[AC] Linton, F. E. "Autonomous Categories and Duality of Functors", *J. Algebra*, 2; 315-349, (1965).

# NON-ABELIAN SHEAF COHOMOLOGY BY DERIVED FUNCTORS

by

R. T. Hoobler*

## INTRODUCTION

Given an arbitrary Grothendieck topology and a sheaf of groups G in that topology, the construction of $H^1(X;G)$, the first cohomology set with coefficients in G evaluated at X, is well known. Its usefulness in algebraic geometry stems from Grothendieck's descent theory. The problem of constructing $H^2(X;G)$ and a connecting map $\delta^1: H^1(X;G'') \longrightarrow H^2(X;G')$ for a central extension of sheaves of groups $G' \longrightarrow G \longrightarrow G''$ giving a nine term exact sequence of pointed sets has been solved by Giraud [6]. He adopted the point of view that $H^1(X;G)$ classified locally trivial (for the given topology) principal homogeneous spaces for G up to isomorphism and then extended this approach to define $H^2(X;G)$ as a set of equivalence classes of gerbes which are essentially local equivalence classes of principal homogeneous spaces. There are of course numerous prerequisites for understanding this approach.

Since the boundary map $\delta^1$ for a central extension of sheaves of groups in the étale topology plays a key role

---

*I wish to express appreciation for support extended by the National Science Foundation through contract NSF GP 8718.

in the theory of Brauer groups of schemes [8] as well as having other applications, we have developed a non-abelian cohomology theory part of which is presented here. Its culmination is Theorem 3.2 which produces a boundary map $\delta^1$ and describes the exactness properties of the corresponding nine term sequence of groups and pointed sets. For group cohomology this map is well known [11] and has even been defined for non-central extensions by Dedecker [3] and Springer [12]. The boundary map is also easily defined for Amitsur cohomology [5,10]. However, it is much harder for sheaf cohomology, although this contains group cohomology as a special case (see Theorem 3.3) since in an appropriate topology non-abelian Amitsur cohomology agrees with non-abelian group cohomology [5, Chapter I, Theorem 7.6].

The essence of our approach is to give a set of axioms on a topology from which a functorial "flask resolution" of $G$, a sheaf of groups, can be produced which resembles Godement's flask resolution of a sheaf of abelian groups. In order to get a "resolution" we define quotients in the category of sheaves of pointed sets with group action. One of the axioms then gives a functorial injection of such sheaves into "flask" sheaves. The other axioms then allow us to copy the procedures of homological algebra and so to produce resolutions of "short exact sequences" and diagrams in which we can "diagram chase" (using group actions on sets instead of multiplication in abelian groups) to get the desired results.

A number of examples are given which include the above applications. This approach to the problem was suggested by Grothendieck's [9] and Artin's [2] results showing that the non-abelian $H^1$ is an effacable functor.

Unfortunately the definitions and techniques developed here do not give an exact nine term cohomology sequence associated to an extension of sheaves of groups $G' \longrightarrow G \longrightarrow G''$ where $G'$ is a normal subgroup sheaf of $G$ and $G'' = G/G'$. The main difficulty is in the definition of $H^2(X,G') \longrightarrow H^2(X,G)$. A description of the relations between the various cohomology groups arising from such a sequence can be given, but it is beyond the scope of this present preliminary report on this approach. Similar problems arise in Giraud's work and are solved there with the aid of the notion of twisting by cocycles. In fact for a central extension $G' \longrightarrow G \longrightarrow G''$, he only has a correspondence $H^2(X;G') \multimap H^2(X;G)$ which makes it unlikely that his $H^2$ is the same as the one defined here. It would be interesting to find a universal mapping property for non-abelian cohomology so that the various definitions could be compared more easily. This exists for $H^1$ and is the basis of the proof of Theorem 3.3.

We have adopted the notation and basic definitions of [1] in this work. Those readers familiar with the material on general Grothendieck topologies in [2] can readily translate the statements and results into the language used there.

The increase in obscurity does not seem to justify using it here. We will avoid all set theoretic questions by assuming that all categories are small. The reader can justify this by using the theory of universes. Finally all functors will be covariant, and given a category $\underline{C}$ , $\underline{C}^0$ will denote the dual category.

## §1 PRELIMINARIES

We will be interested in the following categories:

$\underline{S}$: Category of sets and set maps.

$\underline{S}^{\bullet}$: Category of pointed sets and point preserving maps.

$\underline{S}^G$: Category whose objects are pointed sets with a right group action not necessarily preserving the point, $(S,G)$, where $S \times G \longrightarrow S$ satisfies $s \cdot (g_1 g_2) = (s \cdot g_1) \cdot g_2$ and $s \cdot 1 = s$ for all $s \in S$, $g_1, g_2 \in G$ and whose morphisms are pairs $(f,\theta)$, $f \in \mathrm{Mor}\ \underline{S}^{\bullet}$, $\theta \in \mathrm{Mor}\ \underline{G}$, with $f(s \cdot g) = f(s) \cdot \theta(g)$ for all $s \in S$, $g \in G$.

${}^G\underline{S}$: Category whose objects are pointed sets with a left group action not necessarily preserving the point, $(G,S)$, where $G \times S \longrightarrow S$ satisfies $(g_1 g_2) \cdot s = g_1 \cdot (g_2 \cdot s)$ and $1 \cdot s = s$ for all $s \in S$, $g_1, g_2 \in G$ and whose morphisms are pairs of maps as above.

${}^G\underline{S}^G$: Category of pointed sets with a left and right group action, $(G_1,S,G_2)$, where $(G_1,S) \in {}^G\underline{S}$, $(S,G_2) \in \underline{S}^G$ and $(g_1 \cdot s) \cdot g_2 = g_1 \cdot (s \cdot g_2)$ for all $g_1 \in G_1$, $s \in S$, $g_2 \in G_2$ and whose morphisms are triples of maps satisfying conditions analogous to the above.

$\underline{G}$: Category of groups and group homomorphisms.

$\underline{Ab}$: Category of abelian groups and group homomorphisms.

The components of a morphism in $\underline{S}^G$, ${}^G\underline{S}$, or ${}^G\underline{S}^G$ will be denoted by the same letter since the context will show whether it is a group homomorphism or a pointed set map. The distinguished point in a pointed set will always be written as $e$ and the identity of a group $G$ as $1$. Given $M \in {}^G\underline{S}^G$, let $M_G$ or ${}_GM$ denote the group acting on the right or on the left respectively, and let $M_S$ be the pointed set on which $M_G$ and ${}_GM$ act. Similar notation will be used for objects in $\underline{S}^G$ and ${}^G\underline{S}$. Note that there are several obvious forgetful functors connecting these categories. These will not be given explicit names in order to simplify notation. We will rely on the context to show which category we are working in. As an important example there is a functor $\underline{G} \longrightarrow {}^G\underline{S}^G$ gotten by allowing a group to act on itself by left and right translation.

Now fix once and for all a Grothendieck topology $\underline{T}$. Recall from [1] that this means giving a category $\underline{T}$ with fibred products and a set Cov $\underline{T}$ of families of maps in $\underline{T}$, $\{U_i \xrightarrow{\varphi_i} U\}_{i \in I}$, where in each family of maps $U$ is fixed and which satisfy:

(1) If $\varphi$ is an isomorphism, $\{\varphi\} \in \mathrm{Cov}\ \underline{T}$ .

(2) If $\{U_i \xrightarrow{\varphi_i} U\}$ , $\{V_{ij} \xrightarrow{\psi_{ij}} U_i\} \in \mathrm{Cov}\ \underline{T}$ , then

$\{V_{ij} \xrightarrow{\varphi_i \psi_{ij}} U\} \in \mathrm{Cov}\ \underline{T}$ .

(3) If $\{U_i \longrightarrow U\} \in \mathrm{Cov}\ \underline{T}$ , $V \longrightarrow U \in \mathrm{Mor}\ \underline{T}$ , then

$\{U_i \underset{U}{\times} V \longrightarrow V\} \in \mathrm{Cov}\ \underline{T}$ .

The category of presheaves on $\underline{T}$ with values on a category $\underline{C}$ with products is the category of contravariant functors from $\underline{T}$ to $\underline{C}$ and is denoted by $P(\underline{C})$. The category of sheaves on $\underline{T}$ with values in $\underline{C}$ , denoted by $S(\underline{C})$, is the full subcategory of $P(\underline{C})$ consisting of presheaves $F$ such that for any $\{U_i \longrightarrow U\} \in \mathrm{Cov}\ \underline{T}$ ,

$F(U) \longrightarrow \prod_I F(U_i) \overset{p_1^*}{\underset{p_2^*}{\rightrightarrows}} \prod_{I\times I} F(U_{i_1} \underset{U}{\times} U_{i_2})$ is exact; that is,

$F(U) \longrightarrow \prod_I F(U_i)$ is the equalizer of

$\prod_I F(U_i) \overset{p_1^*}{\underset{p_2^*}{\rightrightarrows}} \prod_{I\times I} F(U_{i_1} \underset{U}{\times} U_{i_2})$ where $p_i^*$ comes from the projection onto the $i^{\underline{th}}$ factor for $i = 1$ or $2$ .

Note that the objects and morphisms in $S(\underline{C})$, $P(\underline{C})$, or $\underline{C}$ can be described by finite products of objects, the "one point" or final object, and morphisms in $S(\underline{S})$, $P(\underline{S})$, or $\underline{S}$ respectively such that various diagrams commute where $\underline{C} = \underline{Ab}$, $\underline{G}$, ${}^G\underline{S}^G$, $\underline{S}^G$, ${}^G\underline{S}$, or $\underline{S}^{\cdot}$ . Alternatively the various functors between these values of $\underline{C}$ and $\underline{S}$ give identifications of $P(\underline{C})$ and $S(\underline{C})$ with group objects and set with group action objects in $P(\underline{S})$ and $S(\underline{S})$. Thus functors from $\underline{S}$, $P(\underline{S})$, or $S(\underline{S})$ to $\underline{S}$, $P(\underline{S})$ or $S(\underline{S})$ which preserve equalizers and finite products, i.e., finite inverse limits, and the final object correspond to functors from $\underline{C}$, $P(\underline{C})$, or $S(\underline{C})$ to $\underline{C}$, $P(\underline{C})$, or $S(\underline{C})$ respectively. Moreover, if $i: S(\underline{S}) \longrightarrow P(\underline{S})$ is the inclusion functor, then $i$ preserves inverse limits and so products and equalizers. These observations will often be used without referring to them. For instance the formation of equalizers in $S(\underline{S})$, $P(\underline{S})$, or $\underline{S}$ commutes with finite inverse limits and the final object. Hence equalizers in $S(\underline{C})$, $P(\underline{C})$, or $\underline{C}$ may be computed in $S(\underline{S})$, $P(\underline{S})$, or $\underline{S}$ respectively where $\underline{C}$ is any of the above categories.

Artin's construction [1] of the sheafification functor, $\#: P(\underline{Ab}) \longrightarrow S(\underline{Ab})$, carries over directly to give a left adjoint $\#: P(\underline{S}) \longrightarrow S(\underline{S})$ to $i: S(\underline{S}) \longrightarrow P(\underline{S})$ which preserves finite inverse limits and the final object. Thus it also gives a left adjoint $\#: P(\underline{C}) \longrightarrow S(\underline{C})$ where $\underline{C}$ is any of the above categories. We will give a brief description of this procedure in order to fix notation.

Given $U \in \underline{T}$, let $J_U$ be the category of coverings of $U$. Thus the objects of $J_U$ are coverings $\{U_i \longrightarrow U\}_{i \in I}$ and a morphism $\varphi: \{U_i \longrightarrow U\}_{i \in I} \longrightarrow \{V_j \longrightarrow U\}_{j \in J}$ is a function $\bar{\varphi}: I \longrightarrow J$ and maps $\varphi_i: U_i \longrightarrow V_{\bar{\varphi}(i)} \in \text{Mor}\, \underline{T}$

such that $U_i \xrightarrow{\varphi_i} V_{\overline{\varphi}(i)}$ commutes.

$$\begin{array}{ccc} U_i & \xrightarrow{\varphi_i} & V_{\overline{\varphi}(i)} \\ & \searrow \quad \swarrow & \\ & U & \end{array}$$

Let $\overline{J}_U$ be the corresponding partially ordered category; that is, ob $\overline{J}_U$ = ob $J_U$ and $\mathrm{Mor}_{\overline{J}_U}(\{U_i \to U\},\{V_j \to U\}) = \phi$ if $\mathrm{Mor}_{J_U}(\{U_i \to U\},\{V_j \to U\}) = \phi$ and otherwise it is a one element set. Note that $\overline{J}_U$ is a connected directed category, i.e., it satisfies L1, L2, and L3 of [1], and so $\varinjlim_{\overline{J}^0_U}$ preserves monomorphisms. Moreover, given $V \xrightarrow{\alpha} U \in \mathrm{Mor}\, \underline{T}$, we have a functor $\alpha^+ : J_U \to J_V$ and $\overline{\alpha}^+ : \overline{J}_U \to \overline{J}_V$ given by $\alpha^+(\{U_i \to U\}) = \{U_i \underset{U}{\times} V \to V\}$ and
$\overline{\alpha}^+(\{U_i \to U\}) = \{U_i \underset{U}{\times} V \to V\}$ .

Now suppose $\underline{C}$ is a left complete category (possesses arbitrary products and equalizers) such that direct limits over categories satisfying L1, L2, and L3 of [1] exist. For instance $\underline{C}$ might be any of the categories at the beginning of §1. For $F \in P(\underline{C})$ and $\{U_i \to U\} \in \mathrm{Cov}\, \underline{T}$,

let $\check{H}^0(\{U_i \to U\};F) = \mathrm{Equalizer}\, (\prod_I F(U_i) \underset{p_2^*}{\overset{p_1^*}{\rightrightarrows}} \prod_{I\times I} F(U_{i_1} \underset{U}{\times} U_{i_2}))$ .

Since the formation of equalizers is functorial, $\check{H}^0(\ ;F) : J^0_U \to \underline{C}$ . Suppose we are given
$\varphi, \psi : \{U_i \to U\}_I \to \{V_j \to U\}_J \in \mathrm{Mor}\, J_U$ . Since
$\{V_{j_1} \underset{U}{\times} V_{j_2} \to U\}_{J\times J}$ is the product of $\{V_j \to U\}$ with itself

in $J_U$, there is $\Phi\colon \{U_i \longrightarrow U\}_I \longrightarrow \{V_{j_1} \underset{U}{\times} V_{j_2} \longrightarrow U\}_{J\times J}$ such that $p_1\Phi = \varphi$ and $p_2\Phi = \psi$ where
$p_i\colon \{V_{j_1} \underset{U}{\times} V_{j_2} \longrightarrow U\}_{J\times J} \longrightarrow \{V_j \longrightarrow U\}_J$ comes from the projection maps onto the $i^{\underline{th}}$ factor. This gives a diagram

$$\begin{array}{ccccc}
\check{H}^0(\{V_j \longrightarrow U\};F) & \xrightarrow{i_V} & \prod\limits_J F(V_j) & \overset{p_1^*}{\underset{p_2^*}{\rightrightarrows}} & \prod\limits_{J\times J} F(V_{j_1} \underset{U}{\times} V_{j_2}) \\
\overline{\varphi}\downarrow\ \downarrow\overline{\psi} & & \varphi^*\downarrow\ \downarrow\psi^* & \swarrow \Phi^* \quad (\varphi\times\varphi)^*\downarrow & \downarrow(\psi\times\psi)^* \\
\check{H}^0(\{U_i \longrightarrow U\};F) & \xrightarrow{i_U} & \prod\limits_I F(U_i) & \overset{p_1^*}{\underset{p_2^*}{\rightrightarrows}} & \prod\limits_{I\times I} F(U_{i_1} \underset{U}{\times} U_{i_2})
\end{array}$$

where $\varphi^* = \Pi F(\varphi_i)$, $\psi^* = \Pi F(\psi_i)$, etc., $i_V$ and $i_U$ are the canonical monomorphisms of the equalizers, and $\overline{\varphi}$, $\overline{\psi}$ are the induced maps between the equalizers. Now
$i_U\overline{\varphi} = \varphi^* i_V = \Phi^* p_1^* i_V = \Phi^* p_2^* i_V = \psi^* i_V = i_U\overline{\psi}$ and so $\overline{\varphi} = \overline{\psi}$.
Thus $\check{H}^0(\{\ \};F)\colon \overline{J}_U^0 \longrightarrow \underline{C}$. Define $\check{H}^0(U;F) = \varinjlim_{\overline{J}_U^0} \check{H}^0(\{\ \};F)$.

This construction is functorial in $U$ and $F$, and so gives a functor $\check{H}^0(\ ;\ )\colon \underline{T}^0 \times P(\underline{C}) \longrightarrow \underline{C}$. Let
$+ : P(\underline{C}) \longrightarrow P(\underline{C})$ be the functor $(+F)(U) = \check{H}^0(U;F)$.
Note that there is a natural transformation $\overline{I}\colon I \longrightarrow +$ coming from $\prod\limits_I \varphi_i^* = \prod\limits_I F(\varphi_i)\colon F(U) \longrightarrow \prod\limits_I F(U_i)$ for any $F \in P(\underline{C})$ and $\{U_i \xrightarrow{\varphi_i} U\} \in \mathrm{Cov}\ \underline{T}$.

Following Artin we introduce the condition

(+) on $F \in P(\underline{C})$:

(+): For all $\{U_i \longrightarrow U\}_{i\in I} \in \mathrm{Cov}\ \underline{T}$, $F(U) \longrightarrow \prod\limits_I F(U_i)$

is a monomorphism. Then copying Artin's proof [1], we see that for $\underline{C} = \underline{S}$, $(+F)$ satisfies $(+)$ for $F \in P(\underline{S})$ and that if $F \in P(\underline{S})$ satisfies $(+)$, then $(+F)$ is a sheaf. Let $\#: P(\underline{S}) \longrightarrow S(\underline{S})$ be given by $\#(F) = +(+F)$. It is easily verified that $\#$ is a left adjoint to $i$ as desired. Moreover, since $\underrightarrow{\lim}$ over connected, directed categories preserves finite products and equalizers, the sheafification functor preserves finite inverse limits. Since the one point presheaf is already a sheaf, there is an induced sheafification functor $\#: P(\underline{C}) \longrightarrow S(\underline{C})$ which is left adjoint to $i$ for any of the above values of $\underline{C}$.

We will construct resolutions by embedding an object in a larger one, taking a quotient, and repeating the process. However, the definition of quotients requires a number of preliminary concepts which are given below along with some of their properties which are easily proven and will be left to the reader. For the remainder of this section $\underline{C}$ will denote one of the categories $\underline{Ab}$, $\underline{G}$, ${}^G\underline{S}^G$, ${}^G\underline{S}$, $\underline{S}^G$, or $\underline{S}^{\cdot}$, $e$ will be the trivial map or final object in $S(\underline{C})$, $P(\underline{C})$, or $\underline{C}$, and $p_i: M_1 \times M_2 \longrightarrow M_i$ will be the canonical projection map onto the $i^{\underline{th}}$ factor of a product. Where there is any ambiguity in notation, the context will indicate which category the object is being constructed in.

Let $\alpha: M \longrightarrow M''$ be a map in $S(\underline{C})$, $P(\underline{C})$, or $\underline{C}$. The kernel of $\alpha$, $\mathrm{Ker}\,(\alpha)$, $\mathrm{Ker}^P(\alpha)$, or $\mathrm{Ker}\,(\alpha)$ respectively, is the equalizer of $\alpha$ and $e$. Note that $i\mathrm{Ker}\,(\alpha) = \mathrm{Ker}^P(i\alpha)$.

$\alpha \in$ Mor $S(\underline{C})$ or Mor $P(\underline{C})$ is of course a monomorphism if and only if $\alpha(U)$ is one-to-one for all $U \in \underline{T}$. The image of $\alpha$, $\alpha(M)$, $\alpha^P(M)$, or $\alpha(M)$, is the smallest subobject of $M''$ through which $\alpha$ factors. Thus $\alpha^P(M)(U) = \alpha(U)(M(U))$ for $U \in \underline{T}$, and $\alpha(M) = \#(i\alpha^P(iM))$. $\alpha$ is onto if the image of $\alpha$ equals $M''$. This says that $\alpha$ is an epimorphism in $S(\underline{S})$, $P(\underline{S})$, or $\underline{S}$ respectively. For notational purposes, we will write $M \overset{\alpha}{\rightarrowtail} M''$ or $M \overset{\alpha}{\twoheadrightarrow} M''$ if $\alpha$ is monic or onto respectively. A sequence $M_1 \xrightarrow{\alpha} M_2 \xrightarrow{\beta} M_3$ in $S(\underline{S}^{\cdot})$, $P(\underline{S}^{\cdot})$ or $\underline{S}^{\cdot}$ is exact at $M_2$ if the image of $\alpha$ equals the kernel of $\beta$. If $M \in S(\underline{S}^G)$, $P(\underline{S}^G)$, or $\underline{S}^G$, $N$ is a subobject of $M_S$ and $G$ is a subgroup object of $M_G$, then the orbit of $N$ under $G$, $O_N(G)$, $O_N^P(G)$, or $O_N(G)$, is the image of $N \times G$ in $M_S$ under the group action. $M_G$ acts transitively on $M_S$ if $O_e(M_G)$, $O_e^P(M_G)$, or $O_e(M_G)$ equals $M_S$. Orbits of subobjects with respect to left group actions are defined similarly and will be denoted as above since the context will determine which side the group acts on.

Let $G \in S(\underline{G})$, $P(\underline{G})$ or $\underline{G}$. $H$ is a normal subgroup sheaf, normal subgroup presheaf, or normal subgroup - in general, a normal subobject - if $H$ is the kernel of $\alpha$ for some $\alpha$ a map in $S(\underline{G})$, $P(\underline{G})$, or $\underline{G}$ respectively. For $G \in S(\underline{G})$ or $P(\underline{G})$, a subsheaf or subpresheaf $H$ of $G$ is normal if and only if $H(U)$ is a normal subgroup of $G(U)$ for all $U \in \underline{T}$. If $H$ is a subgroup object of $G$

and $G_1$, $G_2$ are subgroup objects of $G$ containing $H$ as a normal subobject, then $H$ is a normal subobject of $G_1 \cdot G_2$, the subgroup object of $G$ generated by the image of $G_1 \times G_2$ in $G$ under the multiplication map. Let the normalizer subgroup sheaf, normalizer subgroup presheaf, or normalizer subgroup, $N_G(H)$, $N_G^p(H)$, $N_G(H)$, be the largest subobject $\overline{G}$ of $G$ containing $H$ such that $H$ is a normal subobject of $\overline{G}$. This always exists since the set of subobjects of $G$ containing $H$ as a normal subobject forms a directed system by the above remark and so $\varinjlim$ over this system is a subgroup object of $G$ containing $H$ as a normal subobject. Note that for any $U \in \underline{T}$, $N_{G(U)}(H(U)) \supseteq N_G^p(H)(U)$ or $N_G(H)(U)$, and for $G \in S(\underline{G})$, $iN_G(H) = N_{iG}^p(iH)$.

If $M \in S(\underline{S}^G)$, $P(\underline{S}^G)$ or $\underline{S}^G$, $\sigma\colon M_S \times M_G \longrightarrow M_S$ defines the group action on the right, and $i\colon N \longrightarrow M_S$ is a subobject, then the stabilizer of $N$ in $M$, $St_{M_G}(N)$, $St_{M_G}^p(N)$ or $St_{M_G}(N)$, is the largest subgroup object $j\colon H \longrightarrow M_G$ such that $N \times H$ is the equalizer of $N \times H \underset{ip_1}{\overset{\sigma(i\times j)}{\rightrightarrows}} M_S$.

As above, the subgroup objects $H \subseteq M_G$ satisfying this condition form a directed system, and so the stabilizer always exists. If $\overline{G}$ is a subgroup object of $M_G$ containing the stabilizer of $N$ in $M$ and $N \times \overline{G} \longrightarrow M_S$, defined by the restriction of $\sigma$ , factors through N , then

$St_{M_G}(N)$, $St^p_{M_G}(N)$, or $St_{M_G}(N)$ is a normal subobject of $\overline{G}$ .

Stabilizers of group actions on the left are defined by interchanging right and left. They have similar properties which we leave to the reader to prove and will be denoted in the same way since the context will determine which side the group acts on.

We are now ready to define quotients. Let $\alpha: M \longrightarrow M"$ be a map in $S(\underline{S}^G)$, $P(\underline{S}^G)$, or $\underline{S}^G$, $\sigma: M"_S \times M_G \longrightarrow M"_S$ be the map defining the group action of $M_G$ on $M"_S$ via $\alpha$ , and suppose $M" \in S({}^G\underline{S}^G)$, $P({}^G\underline{S}^G)$ or ${}^G\underline{S}^G$ respectively. Define the set component of the right quotient of M" by M, $\underline{Q}_r(\alpha)_S$, $Q^p_r(\alpha)_S$, or $Q_r(\alpha)_S$, to be the coequalizer of $\sigma$ and e in $S(\underline{S}^{\bullet})$, $P(\underline{S}^{\bullet})$, or $\underline{S}^{\bullet}$ respectively, and let $\pi$ be the corresponding epimorphism of pointed sets. Temporarily let this set component be denoted by $Q_S$ . Since the left action of ${}_GM"$ commutes with the action of $M"_G$ , ${}_GM"$ has an induced action on $Q_S$ . Let ${}_G\underline{M}"(\alpha)$, ${}^p_GM"(\alpha)$, or ${}_GM"(\alpha)$ be ${}_GM"$ with this action on $Q_S$ . Temporarily let ${}_GM"(\alpha)$ be these groups with their action on $Q_S$ and St be $St_{{}_G\underline{M}"(\alpha)}(\underline{Q}_r(\alpha)_S)$, $St^p_{{}^p_GM"(\alpha)}(Q^p_r(\alpha)_S)$, or $St_{{}_GM"(\alpha)}(Q_r(\alpha)_S)$. Define ${}_G\underline{Q}_r(\alpha)$, ${}_GQ^p_r(\alpha)$, or ${}_GQ_r(\alpha)$ to be the sheaf of groups, presheaf of groups, or group ${}_GM"(\alpha)/St$ (depending on the category M" is in) together with the corresponding left group action on $Q_S$ . Note that if $M" \in S({}^G\underline{S}^G)$, then

${}_G\underline{Q}_r(\alpha) = \#({}_GQ^p_r(i\alpha))$. Suppose that $H_1, H_2$ are subgroup objects of $M''_G$ such that there are right group actions $\sigma_1, \sigma_2$ of $H_1, H_2$ on $Q_S$ giving commutative diagrams

$$\begin{array}{ccc} M''_S \times H_i & \longrightarrow & M''_S \\ {\scriptstyle \pi \times H_i}\downarrow & & \downarrow{\scriptstyle \pi} \\ Q_S \times H_i & \xrightarrow{\sigma_i} & Q_S \end{array}$$

for $i = 1$ or $2$. Since $\pi \times H_i$ is an epimorphism, $\sigma_i$ is uniquely determined by $H_i$ and the group action of $M''_G$ on $M''_S$. In particular $\sigma_1 = \sigma_2$ when restricted to $H_1 \cap H_2 \subseteq M''_G$. Thus $\sigma_1, \sigma_2$ define a group action on the right of $H_1 \cdot H_2$ on $Q_S$. Let $\underline{M}''_G(\alpha)$, ${}^pM''_G(\alpha)$, or $M''_G(\alpha)$ be the largest subgroup object of $M''_G$ for which such a group acton on $Q_S$ can be defined. If we temporarily denote this by $M''_G(\alpha)$, then it is $\varinjlim$ over the directed system of all subgroup objects of $M''_G$ satisfying the above condition. Note that $M''_G(\alpha)$ contains $N_{M''_G}(\alpha(M_G))$, $N^p_{M''_G}(\alpha^p(M_G))$, or $N_{M''_G}(\alpha(M_G))$ respectively. As above let St temporarily stand for $St_{M''_G(\alpha)}(\underline{Q}_r(\alpha)_S)$, $St^p_{{}^pM''_G(\alpha)}(Q^p_r(\alpha)_S)$, or $St_{M''_G(\alpha)}(Q_r(\alpha)_S)$, and define $\underline{Q}_r(\alpha)_G$, $Q^p_r(\alpha)_G$, or $Q_r(\alpha)_G$ to be $M''_G(\alpha)/St$ in $S(\underline{G})$, $P(\underline{G})$, or $\underline{G}$ respectively, together with their induced action on $Q_S$. Putting this together gives $\underline{Q}_r(\alpha)$, $Q^p_r(\alpha)$, or $Q_r(\alpha)$ in $S({}^G\underline{S}^G)$, $P({}^G\underline{S}^G)$, ${}^G\underline{S}^G$. Let $Q$, ${}_GQ$, $Q_S$, and $Q_G$ temporarily stand for these right quotients and their components respectively. Then $\pi: M'' \longrightarrow Q$ preserves the left group action, and if the image of $M_G$ is a normal subobject of $M''_G$, then $\pi$ also preserves the right group action. In any case $\pi$ preserves the right group action if $M''_G$ is restricted to $M''_G(\alpha)$. Moreover, $\pi$ is onto in $\mathcal{S}({}^G\underline{S})$, $\mathcal{P}({}^G\underline{S})$, or ${}^G\underline{S}$. In particular, if ${}_GM''$ acts transitively on $M''_S$, then ${}_GQ$ acts transitively on $Q_S$. If $M$, $M''$ are in $\mathcal{S}(\underline{G})$, $\mathcal{P}(\underline{G})$, or $\underline{G}$ and the image of $M$ is a normal subobject of $M''$, then $Q$ is the quotient group in $\mathcal{S}(\underline{G})$, $\mathcal{P}(\underline{G})$, or $\underline{G}$ regarded as a set with left and right group actions coming from left and right translation because of the definition of ${}_GQ$ and $Q_G$. The operation of taking right quotients is functorial on the pointed set component. In particular $Q^p_r(\alpha)_S(U) = Q_r(\alpha(U))_S$ for all $U \in \underline{T}$. The functorial behavior of the group components is not as nice however and will be discussed later. Finally since $\#$ is a left adjoint to $i$, it preserves coequalizers. Thus $\underline{Q}_r(\alpha)_S = \#(Q^p_r(i\alpha)_S)$, and so

$$e \longrightarrow Or \longrightarrow M''_S \longrightarrow Q_S \longrightarrow e$$

is exact in $\mathcal{S}(\underline{S}')$, $\mathcal{P}(\underline{S}')$, or $\underline{S}'$ where $Or$ is the orbit of $e$ under the image of $M_G$ in $\mathcal{S}(\underline{S}')$, $\mathcal{P}(\underline{S}')$, or $\underline{S}'$. (See Proposition 2.2.) In particular the image of $M_S$ in $Q_S$ is trivial if and only if the image of $M_G$ acts transitively on the image of $M_S$.

Left quotient objects are defined similarly and

will be denoted by replacing $r$ with $\ell$ in the notation above. The above statements with left and right interchanged hold for left quotients. We leave their statement and verification up to the reader.

## §2 FLASK RESOLUTIONS AND NON-ABELIAN HOMOLOGICAL ALGEBRA

In this section we first investigate the exactness properties of #, i, and the formation of quotients. Then we introduce some additional assumptions on $\underline{T}$ which enable us to define a canonical "flask" resolution à la Godement [7]. Several examples are then given. The results of the last section applied to them enable us to define the connecting homomorphism for central extensions of sheaves of groups on non paracompact topological spaces, a non-abelian cohomology theory for groups and profinite groups, and the connecting map Grothendieck used in his study of the Brauer group of a scheme. The remainder of this section is devoted to developing the analogue of the $3 \times 3$ lemma in our setting.

Definition 2.1

Given $M', M, M'' \in S({}^G\underline{S}^G)$, $P({}^G\underline{S}^G)$, or ${}^G\underline{S}^G$, $M' \xrightarrow{\alpha} M \xrightarrow{\beta} M''$ is a short exact sequence in $S(\underline{S}^G)$, $P(\underline{S}^G)$, or $\underline{S}^G$ if $\alpha$ is a map in $S(\underline{S}^G)$, $P(\underline{S}^G)$, or $\underline{S}^G$ which is monic and $M'' \cong \underline{Q}_r(\alpha)$, $Q^p_r(\alpha)$, or $Q_r(\alpha)$ in $S({}^G\underline{S})$, $P({}^G\underline{S})$, or ${}^{G}\underline{S}$ respectively. It is a short exact sequence of sets in $S(\underline{S}^G)$, $P(\underline{S}^G)$, or $\underline{S}^G$ if $\alpha$ is a map in $S(\underline{S}^G)$, $P(\underline{S}^G)$, or $\underline{S}^G$ which is monic and $M''_S \cong \underline{Q}_r(\alpha)_S$, $Q^p_r(\alpha)_S$, or $Q_r(\alpha)_S$. It is an exact sequence of sets in $S(\underline{S}^G)$, $P(\underline{S}^G)$, or $\underline{S}^G$ at $M$ if $\alpha(M') \rightarrowtail M \twoheadrightarrow \beta(M)$ is a short exact sequence of sets in $S(\underline{S}^G)$, $P(\underline{S}^G)$, or $\underline{S}^G$. Replacing right by left gives the definition of a short exact sequence (of sets) and an exact sequence of sets at $M$ in $S({}^{G}\underline{S})$, $P({}^{G}\underline{S})$, or ${}^{G}\underline{S}$.

The analogue of the exactness properties of # and i for $S(\underline{Ab})$ is contained in the following two propositions.

Proposition 2.2

1) If $M' \overset{\alpha}{\rightarrowtail} M \overset{\beta}{\to} M''$ is an exact sequence in $S(\underline{S}^{\cdot})$ with $\alpha$ monic, then $iM' \overset{\alpha}{\rightarrowtail} iM \overset{\beta}{\to} iM''$ is an exact sequence in $P(\underline{S}^{\cdot})$ with $\alpha$ monic in $P(\underline{S}^{\cdot})$. If $\beta$ is onto in $S(\underline{C})$, $\underline{C} = $ Ab, G, ${}^{G}\underline{S}^{G}$, ${}^{G}\underline{S}$, $\underline{S}^G$, or $\underline{S}^{\cdot}$, then for any $U \in \underline{T}$, $x \in M''(U)$, there is $\{U_i \xrightarrow{\varphi_i} U\}_I \in J_U$ and $y \in \prod_I M(U_i)$ with $(\prod_I \beta(U_i))(y) = (\prod_I \varphi_i^*)(x)$.

2) If $M' \overset{\alpha}{\rightarrowtail} M \overset{\beta}{\twoheadrightarrow} M''$ is a short exact sequence of sets in $S(\underline{S}^G)$, then $i\mathcal{O}_e(M'_G) \xrightarrow{\alpha|} iM \xrightarrow{\beta} iM''$ is an exact sequence in $P(\underline{S}^{\cdot})$ with $\alpha|$ monic and $\beta$ factors through $Q^p_r(\alpha)_S$. If $M' \in S(\underline{G})$, and ${}_{G}M$ acts transitively on $M_S$ or $\alpha(M')$ is contained in the center of $M_G$ and $M_G$ acts transitively on $M_S$, then $iM' \overset{\alpha}{\rightarrowtail} iM \overset{\beta}{\to} iM''$ is an exact sequence of sets in $P(\underline{S}^G)$ with $\alpha$ monic.

Proof. The first part of 1) is straightforward ( $\alpha$ monic implies $\alpha^{p}(M') = i\alpha(M')$). The rest follows immediately from the fact that for $M \in P(\underline{C})$, $x \in (+M)(U)$, there is $\{U_i \xrightarrow{\varphi_i} U\}_I \in J_U$ and $y \in \prod_I M(U_i)$ with $(\prod_I \bar{i}(M)(U_i))(y) = (\prod_I \varphi_i^*)(x)$ for the desired values of $\underline{C}$. Suppose $\underline{C} = \underline{S}^{\cdot}$ (the argument for the other categories being similar). Then $x \in \varinjlim_{\bar{J}_U^0} \check{H}^0(\{U_i \to U\};M)$ can be represented by $y \in \prod_I M(U_i)$ with $p_1^*(y) = p_2^*(y) \in \prod_{I\times I} M(U_{i_1} \times_U U_{i_2})$ for some $\{U_i \xrightarrow{\varphi_i} U\} \in \bar{J}_U^0$.

But now $(\prod_I \bar{i}(M)(U_i))(y) \in \prod_I \varinjlim_{\bar{J}_{U_i}^0} \check{H}^0(\{V_{j,i} \to U_i\};M)$ is represented by $p_1^*(y) \in \prod_{i_2\in I} \check{H}^0(\{U_{i_1} \times_U U_{i_2} \to U_{i_2}\};M)$ and $(\prod_I \varphi_i^*)(x)$ is represented by $p_2^*(y) \in \prod_{i_2\in I} \check{H}^0(\{U_{i_1} \times_U U_{i_2} \to U_{i_2}\};M)$ which gives the desired result.

For 2), $\alpha|$ is the inclusion the orbit sheaf of $e$ under $\alpha(M'_G)$ where $i\alpha(M'_G) = \alpha^{p}(M'_G)$. It is clear that $\alpha|$ is monic, $\beta(\alpha|)$ is the trivial map and (by definition) $\beta$ factors through $Q_r^{p}(\alpha)_S$. If $\beta(U)(x) = e$, $x \in M_S(U)$, $U \in \underline{T}$, then there is $\{U_i \xrightarrow{\varphi_i} U\} \in J_U$ and $g \in \prod_I M'_G(U_i)$ with $(\prod_I \alpha(U_i))(e \cdot g) = (\prod_I \varphi_i^*)(x)$. Hence $x \in O_e(\alpha(M'_G))(U)$ as desired. Finally under the additional hypotheses we must show that given $x,y \in M_S(U)$ with $\beta(U)(x) = \beta(U)(y)$, there is $g \in M'_G(U)$ with $x \cdot \alpha(U)(g) = y$. First let us show that for any $U \in \underline{T}$, given $x \in M_S(U)$ and $h \in M'_G(U)$ with $x = x \cdot \alpha(U)(h)$, then $h = e$. There is $\{U_i \xrightarrow{\varphi_i} U\}_I \in J_U$

$g \in \prod_{I} {}_G M(U_i)$ or $g \in \prod_I M_G(U_i)$ with $(\prod_I \varphi_i^*)(x) = g \cdot e$ or $e \cdot g$ respectively. Then in the first case

$(g \cdot e) \cdot ((\prod_I \varphi_i^*)(\alpha(U)(h))) = g \cdot e$ and so

$e \cdot \alpha(U)(h) = e = \alpha(U)(e \cdot h)$. Hence $h = e$ as stated. The proof of the other case using the central assumption is essentially the same. Returning to the original problem and using the notation there, we can find a covering $\{U_i \xrightarrow{\varphi_i} U\}$ and $\tilde{g} \in \prod_I M'_G(U_i)$ with $(\prod_I \varphi_i^*)(x) \cdot (\prod_I \alpha(U_i))(\tilde{g}) = (\prod_I \varphi_i^*)(y)$.

Hence $p_1^*(\prod_I \varphi_i^*)(x) \cdot \prod_{I\times I} \alpha(U_{i_1} \underset{U}{\times} U_{i_2})[(p_1^*(\tilde{g})) \cdot (p_2^*(\tilde{g})^{-1})]$

$= p_1^*(\prod_I \varphi_i^*)(x)$, and so $p_1^*(\tilde{g}) = p_2^*(\tilde{g})$ by the above observation. Hence $\tilde{g} = (\prod \varphi_i^*)(g)$ for some $g \in M'_G(U)$. Since $M$ is a sheaf we get $x \cdot \alpha(U)(g) = y$ as desired. ∎

<u>Corollary 2.3</u>

Let $U \in \underline{T}$ and $M' \overset{\alpha}{\rightarrowtail} M \overset{\beta}{\rightarrow} M''$ be a short exact sequence of sets in $S(\underline{S}^G)$. If $M' \in S(\underline{Ab})$ and $\alpha(M')$ is contained in the center of $M_G$, then

$M'(U) \overset{\alpha(U)}{\rightarrowtail} \mathcal{O}_e(M_G)(U) \xrightarrow{\beta(U)} M''(U)$ is exact in $\underline{S}^G$ and $\underline{S}^{\bullet}$ where $\mathcal{O}_e(M_G)_G = M_G$.

<u>Proposition 2.4</u>

$\#: P(\underline{C}) \longrightarrow S(\underline{C})$ preserves monomorphisms and maps which are onto for $\underline{C} = \underline{G}, {}^{G}\underline{S}^{G}, \underline{S}^{G}, {}^{G}\underline{S}, \underline{S}^{\cdot}$ as well as exactness in $\underline{S}^{\cdot}$, short exact sequences of sets, and transitivity of group actions.

Proof: Since $\varinjlim_{\overline{J}^0_U}$ preserves monomorphisms, it is clear that # preserves monomorphisms as well as maps which are onto for the above categories. Given $M' \xrightarrow{\alpha} M \xrightarrow{\beta} M''$ an exact sequence of presheaves of pointed sets, we have $\#(\alpha^{p}(M')) = \#(\mathrm{Ker}^{p}(\beta)) = \mathrm{Ker}(\#(\beta))$ since $\varinjlim_{\overline{J}^0_U}$ preserves the exactness of the sequence in $P(\underline{S}^{\cdot})$ if $\alpha$ is monic. This also shows that # preserves short exact sequences of sets since $\#(Q^{p}_{r}(\alpha)_{S}) = \underline{Q}_r(\#(\alpha))_{S}$ by universal mapping properties. Finally # preserves transitivity since $\#(\sigma)$ is onto if $\sigma$ is onto where $\sigma\colon e \times M_G \longrightarrow M_S$ comes from the right group action. ∎

Returning to the functoriality of ${}_{G}\underline{Q}_r(\alpha)$ and $\underline{Q}_r(\alpha)_G$, there are two cases of interest where this behavior can be described. The statements are based on the diagram below where all objects are in $S({}^{G}\underline{S}^{G})$, $\alpha, \overline{\alpha} \in \mathrm{Mor}\, S({}^{G}\underline{S}^{G})$, and $f, f'' \in \mathrm{Mor}\, S(\underline{S}^{G})$.

$$\begin{array}{ccccc} M & \xrightarrow{\alpha} & M'' & \xrightarrow{\pi} & \underline{Q}_r(\alpha) \\ {\scriptstyle f}\downarrow & & {\scriptstyle f''}\downarrow & & {\scriptstyle \overline{f}}\downarrow \\ N & \xrightarrow{\overline{\alpha}} & N'' & \xrightarrow{\overline{\pi}} & \underline{Q}_r(\overline{\alpha}) \end{array}$$

The two cases are:

(A) - $(f''|): \alpha(M_G) \longrightarrow \alpha(N_G)$ and $f'': M''_S \longrightarrow N''_S$ are onto

(B) - $f''(M''_G)$ is contained in the center of $N''_G$.

To simplify notation, if $M \in P(\underline{C})$, $\{U_i \xrightarrow{\varphi_i} U\} \in \mathrm{Cov}\ \underline{T}$, let $\varphi^*_I$ denote $\prod_I \varphi^*_i : M(U) \longrightarrow \prod_I M(U_i)$. If $\alpha: M' \longrightarrow M \in \mathrm{Mor}\ P(\underline{C})$, let $\alpha$ denote $\prod_I \alpha(U_i): \prod_I M'(U_i) \longrightarrow \prod_I M(U_i)$, the context determining $I$ and $\{U_i\}$.

<u>Proposition 2.5</u>

1) In either (A) or (B), $f''(M''_G(\alpha)) \subseteq N''_G(\overline{\alpha})$. In (A), $\overline{f} \in \mathrm{Mor}\ S(\underline{S}^G)$.

2) In (A), if $f'' \in \mathrm{Mor}\ S({}^G\underline{S}^G)$, then $\overline{f} \in \mathrm{Mor}\ S({}^G\underline{S}^G)$.

<u>Proof</u>: Suppose $f''(M''_G)$ is contained in the center of $N''_G$. Then $f''(M''_G(\alpha)) \subseteq f''(M''_G) \subseteq N_{N''_G}(\overline{\alpha}(N_G)) \subseteq N''_G(\overline{\alpha})$ as desired. In (B), if $\overline{g} \in f''(M''_G(\alpha))(U)$ and $\overline{x}, \overline{y} \in N''_S(U)$ with $\overline{\pi}(\overline{x}) = \overline{\pi}(\overline{y})$, then there is a covering of $U$, $\{U_i \xrightarrow{\varphi_i} U\}$, $g \in \prod_I M''_G(\alpha)(U_i)$ with $f''(g) = \varphi^*_I(\overline{g})$, $\overline{g}_1 \in \prod_I N_G(U_i)$ with $\varphi^*_I(\overline{x}) = \varphi^*_I(\overline{y}) \cdot \overline{\alpha}(\overline{g}_1)$, $g_1 \in \prod_I \alpha(M_G)(U_i)$ with $f''(g_1) = \overline{\alpha}(\overline{g}_1)$, and $y \in \prod_I M''_S(U_i)$ with $f''(y) = \varphi^*_I(\overline{y})$. Then $f''(y \cdot g_1) = \varphi^*_I(\overline{x})$ and $\pi((y \cdot g_1) \cdot g) = \pi(y \cdot g)$. Hence $\varphi^*_I \overline{\pi}(\overline{x} \cdot \overline{g}) = \overline{\pi}(f''(y \cdot g_1) \cdot f''(g)) = \overline{f}\pi((y \cdot g_1) \cdot g) = \overline{f}\pi(y \cdot g) = \varphi^*_I \overline{\pi}(\overline{y} \cdot \overline{g})$. Since $\underline{Q}_r(\overline{\alpha})_S$ is a sheaf, this shows that $f''(M''_G(\alpha))(U) \subseteq N''_G(\alpha)(U)$.

Finally, in case (B), $\overline{f} \times f''$ is an epimorphism in the commutative diagram below, and so $f''(St_{M''_G(\alpha)}(\underline{Q}_r(\alpha)_S)) \subseteq St_{N''_G(\alpha)}(\underline{Q}_r(\alpha)_S)$.

$$\begin{array}{ccc} \underline{Q}_r(\alpha)_S \times St_{M''_G(\alpha)}(\underline{Q}_r(\alpha)_S) & \overset{p_1}{\rightrightarrows} & \underline{Q}_r(\alpha)_S \\ \Big\downarrow \overline{f} \times f'' & & \Big\downarrow \overline{f} \\ \underline{Q}_r(\overline{\alpha})_S \times f''(St_{M''_G(\alpha)}(\underline{Q}_r(\alpha)_S)) & \underset{p_1}{\rightrightarrows} & \underline{Q}_r(\overline{\alpha})_S \end{array}$$

Thus $\overline{f} \in \operatorname{Mor} S(\underline{S}^G)$. If $f'' \in \operatorname{Mor} S({}^G\underline{S}^G)$, a similar diagram shows that $\overline{f} \in \operatorname{Mor} S({}^G\underline{S}^G)$. ∎ A similar proposition holds for $\underline{Q}_\ell(\alpha)$ and $\underline{Q}_\ell(\overline{\alpha})$ which we leave to the reader to state and prove.

We are now ready to begin the construction of a canonical "flask" resolution. First we need a canonical embedding of a sheaf into a "flask" sheaf. The following definition gives the necessary axioms.

<u>Definition 2.6</u>

Godement resolutions can be constructed in $S(\underline{S})$ if there is a functor $C\colon S(\underline{S}) \longrightarrow S(\underline{S})$ and a natural transformation $j\colon I \longrightarrow C$ with the following properties:

GR1: $C$ preserves finite inverse limits and monomorphisms.

GR2: For all $\alpha\colon M \longrightarrow M'' \in \operatorname{Mor} S(\underline{S}^G)$, $M'' \in S({}^G\underline{S}^G)$, the canonical map $Q_r^p(C(\alpha))_S \longrightarrow iC\underline{Q}_r(\alpha)_S$ is an isomorphism, and similarly for $\alpha \in \operatorname{Mor} S({}^G\underline{S})$.

Moreover, $iC$ preserves maps which are onto.

GR3: If $M_1$ and $M_2$ are subsheaves of $M$, then $CM_1 = CM_2$ if and only if $M_1 = M_2$. $Ce = e$.

GR4: The maps $j(CM): CM \longrightarrow C^2M$ and $C(j(M)): CM \longrightarrow C^2M$ have left inverses for all $M \in \mathcal{S}(\underline{S})$.

GR5: Let $U \in \underline{T}$, $\mathcal{S}_U(\underline{S})$ be the category of sheaves in $\underline{T}$ regarded as sheaves in the induced topology on the category $\underline{T}/U$ of objects over $U$ [1, Chap. II, 4.12]. Let $C_U: \mathcal{S}_U(\underline{S}) \longrightarrow \mathcal{S}_U(\underline{S})$ be defined by $C_U(M)(V) = CM(V)$. If $\{U_i \xrightarrow{\varphi_i} U\} \in \text{Cov } \underline{T}/U$, then the map $C_U(M) \longrightarrow \prod_I \varphi_{i*}\varphi_i^* C_U(M)$ has a left inverse for any $M \in \mathcal{S}_U(\underline{S})$ which is natural in $M$.

Note that GR1 and GR3 imply that $C: \mathcal{S}(\underline{C}) \longrightarrow \mathcal{S}(\underline{C})$ where $\underline{C}$ is any of the categories at the beginning of §1. This, the functoriality of $C$, and the definition of $Q_r^p(C(\alpha))_S$ produce the canonical map in GR2. We give below several examples of topologies in which Godement resolutions can be defined.

<u>Examples</u>: 1) Let $X$ be a topological space, $\underline{T}$ the category of open subsets of $X$ with inclusion maps for morphisms, $\{U_i \subseteq U\} \in \text{Cov } \underline{T}$ if $\cup U_i = U$. Let $X_{dis}$ be the space $X$ with the discrete topology, $f: X_{dis} \longrightarrow X$ the canonical continuous map. Define a Grothendieck topology on $X_{dis}$ as above, and let $\mathcal{S}_X(\underline{S})$ and $\mathcal{S}_{X_{dis}}(\underline{S})$ be the category

of sheaves of sets on $X$ and $X_{dis}$ respectively. Then it is well known that there is a left adjoint $f^*: S_X(\underline{S}) \longrightarrow S_{X_{dis}}(\underline{S})$ to the direct image functor $f_*: S_{X_{dis}}(\underline{S}) \longrightarrow S_X(\underline{S})$. The trivial triple on $S_{X_{dis}}(\underline{S})$ gives a triple $(C,j,k) = (f_*f^*, \beta, f_*\alpha f^*)$ where $\beta: I \longrightarrow f_*f^*$ and $\alpha: f^*f_* \longrightarrow I$ are the respective adjunction maps [7]. Moreover, $(f_*f^*)(M)(U) = f^*(M)(U_{dis})$ where $U_{dis}$ has the discrete topology on it, and so $CM(U) = \prod_{y \in U_{dis}} f^*(M)_y \cong \prod_{y \in U} M_y$. $j(M): M \longrightarrow CM$ is, of course, the map restricting a section of $M$ over $U$ to its value at all of the stalks. This construction is the one Godement originally gave [7]. The verifications of GR1 - GR5 are either trivial or simpler versions of the arguments in the next example.

2) Let $X$ be a prescheme, $\underline{T}$ the etale site on $X$ [2, Exposé VII]. This is the full subcategory of schemes $U$ over $X$ belonging to a fixed universe such that the structure map $U \longrightarrow X$ is etale (Case (2) of [1]), with $\{U_i \xrightarrow{\varphi_i} U\} \in \text{Cov}\,\underline{T}$ if $U = \cup\, \varphi_i(U_i)$ ($\varphi_i$ is necessarily etale). Following [2, Exposé VIII] a geometric point $\overline{y}$ of $X$ is an $X$-scheme which is the spectrum of a separably closed field. For each $y \in X$, choose a separable closure $\overline{k(y)}$ of $k(y)$, and let $GP(X)$ be the set of the corresponding geometric points. Let $\overline{X} = \coprod_{\overline{y} \in GP(X)} \overline{y}$, the disjoint union of

the preschemes Spec $(\overline{k(y)})$, $y \in X$, and let $\amalg i_y : \overline{X} \longrightarrow X$ be the canonical map. Note that for $U \xrightarrow{\varphi} X \in \underline{T}$,

$$U \underset{X}{\times} \overline{X} = \coprod_{y \in \varphi(U)} \Big( \coprod_{y_i \in \varphi^{-1}(y)} \overline{y}_i \Big) = \coprod_{\overline{y} \in GP(U)} \overline{y} \quad \text{where } \overline{y}_i = \overline{y} \text{ since}$$

$U \underset{X}{\times} \overline{X}$ is etale over $\overline{X}$ which implies that each fibre over $\overline{y} \in GP(X)$ is a finite disjoint union of copies of $\overline{y}$.

Let $S_X(\underline{S})$ and $S_{\overline{X}}(\underline{S})$ be the category of sheaves of sets on the etale sites over $X$ and $\overline{X}$ respectively. As in the above example, the trivial triple in $S_{\overline{X}}(\underline{S})$ induces a triple $(C,j,k) = ((\amalg i_y)_*(\amalg i_y)^*, \beta, (\amalg i_y)_*\alpha(\amalg i_y)^*)$ in $S_X(\underline{S})$ where $\alpha$, $\beta$ are the respective adjunction maps. For $U \xrightarrow{\varphi} X \in \underline{T}$,

$$CM(U) = (\amalg i_y)^*M\Big( \coprod_{y \in \varphi(U)} \Big( \coprod_{y_i \in \varphi^{-1}(y)} \overline{y}_i \Big)\Big).$$

Since $(\amalg i_y)^*M$ is a sheaf on the etale site over $\overline{X}$ and $\{\overline{y}_i \longrightarrow \coprod_{y \in \varphi(U)} ( \coprod_{y_i \in \varphi^{-1}(y)} \overline{y}_i)\}$ is a covering family,

$$CM(U) = \prod_{y \in \varphi(U)} \Big( \prod_{y_i \in \varphi^{-1}(y)} (\amalg i_y)^*M(\overline{y}_i) \Big) = \prod_{\overline{y} \in GP(U)} (\amalg i_y)^*M(\overline{y}).$$

Essentially by definition $(\amalg i_y)^*M(\overline{y}) = M_{\overline{y}}$, the stalk of $M$ at $\overline{y}$, where $M_{\overline{y}} = \varinjlim_{C^0_{\overline{y}}} M(X')$, $C_{\overline{y}}$ being the category of preschemes $X'$ etale over $X$ with a map $\overline{y} \longrightarrow X'$ over $\overline{y} \longrightarrow X$ [2, Chapter III; 1, Exposé VIII]. If $V \xrightarrow{\psi} U \in \text{Mor } \underline{T}$, then the map

$$CM(U) = \prod_{\overline{y} \in GP(U)} M_{\overline{y}} \longrightarrow CM(V) = \prod_{\overline{y} \in GP(\psi(V))} \Big( \prod_{\overline{y}_i \in \psi^{-1}(\overline{y})} M_{\overline{y}_i} \Big) \quad \text{comes}$$

from composing the projection $\prod_{\overline{y} \in GP(U)} M_{\overline{y}} \longrightarrow \prod_{\overline{y} \in GP(\psi(V))} M_{\overline{y}}$

with the product of the diagonal maps $M_{\bar{y}} \longrightarrow \prod_{\bar{y}_i \in \psi^{-1}(\bar{y})} M_{\bar{y}_i}$ (remember $\bar{y} = \bar{y}_i$ ). In particular for $\alpha\colon M \longrightarrow N \in \operatorname{Mor} S(\underline{S})$, $C(\alpha)$ is determined by the maps $\alpha_{\bar{y}}\colon M_{\bar{y}} \longrightarrow N_{\bar{y}}$ for all $\bar{y} \in GP(X)$. The map $j(M)$ may be thought of as sending a section to its value in each stalk of $M$ .

Since $C$ and $j$ come from a triple, GR4 is satisfied. GR1 is immediate since the product of equalizers is the equalizer of the product and $\varinjlim$ over $C^0_{\bar{y}}$, $C^0_{\bar{y}}$ being a connected directed category, preserves equalizers and monomorphisms. GR3 is Proposition 1.8 of Chapter II of [1]. For GR2, recall that if $M \xrightarrow{\alpha} M'' \in \operatorname{Mor} \underline{S}^G$, $\sigma\colon M''_S \times M_G \longrightarrow M''_S$ is the map defining the right group action of $M_G$ on $M''_S$ via $\alpha$ , then the coequalizer of $M''_S \times M_G \overset{\sigma}{\underset{p_1}{\rightrightarrows}} M''_S$ is $M''_S/\sim$ where $\sim$ is the equivalence relation $x \sim y$ if there is a $g \in M_G$ with $x \cdot \alpha(g) = y$ . Now a straightforward calculation using the definition of $(M''_{\bar{y}})_S$ and the description above of $Q_r(\alpha_{\bar{y}})_S$ shows that $Q_r(\alpha_{\bar{y}})_S \cong (\underline{Q}_r(\alpha)_S)_{\bar{y}}$ . Thus for $M \xrightarrow{\alpha} M'' \in \operatorname{Mor} S(\underline{S}^G)$ we see from the structure of $CM_G$ and $CM''_S$ that the coequalizer of

$$CM''_S(U) \times CM_G(U) \overset{\sigma(U)}{\underset{p_1}{\rightrightarrows}} CM''_S(U) \text{ is } \prod_{\bar{y} \in GP(U)} Q_r(\alpha_{\bar{y}})_S .$$

This and the above description of the restriction map $CM(U) \longrightarrow CM(V)$ for $V \longrightarrow U \in \operatorname{Mor} \underline{T}$ shows that GR2 holds for right quotients.

The argument for left quotients is identical, and $iC$ preserves maps which are onto by Proposition 2.2. Finally, if $\{U_i \xrightarrow{\varphi_i} U\} \in \mathrm{Cov}\ \underline{T}/U$, we must show that

$C_U(M)(V) = CM(V) \longrightarrow \prod_I \varphi_{i*}\varphi_i^* C_U M(V) = \prod_I M(V \times_U U_i)$ has a left

inverse which is natural with respect to $V$ and $M$. A section $\psi$ of $\coprod_I(\coprod_{\bar{y}_i \in GP(U_i)} \bar{y}_i) \longrightarrow \coprod_{\bar{y} \in GP(U)} \bar{y}$ can be defined by

choosing a point in the fibre over $\bar{y}$ for each $\bar{y} \in GP(U)$ (the fibre is non-empty since $\{\varphi_i\} \in \mathrm{Cov}\ \underline{T}/U$) since the structure sheaves of two such points are isomorphic. Then this induces $((\coprod i_y)^*M)(V \times_U \psi)\colon \prod_I CM(V \times_U U_i) \longrightarrow CM(V)$ which is clearly natural and left inverse to $\prod_I CM(\varphi_i)$ since $\prod_I CM(V \times_U U_i) = CM(\coprod_I V \times_U U_i)$. Hence GR5 is satisfied.

3) Let $G$ be a group, $\underline{T}_G$ the category of left $G$-sets with the canonical topology [1, Chapter I, Example 0.6]. Thus $\{U_i \xrightarrow{\varphi_i} U\}_I \in \mathrm{Cov}\ \underline{T}_G$ if and only if $\bigcup_I \varphi_i(U_i) = U$. In this topology all sheaves are representable. If $M$ is a sheaf in $\underline{T}_G$, then the object representing $M$ is $M(G)$ which has a left $G$-structure coming from functoriality via right translation by $g^{-1}$ for $g \in G$. Let $\underline{T}_{dis}$ be the category of left $e$-sets, where $e$ is the trivial group with the corresponding topology. The forgetful functor $f\colon \underline{T}_G \longrightarrow \underline{T}_{dis}$ gives a morphism of topologies. Let $S(\underline{S})$ and $S_{dis}(\underline{S})$ be the category of sheaves of sets on $\underline{T}_G$ and $\underline{T}_{dis}$

respectively. Repeating the above process (and using notation which conforms with it) there is a left adjoint $f^*\colon \mathcal{S}(\underline{S}) \longrightarrow \mathcal{S}_{dis}(\underline{S})$ to the direct image functor $f_*$. This defines a triple $(C,j,k) = (f_*f^*,\alpha,f_*\beta f^*)$ where $\alpha$ and $\beta$ are the adjunction morphisms. Now $CM(U) = f^*M(U_{dis})$ where $U_{dis}$ is $U$ regarded as a set. Since $f^*M \in \mathcal{S}_{dis}(\underline{S})$ and $U$ is the disjoint union of its points in $\underline{T}_{dis}$, $CM(U) = \prod_{y\in U} M_y = \mathrm{Hom}_{\underline{T}_G}(\coprod_{y\in U} G_y, M)$ where $M_y = M(G)$ and $G_y = G$. For $V \xrightarrow{\varphi} U \in \mathrm{Mor}\,\underline{T}_G$, $CM(U) \longrightarrow CM(V)$ is the composition of the projection $\prod_{y\in U} M_y \longrightarrow \prod_{y\in\varphi(V)} M_y$ followed by the product of the diagonal maps $M_y \longrightarrow \prod_{y_i\in\varphi^{-1}(y)} M_{y_i}$. The object representing $CM$ is $\prod_{g\in G} M_g$ and the left $G$-action is induced from right translation by $g^{-1}$ on the index set $G$. The functorial behavior of $CM$ and the definition of equalizers in $\underline{T}_G$ shows that GR1 holds. GR3 is clear, and GR4 follows since $C$ was defined by a triple. Suppose that $M \xrightarrow{\alpha} M''$ is a map in $\mathcal{S}(\underline{S}^G)$. The definition of the coequalizer of

$$\prod_{y\in U}(M''_S)_y \times \prod_{y\in U}(M_G)_y \underset{p_1}{\overset{\Pi\sigma_y}{\rightrightarrows}} \prod_{y\in U}(M''_S)_y$$

where $\sigma_y$ comes from the group action of $(M_G)_y$ on $(M''_S)_y$ shows that it is $\prod_{y\in U} Q_r(\alpha_y)_S$. But $Q_r(\alpha_y)_S$, once the notation and identifications are untangled, is just the object representing $\underline{Q}_r(\alpha)_S$. Hence since the only coverings of G are disjoint sums of G, this and 2.2

show that GR2 is satisfied. Finally, as in 2), to verify GR5, choose a section $\psi: U \longrightarrow \coprod_I U_i$ to the left G-map $\coprod_I U_i \longrightarrow U$. While $\psi$ cannot usually be chosen as a G-map, it does define a map $\coprod_{y\in V} G_y \longrightarrow \coprod_I (\coprod_{y \in V \times_U U_i} G_y)$ and so a map

$\prod_I \mathrm{Hom}_{\underline{T}_G}(\coprod_{y\in V\times_U U_i} G_y, M) \longrightarrow \mathrm{Hom}_{\underline{T}_G}(\coprod_{y\in V} G_y, M)$ which is natural in $V$ since the G-action on $\coprod_{y\in V} G_y$ comes from left translation by $g$ in each factor $G_y$. Thus GR5 is also satisfied.

The Grothendieck topology which defines the Tate cohomology groups of a profinite group [1, Chapter 1, Example (0.6 bis)] also has Godement resolutions. The construction is essentially the one given above with appropriate continuity restrictions.

The next theorem summarizes the properties of $C$ and $j$. In particular it will be used to show that we get "flask" resolutions and that these resolutions may be used to resolve short exact sequences.

Let $M \in S(\underline{C})$ where $\underline{C} = \underline{Ab}$, $\underline{G}$, ${}^G\underline{S}^G$, $\underline{S}^G$, ${}^G\underline{S}$, or $\underline{S}^{\cdot}$ and fix $U \in \underline{T}$. Define a functor $S^*(M)(\ )$ from $J^0_U$ to the category of augmented co-semi-simplicial objects in $\underline{C}$ by

$$S^*(M)(\{U_i \xrightarrow{\varphi_i} U\}_I) : M(U) \xrightarrow{\varphi^*} \prod_I M(U_i) \overset{p_1^*}{\underset{p_2^*}{\rightrightarrows}} \prod_{I\times I} M(U_{i_1} \underset{U}{\times} U_{i_2})$$

$$\overset{p_1^*}{\underset{p_3^*}{\Rrightarrow}} \prod_{I^3} M(U_{i_1} \underset{U}{\times} U_{i_2} \underset{U}{\times} U_{i_3}) \overset{p_1^*}{\underset{p_4^*}{\longrightarrow}}$$

where for simplicity we will write $\varphi^*$ for $\varphi_I^*$ and

$p_j^*: \prod_{I^n} M(U_{i_1} \underset{U}{\times} \cdots \underset{U}{\times} U_{i_n}) \longrightarrow \prod_{I^{n+1}} M(U_{i_1} \underset{U}{\times} \cdots \underset{U}{\times} U_{i_{n+1}})$ ,

$1 \leq j \leq n+1$ , for the map coming from the projection maps

$p_{i_1,\ldots,\hat{i}_j,\ldots,i_{n+1}} : U_{i_1} \underset{U}{\times} \cdots \underset{U}{\times} U_{i_{n+1}}$

$\longrightarrow U_{i_1} \underset{U}{\times} \cdots \underset{U}{\times} U_{i_{j-1}} \underset{U}{\times} U_{i_{j+1}} \underset{U}{\times} \cdots \underset{U}{\times} U_{i_{n+1}}$

which excludes the $j^{\underline{th}}$ factor. (The degeneracy maps come from the diagonal maps, but these won't be needed.)

<u>Theorem 2.7</u>

Let $\underline{C}$ be any of the above categories, $M \in S(\underline{C})$.

1) For any $\{U_i \xrightarrow{\varphi_i} U\} \in \text{Cov } \underline{T}$ , there are maps

$\varphi_*: \prod_I CM(U_i) \longrightarrow CM(U)$ and

$p_{1*}^n: \prod_{I^n} CM(U_{i_1} \underset{U}{\times} \cdots \underset{U}{\times} U_{i_n}) \longrightarrow \prod_{I^{n-1}} CM(U_{i_1} \underset{U}{\times} \cdots \underset{U}{\times} U_{i_{n-1}})$, $n > 1$ ,

such that

$$\text{(a)} \quad \varphi_*\varphi^* = p_{1*}^n p_1^* = 1 \quad \text{and}$$

$$\text{(b)} \quad p_{1*}^n p_k^* = p_{k-1}^* p_{1*}^{n-1}$$

for $n > 1$ and $p_{1*}^2 p_2^* = \varphi^*\varphi_*$ .

2) Let $\alpha: M \longrightarrow M'' \in \operatorname{Mor} S(\underline{C})$. $\alpha$ is monic if and only if $C(\alpha)$ is monic. In particular $j(M): M \longrightarrow CM$ is monic. Moreover, $iC\operatorname{Ker}(\alpha) = \operatorname{Ker}^P(C(\alpha))$ and $iC(\alpha(M)) = C(\alpha)^P(CM)$.

3) If $M' \xrightarrow{\alpha} M \xrightarrow{\beta} M''$ is an exact sequence in $S(\underline{S}^{\cdot})$, then $iCM' \xrightarrow{C(\alpha)} iCM \xrightarrow{C(\beta)} iCM''$ is an exact sequence in $P(\underline{S}^{\cdot})$.

4) If $M \in S(\underline{S}^G)$, then $iC\mathcal{O}_{M_G}(e) = \mathcal{O}^P_{CM_G}(e)$. In particular if $M_G$ acts transitively on $M_S$, then $iC(M_G)$ acts transitively on $iC(M_S)$. If $H$ is a normal subgroup sheaf of $G$, then $CH$ is a normal subgroup sheaf of $CG$, $iC(\#(iG/iH)) = iC(G)/iC(H)$, and so $iCH \rightarrowtail iCG \twoheadrightarrow iC(\#(G/H))$ is a short exact sequence in $P(\underline{G})$.

5) Let $\alpha: M \longrightarrow M'' \in \operatorname{Mor} S(\underline{S}^G)$, $M'' \in S({}^G\underline{S}^G)$. Then there is a map $\theta: iC(\underline{Q}_r(\alpha)) \longrightarrow Q^P_r(C(\alpha)) \in \operatorname{Mor} P({}^G\underline{S}^G)$ which is an isomorphism on the pointed set components and onto in $P({}^G\underline{S})$. If moreover $\alpha(M_G)$ is a normal subgroup sheaf of $M''_G$, then $\theta$ is onto in $P({}^G\underline{S}^G)$. A similar statement holds for left quotients.

<u>Proof</u>: 1) Note that $\varphi_{i*}\varphi_i^* C_U(M)(V) = CM(U_i \underset{U}{\times} V)$ for any $M \in S(\underline{C})$ and $\{U_i \xrightarrow{\varphi_i} U\} \in \operatorname{Cov} \underline{T}/U$ by [1, I, 2.8]. Now by GR5, $(\varphi \times V)^*: C(M)(V) \longrightarrow \prod_I CM(U_i \underset{U}{\times} V)$ has a left inverse $\overline{\varphi}(V)$ natural in $V$ and $M$. The naturality in $M$ shows that $\overline{\varphi}(V) \in \operatorname{Mor} \underline{C}$ for all $V$ ($\varphi_{i*}\varphi_i^*$ commutes with

finite inverse limits [1, Chapter II, 4.14]). Let $\varphi_* = \bar{\varphi}(U): \prod_I CM(U_i) \longrightarrow CM(U)$. Now $\bar{\varphi}(U_{i_1} \times_U \cdots \times_U U_{i_n})$ is a left inverse of

$CM(U_{i_1} \times_U \cdots \times_U U_{i_n}) \longrightarrow \prod_I CM(U_i \times_U U_{i_1} \times_U \cdots \times_U U_{i_n})$ for each n-tuple $(i_1, \cdots, i_n) \in I^n$. Let

$p^n_{1*} = \prod_{I^n} \bar{\varphi}(U_{i_1} \times_U \cdots \times_U U_{i_n})$. Then $p^n_{1*}$ is a left inverse to

$$p_1^*: \prod_{I^n} CM(U_{i_1} \times_U \cdots \times_U U_{i_n}) \longrightarrow \prod_{I^{n+1}} CM(U_i \times_U U_{i_1} \times_U \cdots \times_U U_{i_n})$$

which gives (a). Since $\bar{\varphi}$ is a sheaf map, the diagram below commutes which gives (b):

$$\begin{array}{ccc}
\prod_{I\times I^{n-1}} \varphi_{i_1*}\varphi^*_{i_1} C_U M(U_{i_2} \times_U \cdots \times_U U_{i_n}) & \xrightarrow{p^*_{k-1}} & \prod_{I\times I^n} \varphi_{i_1*}\varphi^*_{i_1} C_U M(U_{i_2} \times_U \cdots \times_U U_{i_{k-1}} \times_U U_i \times_U \cdots \times_U U_{i_n}) \\
\Vert & & \Vert \\
\prod_{I^n} CM(U_{i_1} \times_U U_{i_2} \times_U \cdots \times_U U_{i_n}) & \xrightarrow{p^*_k} & \prod_{I^{n+1}} CM(U_{i_1} \times_U \cdots \times_U U_{i_{k-1}} \times_U U_i \times_U \cdots \times_U U_{i_n}) \\
\downarrow p^n_{1*} & & \downarrow p^{n+1}_{1*} \\
\prod_{I^{n-1}} CM(U_{i_2} \times_U \cdots \times_U U_{i_n}) & \xrightarrow{p^*_{k-1}} & \prod_{I^n} CM(U_{i_2} \times_U \cdots \times_U U_{i_{k-1}} \times_U U_i \times_U \cdots \times_U U_{i_n})
\end{array}$$

2) Let $\alpha: M \longrightarrow M''$ be a map of sheaves with $C(\alpha)$ monic. Let $f_1, f_2: L \longrightarrow M$ with $\alpha f_1 = \alpha f_2$. Let $F_j = \text{Equalizer } (L \times M \overset{p_2}{\underset{f_j p_1}{\rightrightarrows}} M)$ be the graph of $f_j$,

$j = 1$ or 2. Since $C$ preserves finite inverse limits $CF_j$ is the graph of $C(f_j): CL \longrightarrow CM$ in $CL \times CM$. But $C(\alpha)$ is monic and so $CF_1 = CF_2$. By GR3 this gives $F_1 = F_2$ or $f_1 = f_2$. Since $C(j(M)): CM \longrightarrow C^2M$ has a left inverse, it is a monomorphism. Thus $j(M)$ is a monomorphism. Since $i\mathrm{Ker}\,(C(\alpha)) = \mathrm{Ker}^P(C(\alpha))$ and $C$ preserves equalizers, $iC\mathrm{Ker}\,(\alpha) = \mathrm{Ker}^P(C(\alpha))$. Moreover $C(\alpha(M))$ is a subsheaf of $CM"$, and since $iC$ preserves maps which are onto, $iC(\alpha(M)) = C(\alpha)^P(CM)$,

3) follows immediately from 2) as do the first two assertions of 4). The condition that $H$ is a normal subgroup sheaf of $G$ is equivalent to saying that

$H \times G \xrightarrow{p_2 \times p_1 \times \mathrm{inv}\, p_2} G \times H \times G \longrightarrow G$ factors through $H$

where inv is the inverse map and the last map comes from multiplication. Thus $CH$ is a normal subgroup sheaf of $CG$. Now as a set $iC(\#(iG/iH)) = iCG/iCH$ since $\#(iG/iH) = \underline{Q}_r(j)_S$, $j: H \longrightarrow G$ being the inclusion map. But the stabilizer subgroup in $CG$ of its action on $CG/CH$ is precisely $CH$ which comes from the stabilizer subgroup in $G$ of its action on $G/H$. Hence the argument below for 5) shows that $iC(\#(iG/iH)) = iCG/iCH$ in $P({}^G\underline{S}^G)$ or equivalently in $P(\underline{G})$.

5)

$$i_G C(\underline{Q}_r(\alpha)) = iC({}_G\underline{Q}_r(\alpha)) = iC({}_GM"(\alpha)/St_{{}_GM"(\alpha)}(\underline{Q}_r(\alpha)_S)).$$

Since ${}^P_G CM"(C(\alpha)) = iC({}_GM")$, the definition of

$St^{p}_{{}^{p}_{G}CM''(C(\alpha))}(Q^{p}_{r}(C(\alpha))_{S})$, GR1, and the onto statement of GR2 show that $iC(St_{{}_{G}M''(\alpha)}(\underline{Q}_{r}(\alpha)_{S}))$ is contained in $St^{p}_{{}^{p}_{G}CM''(C(\alpha))}(Q^{p}_{r}(C(\alpha))_{S})$ and that $iC({}_{G}\underline{Q}_{r}(\alpha)) \longrightarrow {}_{G}Q^{p}_{r}(C(\alpha))$ is onto. (In 4), C applied to the first stabilizer equals the second stabilizer which finishes the proof of 4). 4) then shows that the map is onto.) ${}^{p}iCM''_{G}(C(\alpha))$ is the sup of the subgroup presheaves in $iCM''_{G}$ whose actions on $iCM''_{S}$ induce an action on $Q^{p}_{r}(C(\alpha))_{S}$. Since $M''_{G}(\alpha)$ is a subgroup sheaf of $M''_{G}$ whose action on $M''_{S}$ induces an action on $\underline{Q}_{r}(\alpha)_{S}$, GR1 and GR2 show that $iCM''_{G}(\alpha)$ is contained in ${}^{p}iCM''_{G}(C(\alpha))$. Moreover the argument above shows that $iC(St_{M''_{G}(\alpha)}(\underline{Q}_{r}(\alpha)_{S})) \subseteq St^{p}_{{}^{p}iCM''_{G}(C(\alpha))}(Q^{p}_{r}(C(\alpha))_{S})$ where we have identified $C\underline{Q}_{r}(\alpha)_{S}$ with $Q^{p}_{r}(C(\alpha))_{S}$. Hence we get a map $iC(\underline{Q}_{r}(\alpha)_{G}) \longrightarrow Q^{p}_{r}(C(\alpha))_{G}$ which fails to be onto only because $iCM''_{G}(\alpha)$ may not equal ${}^{p}iCM''_{G}(C(\alpha))$. This together with the map between the groups acting on the left and the isomorphism of the pointed set components gives $iC(\underline{Q}_{r}(\alpha)) \longrightarrow Q^{p}_{r}(C(\alpha)) \in \text{Mor}\ P({}^{G}\underline{S}^{G})$ which is onto in $P({}^{G}\underline{S})$. If $\alpha(M_{G})$ is a normal subgroup of $M''_{G}$, then $C(\alpha)^{p}(CM_{G})$ is a normal subgroup of $iCM''_{G}$. Thus $iCM''_{G}(\alpha) = {}^{p}iCM''_{G}(C(\alpha))$ and the map is onto in $P({}^{G}\underline{S}^{G})$. ∎

This gives a reasonably good picture of the functorial properties of C. The last result of this section does the same thing for $\underline{Q}_{r}$. In particular we will need a non-abelian analogue of the $3 \times 3$ lemma which will follow by combining the results below.

<u>Theorem 2.8</u>

1) Given

$$\begin{array}{ccccccc} & & M' & \overset{\alpha}{\rightarrowtail} & M & \overset{\beta}{\longrightarrow} & M'' \\ & & {\scriptstyle j'}\downarrow & & {\scriptstyle j}\downarrow & & {\scriptstyle j''}\downarrow \\ & & N' & \overset{\bar{\alpha}}{\rightarrowtail} & N & \overset{\bar{\beta}}{\longrightarrow} & N'' \\ & & {\scriptstyle \pi'}\downarrow & & {\scriptstyle \pi}\downarrow & & \\ e & \longrightarrow & \underline{Q}_r(j') & \overset{\tilde{\alpha}}{\longrightarrow} & \underline{Q}_r(j) & & \end{array}$$

where all of the sheaves are in $S({}^G\underline{S}^G)$, suppose that $j, j', \alpha, \bar{\alpha} \in \text{Mor}\, S(\underline{S}^G)$, the other maps are in $S(\underline{S}^{\cdot})$, $\alpha, \bar{\alpha}, j', j$, and $j''$ are monic in $S(\underline{S}^{\cdot})$, the first two rows are exact in $S(\underline{S}^{\cdot})$ and $M'_G$ acts transitively on $M'_S$. Then the third row is exact in $S(\underline{S}^{\cdot})$. Moreover, if $N'_G$ or ${}_G N'$ acts transitively on $N'_S$ and $\tilde{\alpha}\pi'$, $\pi' \in \text{Mor}\, S(\underline{S}^G)$ or $S({}^G\underline{S})$ respectively, then $\tilde{\alpha}$ is monic.

2) Consider

$$\begin{array}{ccccc} M' & \overset{\alpha}{\longrightarrow} & M & \overset{\beta}{\longrightarrow} & M'' \\ {\scriptstyle j'}\downarrow & & {\scriptstyle j}\downarrow & & {\scriptstyle j''}\downarrow \\ N' & \overset{\bar{\alpha}}{\longrightarrow} & N & \overset{\bar{\beta}}{\longrightarrow} & N'' \\ {\scriptstyle \pi'}\downarrow & & {\scriptstyle \pi}\downarrow & & {\scriptstyle \pi''}\downarrow \\ \underline{Q}.(j') & \overset{\tilde{\alpha}}{\longrightarrow} & \underline{Q}.(j) & \overset{\tilde{\beta}}{\longrightarrow} & \underline{Q}.(j'') \end{array}$$

where all of the sheaves are in $S({}^G\underline{S}^G)$, $\bar{\alpha}, \tilde{\alpha}, \pi' \in \text{Mor}\, S(\underline{S}^G)$, $N''_S \simeq \underline{Q}_r(\bar{\alpha})_S$, and if $N_G$ is restricted to $\bar{\alpha}(N'_G)$, then $\pi$ becomes a map in $S(\underline{S}^G)$.

If $. = r$, and $j', j, j'', \beta, \overline{\beta} \in \text{Mor } S(\underline{S}^G)$, and

$\overline{\beta}|: j(M)_G \longrightarrow j''(M'')_G$ is onto, then $\underline{Q}_r(j'')_S \cong \underline{Q}_r(\tilde{\alpha})_S$.

If $. = \ell$, and $j', j, j'', \beta, \overline{\beta} \in \text{Mor } S({}^G\underline{S})$, and

$\overline{\beta}|: {}_Gj(M) \longrightarrow {}_Gj''(M'')$ is onto, then $\underline{Q}_\ell(j'')_S \cong \underline{Q}_r(\tilde{\alpha})_S$.

Moreover left and right may be interchanged in the hypotheses if they are interchanged in the conclusion.

<u>Proof</u>: 1) Let $U \in \underline{T}$, $\overline{x} \in \underline{Q}_r(j')_S(U)$ with $\tilde{\alpha}(\overline{x}) = e$. Since $\pi'$ is onto, there is $\{U_i \xrightarrow{\varphi_i} U\}_I \in \text{Cov } \underline{T}$ and $x \in \prod_I N'_S(U_i)$ with $\pi'(x) = \varphi_I^*(\overline{x})$. But $\pi(\overline{\alpha}(x)) = e$, and so by 2.2 (and taking a refinement of $\{U_i \longrightarrow U\}$ if necessary) there is $g \in \prod_I M_G(U_i)$ with $\overline{\alpha}(x) = e \cdot j(g)$. Since $j''$ is monic, $e \cdot g \in \prod_I \alpha(M'_S)(U_i)$. Taking a refinement of $\{U_i \longrightarrow U\}$ if necessary, there is $g_1 \in \prod_I M'_G(U_i)$ such that $\alpha(e \cdot g_1) = e \cdot g$, and so $j'(e \cdot g_1) = x$ since $\overline{\alpha}$ is monic. Hence $\overline{x} = e$ as desired since $\underline{Q}_r(j')$ is a sheaf. For the rest assume that $N'_G$ acts transitively on $N'_S$ and $\tilde{\alpha}\pi' \in \text{Mor } S(\underline{S}^G)$. Then given $\overline{x}, \overline{y} \in \underline{Q}_r(j')_S(U)$ with $\tilde{\alpha}(\overline{x}) = \tilde{\alpha}(\overline{y})$, there is $\{U_i \xrightarrow{\varphi_i} U\}_I \in J_U$, $x, y \in \prod_I N'_S(U_i)$ with $\pi'(x) = \varphi_I^*(\overline{x})$ and $\pi'(y) = \varphi_I^*(\overline{y})$, and $g \in \prod_I N'_G(U_i)$ with $x \cdot g = e$. Then $\tilde{\alpha}(\pi'(x \cdot g)) = \tilde{\alpha}(e) = \tilde{\alpha}(\pi'(y \cdot g))$. Thus the above shows that $\pi'(x \cdot g) = \varphi_I^*(\overline{x}) \cdot \pi'(g)$ $= \pi'(y \cdot g) = \varphi_I^*(\overline{y}) \cdot \pi'(g)$, and so $\overline{x} = \overline{y}$. The other case is similar.

2) The hypotheses on the group actions the maps preserve are, except for $\pi'$, $\beta$, and $\overline{\beta}$ required for the other

hypotheses and the conclusion to make sense. For $\cdot = r$ , the result follows from a diagram chase in the diagram below.

$$\begin{array}{ccccc}
N_S \times M_G \times N'_G & \underset{p_1\times p_2}{\rightrightarrows} & N_S\times M_G & \xrightarrow{\overline{\beta}\times M_G} & N''_S\times M_G \\
{\scriptstyle p_1\times p_3}\downarrow\downarrow & & {\scriptstyle p_1}\downarrow\downarrow & & {\scriptstyle p_1}\downarrow\downarrow\\
N_S\times N'_G & \underset{p_1}{\rightrightarrows} & N_S & \xrightarrow{\overline{\beta}} & N''_S\\
{\scriptstyle \pi\times N'_G}\downarrow & & {\scriptstyle\pi}\downarrow && {\scriptstyle\pi''}\downarrow\\
\underline{Q}_r(j)_S\times N'_G & \underset{p_1}{\rightrightarrows} & \underline{Q}_r(j)_S & \xrightarrow{\tilde{\beta}} & \underline{Q}_r(j'')_S
\end{array} .$$

All unlabeled maps come from the group actions and $p_i$ denotes projection from the product onto the $i^{\underline{th}}$ factor. By definition $\overline{\beta}$ and $\pi$ are coequalizers. Since $\overline{\beta}| : j(M)_G \longrightarrow j''(M'')_G$ is onto, $\pi''$ is also a coequalizer (the coequalizer of $L \longrightarrow M \rightrightarrows N$ is the coequalizer of $M \rightrightarrows N$ if $L \longrightarrow M$ is an epimorphism). A straightforward argument using Proposition 2.2 and the structure of the coequalizer $Q$ of $M_S \times N_G \underset{p_1}{\rightrightarrows} M_S$ where $N_G$ is a sheaf of groups acting on any sheaf of sets shows that $Q \times L$ is the coequalizer of $M_S \times N_G \times L \underset{p_1\times p_3}{\rightrightarrows} M_S \times L$ for any $L \in S(\underline{S})$. Thus $\overline{\beta} \times M_G$ and $\pi \times N'_G$ are also coequalizers. Finally the hypothesis on $N_G$ shows that the whole diagram is commutative. Now a diagram chase shows that $\tilde{\beta}$ is also a coequalizer. Hence $\underline{Q}_r(j'')_S \cong \underline{Q}_r(\tilde{\alpha})_S$ since

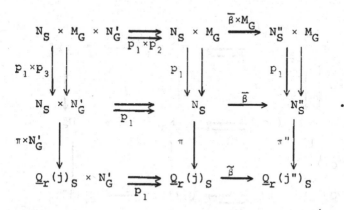

$N'_G \longrightarrow \underline{Q}_r(j')_G$ is onto.

For $. = \ell$ , the result follows from the diagram below.

$$\begin{array}{ccccc}
{}_G M \times N_S \times N'_G & \overset{}{\underset{p_1 \times p_2}{\rightrightarrows}} & {}_G M \times N_S & \xrightarrow{{}_G M \times \overline{\beta}} & {}_G M \times N''_S \\
{\scriptstyle p_2 \times p_3}\downarrow\downarrow & & {\scriptstyle p_2}\downarrow\downarrow & & {\scriptstyle p_2}\downarrow\downarrow \\
N_S \times N'_G & \underset{p_1}{\rightrightarrows} & N_S & \xrightarrow{\overline{\beta}} & N''_S \\
{\scriptstyle \pi \times N'_G}\downarrow & & {\scriptstyle \pi}\downarrow & & {\scriptstyle \pi''}\downarrow \\
\underline{Q}_\ell(j)_S \times N'_G & \underset{p_1}{\rightrightarrows} & \underline{Q}_\ell(j)_S & \xrightarrow{\widetilde{\beta}} & \underline{Q}_\ell(j'')_S
\end{array}$$

The arguments above show that $\pi$, $\pi''$, $\overline{\beta}$, $\pi \times N'_G$ , and ${}_G M \times \overline{\beta}$ are coequalizers. A diagram chase as above then shows that $\widetilde{\beta}$ is a coequalizer as desired. ∎

## §3. COHOMOLOGY WITH NON-ABELIAN COEFFICIENTS

This section is devoted to the definition of $H^n(U;M)$, $M \in S({}^G\underline{S}^G)$, and a description of its properties including the exactness of a 9 term cohomology sequence associated to a central extension of coefficient sheaves. It concludes with a comparison theorem which says that our definition agrees with the usual one for well known cohomology theories.

As before we fix a topology in which Godement resolutions can be constructed, $(\underline{T}, C, j)$, for the entire section. Let $M \in S({}^G\underline{S}^G)$. The canonical resolution of $M$ is the complex:

$$\begin{array}{ccccccccc}
e \longrightarrow & CM & \xrightarrow{d^0} & & C\underline{Q}_\ell(j_0) & \xrightarrow{d^1} & C\underline{Q}_r(j_1) & \xrightarrow{d^2} & \cdots \\
& {\scriptstyle j_0}\nearrow & \searrow{\scriptstyle \pi_0} & {\scriptstyle j_1}\nearrow & \searrow{\scriptstyle \pi_1} & {\scriptstyle j_2}\nearrow & \searrow{\scriptstyle \pi_2} & {\scriptstyle j_3}\nearrow & \\
M & & \underline{Q}_\ell(j_0) & & \underline{Q}_r(j_1) & & \underline{Q}_\ell(j_2) & &
\end{array}$$

where of course $j_0 = j(M)$, $j_1 = j(\underline{Q}_\ell(j_0))$, etc., and $d^i \in \operatorname{Mor} S(\underline{S}^\cdot)$ preserves the left group action if $i$ is odd and the right group action if $i$ is even. For the remainder of this section fix $U \in \underline{T}$ (thus giving $\Gamma_U\colon S({}^G\underline{S}^G) \longrightarrow {}^G\underline{S}^G$ which we will derive). Let $Z^n(M) \in {}^G\underline{S}^G$ be defined by $Z^n(M)_S = \operatorname{Ker}_{\underline{S}}.(C\underline{Q}.(j_{n-1})_S(U) \xrightarrow{d^n} C\underline{Q}.(j_n)_S(U))$ with $Z^n(M)_G$ and ${}_GZ^n(M)$ being the largest subgroups of $\underline{Q}.(j_{n-1})_G(U)$ and ${}_G\underline{Q}.(j_{n-1})(U)$ respectively which stabilize $Z^n(M)_S$ as a set where $. = \ell$ or $r$ depending on $n$. Proposition 2.2 shows that $Z^n(M)_S = O_e({}_G\underline{Q}_r(j_{n-1}))(U) \subseteq \underline{Q}_r(j_{n-1})_S(U)$ and ${}_GZ^n(M) = {}_G\underline{Q}_r(j_{n-1})(U)$ for $n$ even and $Z^n(M)_S = O_e(\underline{Q}_\ell(j_{n-1})_G)(U)$ and $Z^n(M)_G = \underline{Q}_\ell(j_{n-1})_G(U)$ for $n$ odd where $\underline{Q}_r(j_{-1})$ is always to be interpreted as $M$ and $\underline{Q}_\ell(j_{-2})$ as $e$. For $n$ odd, let $O_e(C\underline{Q}_r(j_n)_G) \in S(\underline{S}^G)$ be the sheaf of pointed sets whose right group component is $C\underline{Q}_r(j_n)_G$. Define $O_e({}_GC\underline{Q}_\ell(j_n)) \in S({}^G\underline{S})$ in a similar way for $n$ even. Then by restriction we have a map

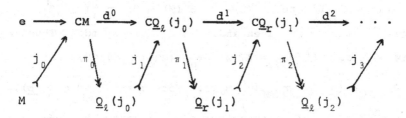

$\pi'_{n-1}: \mathcal{O}_e(C\underline{Q}_r(j_{n-2})_G)(U) \longrightarrow Z^n(M)$ which is in $\underline{S}^G$ for n odd and a map $\pi'_{n-1}: \mathcal{O}_e({}_GC\underline{Q}_\ell(j_{n-2}))(U) \longrightarrow Z^n(M)$ which is in ${}^G\underline{S}$ for n even. Define $H^n(U;M) = H^n(M) = Q.(\pi'_{n-1}) \in {}^G\underline{S}^G$ where $. = \ell$ for n even and $. = r$ for n odd. Thus if n is odd say, $H^n(M)_S = Z^n(M)_S/\pi'_{n-1}(C\underline{Q}_r(j_{n-2})_G(U)$ and ${}_GH^n(M) = {}_GZ^n(M)/St_{{}_GZ^n(M)}(H^n(M)_S)$. Note that if $M \in S(\underline{Ab})$, then $H^n(M)$ is, via a forgetful functor, just the $n^{\underline{th}}$ homology of the complex:

$$e \longrightarrow CM(U) \xrightarrow{d^0} C\underline{Q}_\ell(j_0)(U) \xrightarrow{d^1} C\underline{Q}_r(j_1)(U) \xrightarrow{d^2} \cdots$$

If $x \in Z^n(M)_S$, let $\{x\}$ denote the equivalence class of x in $H^n(M)_S$. Finally observe that $H^n(M)_S$ is a pointed set with neutral elements. The neutral elements, $H^n(M)'_S$, are the image in $H^n(M)_S$ of $\pi_{n-1}(C\underline{Q}.(j_{n-2})_S(U)) \cap Z^n(M)_S$. If ${}_GM$ acts transitively on $M_S$, then $H^0(M)_S = M_S(U)$ where the groups acting on $H^0(M)_S$ are quotients of those acting on $M_S(U)$. Thus if $M \in S(\underline{G})$, $H^0(M) = M(U)$, $Z^1(M)_S = \underline{Q}_\ell(j_0)_S(U)$ (since $\underline{Q}_\ell(j_0)_G$ acts transitively on $\underline{Q}_\ell(j_0)_S$), and $H^1(M)'_S = e$ since $CM_G(U)$ acts transitively on $CM_S(U)$.

Now $H^0(M)_S = \mathcal{O}_e({}_GM)(U)$ defines a functor $H^0(\ )_S: S({}^G\underline{S}) \longrightarrow \underline{S}^{\cdot}$. If we restrict the category of sheaves, then we can enlarge the range category. Thus $H^0(\ ): S(\underline{G}) \longrightarrow \underline{G}$. Moreover, given $\beta: M \longrightarrow M'' \in \text{Mor}\ S({}^G\underline{S})$ if ${}_GM(U)$ acts faithfully on $M_S(U)$ or $\beta$ is onto in $P({}^G\underline{S})$, then $H^0(\beta) \in \text{Mor}\ {}^G\underline{S}$ since in either case

$\beta(\mathrm{St}_{{}_G M(U)}(O_e({}_G M))(U))$ acts trivially on $O_e({}_G M'')(U)$.

The functoriality of $H^1(M)$ and $H^2(M)$ is a little more complicated. For our purposes the following observations for $\beta\colon M \longrightarrow M'' \in \mathrm{Mor}\, S(\underline{G})$ suffice:

1) $H^1(M)_S$ defines a functor $H^1(\ )_S\colon S(\underline{G}) \longrightarrow \underline{S}^{\cdot}$ ,

2) $H^2(\beta)$ exists in $\underline{S}^{\cdot}$ if $\beta$ is onto, and

3) $H^1(\beta)$, $H^2(\beta)$ exist in ${}^G\underline{S}^G$ if $M \in S(\underline{Ab})$ and $\beta(M)$ is contained in the center of $M''$ .

In either 2) or 3) $H^2(\beta)(H^2(M)^{!}_{S}) \subseteq H^2(M'')^{!}_{S}$ , and in 3), given $g \in H^i(M)$, $x \in H^i(M'')$, $H^i(\beta)(g) \cdot x = x \cdot H^i(\beta)(g)$.

The proof is based on the following diagram:

$$\begin{array}{ccccccccc}
M(U) & \overset{j_0}{\rightarrowtail} & CM(U) & \overset{\pi_0}{\longrightarrow} & \underline{Q}_\ell(j_0)(U) & \overset{j_1}{\rightarrowtail} & C\underline{Q}_\ell(j_0)(U) & \overset{\pi_1}{\longrightarrow} & \underline{Q}_r(j_1)(U) \\
\downarrow{\scriptstyle\beta} & & \downarrow{\scriptstyle C\beta} & & \downarrow{\scriptstyle\beta_1} & & \downarrow{\scriptstyle C\beta_1} & & \downarrow{\scriptstyle\beta_2} \\
M''(U) & \overset{j''_0}{\rightarrowtail} & CM''(U) & \overset{\pi''_0}{\longrightarrow} & \underline{Q}_\ell(j''_0)(U) & \overset{j''_1}{\rightarrowtail} & C\underline{Q}_\ell(j''_0)(U) & \overset{\pi''_1}{\longrightarrow} & \underline{Q}_r(j''_1)(U)
\end{array}\ .$$

In general $\beta_1 \in \mathrm{Mor}\, S(\underline{S}^{\cdot})$. However, $\beta_1\pi_0 = \pi''_0 C\beta \in \mathrm{Mor}\, S(\underline{S}^G)$, and so it induces a map $H^1(\beta) = \beta^1_*\colon H^1(M) \longrightarrow H^1(M'') \in \mathrm{Mor}\, \underline{S}^{\cdot}$ . In either 2) or 3), $\beta_1$ (and so $C\beta_1$) is a map in $S({}^G\underline{S}^G)$ by Proposition 2.5 since in 3), $\mathrm{St}_{CM_G(j_0)}(\underline{Q}_\ell(j_0)) = j_0(M) = \mathrm{St}_{{}_G CM(j_0)}(\underline{Q}_\ell(j_0))$,

and $\beta(M)$ is contained in the center of $M''$ . Thus $C\beta$ takes stabilizers into stabilizers as required. Moreover, in this case $H^1(\beta) \in \mathrm{Mor}\, {}^G\underline{S}^G$ since $\pi_0(CM(U))$ is the stabilizer of $H^1(M)_S$ in ${}_G\underline{Q}_\ell(j_0)(U)$ and $\underline{Q}_\ell(j_0)_G(U)$ which are the groups acting on $H^1(M)_S$ . The action of $H^1(M) \in \underline{Ab}$ on $H^1(M")_S$ is independent of the side it acts on since this is true of the action of $\underline{Q}_\ell(j_0)(U) \in \underline{Ab}$ on $\underline{Q}_\ell(j_0")_S(U)$. Since $\beta_1$ and so $C\beta_1$ are onto in $S(\underline{S}^G)$ in 2) and in 3) $\underline{Q}_\ell(j_0) \in S(\underline{Ab})$ with $\beta_1(\underline{Q}_\ell(j_0))$ contained in the center of $\underline{Q}_\ell(j_0")_G$ and ${}_G\underline{Q}_\ell(j_0")$ and its action on $\underline{Q}_\ell(j_0")_S$ is independent of the side, we can apply Proposition 2.5 and the above argument again to finish the proof. Note that if $\beta: M \longrightarrow M" \in \text{Mor}\, S(\underline{G})$ and $\beta(M)$ is contained in the center of $M"$ , then $H^1(\beta)$ and $H^2(\beta)$ exist in $\underline{S}^{\cdot}$ since $\beta$ can be factored as a map which is onto followed by a monomorphism into the center of $M"$ .

The first exactness statement is needed for the comparison theorem.

<u>Proposition 3.1</u>

Let $M' \overset{\alpha}{\rightarrowtail} M \overset{\beta}{\twoheadrightarrow} M"$ be a short exact sequence in $S(\underline{S}^G)$ with $M', M \in S(\underline{G})$. Then there is a natural transformation $\delta^0: H^0(M") \longrightarrow H^1(M')$ in $\underline{S}^{\cdot}$ giving a sequence

$$1 \longrightarrow H^0(M') \xrightarrow{\alpha_*^0} H^0(M) \xrightarrow{\beta_*^0} H^0(M") \xrightarrow{\delta^0} H^1(M') \xrightarrow{\alpha_*^1} H^1(M)$$

which is exact in $\underline{S}^G$ at $H^0(M')$ and $H^0(M)$, exact in ${}^G\underline{S}$ at $H^0(M")$, and exact in $\underline{S}^{\cdot}$ at $H^1(M')$.

<u>Proof</u>: The proof is based on the following diagram where the maps are the obvious ones:

$$\begin{array}{ccccc}
M'(U) & \overset{\alpha}{\rightarrowtail} & M(U) & \overset{\beta}{\longrightarrow} & M''(U) \\
{\scriptstyle j_0'}\downarrow & & {\scriptstyle j_0}\downarrow & & {\scriptstyle j_0''}\downarrow \\
CM'(U) & \overset{\overline{\alpha}}{\rightarrowtail} & CM(U) & \overset{\overline{\beta}}{\twoheadrightarrow} & CM''(U) \\
{\scriptstyle \pi_0'}\downarrow & & {\scriptstyle \pi_0}\downarrow & & {\scriptstyle \pi_0''}\downarrow \\
\underline{Q}_\ell(j_0')(U) & \overset{\alpha_1}{\rightarrowtail} & \underline{Q}_\ell(j_0)(U) & \overset{\beta_1}{\longrightarrow} & \underline{Q}_\ell(j_0'')(U)
\end{array} .$$

The first row is exact in $\underline{S}^G$ by 2.2. The second row is a short exact sequence of sets in $\underline{S}^G$ by 2.7. The third row comes from a short exact sequence of sets in $S(\underline{S}^G)$ and $\alpha_1$ is monic by 2.8 since $\alpha_1\pi_0' \in \text{Mor}\, S(\underline{S}^G)$. Since $\underline{Q}_\ell(j_0')_G$ acts transitively, 2.2 shows that the third row and all of the columns are exact in $\underline{S}^{\cdot}$, and $Z^1(M')_S = \underline{Q}_\ell(j_0')(U)$.

Define $\delta^0 : M''(U) \longrightarrow H^1(M') \in \text{Mor}\, \underline{S}^{\cdot}$ in the usual way. Thus if $x \in M''_S(U) = H^0(M'')$, choose $y \in CM(U)$ with $\overline{\beta}(y) = j_0''(x)$. Since $\beta_1(\pi_0(y)) = e$ , there is a unique $z \in Z^1(M')_S$ with $\alpha_1(z) = \pi_0(y)$. Let $\delta^0(x) = \{z\}$. It is immediate that $\delta^0$ is independent of the choice of $y$ and is natural for maps

$$\begin{array}{ccccc}
M' & \overset{\alpha}{\rightarrowtail} & M & \overset{\beta}{\twoheadrightarrow} & M'' \\
{\scriptstyle f'}\downarrow & & {\scriptstyle f}\downarrow & & {\scriptstyle f''}\downarrow \\
N' & \overset{\overline{\alpha}}{\rightarrowtail} & N & \overset{\overline{\beta}}{\twoheadrightarrow} & N''
\end{array}$$

where $f', f \in \text{Mor}\, S(\underline{G})$ and $f''$ is the induced map. We already have exactness in $\underline{S}^G$ at $H^0(M')$ and $H^0(M)$. Since we are forming left quotients in the third row, if $g \in {}_GH^0(M)$ and $x \in H^0(M'')$, then $\delta^0(x) = \delta^0(\beta(g) \cdot x)$. On the other hand, if $\delta^0(x) = \delta^0(x')$ for $x, x' \in H^0(M'')_S = M''_S(U)$, then we can choose $y, y' \in CM(U)$ with $\pi_0(y) = \pi_0(y')$ and $\overline{\beta}(y) = j_0''(x)$, $\overline{\beta}(y') = j_0''(x')$. Since $\pi_0(y' \cdot y^{-1}) = e$ . $y' \cdot y^{-1} \in M(U)$, and so $\beta(y' \cdot y^{-1}) \cdot x = x'$ as desired. Exactness in $\underline{S}^{\cdot}$ at $H^1(M')$ follows from the transitive action of $CM_G(U)$ on $CM_S(U)$ and the exactness in $\underline{S}^{\cdot}$ of the third column. ∎

The next theorem provides a boundary map $\delta^1\colon H^1(M'') \longrightarrow H^2(M')$ for a central extension of sheaves of groups.*

<u>Theorem 3.2</u>

Let $M' \overset{\alpha}{\rightarrowtail} M \overset{\beta}{\twoheadrightarrow} M''$ be a short exact sequence of sheaves of groups with $\alpha(M')$ contained in the center of $M$ . Then $\delta^0$ is a homomorphism, and there is a natural transformation $\delta^1\colon H^1(M'') \longrightarrow H^2(M')$ in $\underline{S}^{\cdot}$ such that

1) $1 \longrightarrow H^0(M') \xrightarrow{\alpha_*^0} H^0(M) \xrightarrow{\beta_*^0} H^0(M'')$
$\xrightarrow{\delta^0} H^1(M') \xrightarrow{\alpha_*^1} H^1(M) \xrightarrow{\beta_*^1} H^1(M'')$

is an exact sequence of groups at the first three terms and exact in $\underline{S}^G$ at the other terms.

2) $H^1(M) \xrightarrow{\beta_*^1} H^1(M'') \xrightarrow{\delta^1} H^2(M')$ is an exact sequence of pointed sets.

---

* Using a slightly different definition of $\underline{Q}_r(\alpha)$, the existence of $\delta^1$ for a short exact sequence of sheaves of groups can be shown. However, with this definition only 1) - 3) (suitably modified in the non-central case) of the exactness properties below can be proven.

3) $H^1(M'') \xrightarrow{\delta^1} H^2(M') \xrightarrow{\alpha^2_*} H^2(M)$ is an exact sequence of sets with neutral elements, i.e. $\delta^1(H^1(M'')_S) = (\alpha^2_*)^{-1}(H^2(M)^{\prime}_S)$.

4) $H^2(M') \xrightarrow{\alpha^2_*} H^2(M) \xrightarrow{\beta^2_*} H^2(M'')$ is an exact sequence of sets with neutral elements under the right group action; that is, for all $\bar{g} \in H^2(M')_G$, $\bar{x} \in H^2(M)_S$, we have $\beta^2_*(\bar{x}) = \beta^2_*(\bar{x} \cdot \alpha^2_*(\bar{g}))$, and if $\beta^2_*(\bar{x}) \in H^2(M'')^{\prime}_S$, then there is $\bar{g} \in H^2(M')_G$ with $\bar{x} \cdot \alpha^2_*(\bar{g}) \in H^2(M)^{\prime}_S$.

<u>Proof</u>: The proof is based on the following diagram where the maps are the obvious ones:

$$
\begin{array}{ccccc}
M'(U) & \xrightarrow{\alpha} & M(U) & \xrightarrow{\beta} & M''(U) \\
\downarrow{\scriptstyle j'_0} & & \downarrow{\scriptstyle j_0} & & \downarrow{\scriptstyle j''_0} \\
CM'(U) & \xrightarrow{\bar{\alpha}} & CM(U) & \xrightarrow{\bar{\beta}} & CM''(U) \\
\downarrow{\scriptstyle \pi'_0} & & \downarrow{\scriptstyle \pi_0} & & \downarrow{\scriptstyle \pi''_0} \\
\underline{Q}_\ell(j'_0)(U) & \xrightarrow{\alpha_1} & \underline{Q}_\ell(j_0)(U) & \xrightarrow{\beta_1} & \underline{Q}_\ell(j''_0)(U) \\
\downarrow{\scriptstyle j'_1} & & \downarrow{\scriptstyle j_1} & & \downarrow{\scriptstyle j''_1} \\
C\underline{Q}_\ell(j'_0)(U) & \xrightarrow{\bar{\alpha}_1} & C\underline{Q}_\ell(j_0)(U) & \xrightarrow{\bar{\beta}_1} & C\underline{Q}_\ell(j''_0)(U) \\
\downarrow{\scriptstyle \pi'_1} & & \downarrow{\scriptstyle \pi_1} & & \downarrow{\scriptstyle \pi''_1} \\
\underline{Q}_r(j'_1)(U) & \xrightarrow{\alpha_2} & \underline{Q}_r(j_1)(U) & \xrightarrow{\beta_2} & \underline{Q}_r(j_1)(U)
\end{array}
$$

The first row is an exact sequence of groups, and the second row is a central extension of $CM''(U)$ by $CM'(U)$. The third row comes from short exact sequence of sets in $S(\underline{S}^G)$ by 2.8 (regard $M'' = \underline{Q}_r(\alpha)$) and so is an exact sequence of sets in $\underline{S}^G$ by 2.2. Moreover, $\alpha_1, \beta_1 \in \mathrm{Mor}\, S({}^G\underline{S}^G)$,

$\beta_1 \in \mathrm{Mor}\ S(\underline{S}^G)$ is onto, and the sheaves of groups acting on the right act transitively. Thus by 2.7 the fourth row is a short exact sequence of sets in $\underline{S}^G$, $\bar{\alpha}_1$, $\bar{\beta}_1 \in \mathrm{Mor}\ {}^G\underline{S}^G$, the groups acting on the right act transitively, and $\bar{\beta}_1 \in \mathrm{Mor}\ \underline{S}^G$ is onto. Then 2.8 shows that the fifth row comes from a short exact sequence of sets in $S(\underline{S}^G)$ (since $\underline{Q}_r(j_1') \in S(\underline{Ab})$) and $\alpha_2$, $\beta_2 \in \mathrm{Mor}\ S({}^G\underline{S}^G)$ by earlier remarks. So by Proposition 2.2 it is an exact sequence of sets in $\underline{S}^G$. Moreover, the first two arrows in each of the three columns define exact sequences of sets in ${}^G\underline{S}$ as do the last two arrows in the first column.

Since the sheaves of groups acting on the left, right act transitively on the corresponding sheaves of sets defining the first, third row respectively, the sets in these rows are the 0 and 1 cocyles. Moreover, the last two arrows in the two columns on the right form exact sequences in $\underline{S}^{\cdot}$ by 2.2. Finally for convenience we will identify $H^1(M')$, $H^2(M') \in \underline{Ab}$ with $H^1(M')_S$, $H^2(M')_S$ or $H^1(M')_G$, $H^2(M')_G$, etc.

Since $j_0''$, $\bar{\beta} \in \mathrm{Mor}\ \underline{G}$ and $\pi_0 \in \mathrm{Mor}\ \underline{S}^G$, it is clear that $\delta^0$ is a homomorphism. A straightforward diagram chase using $\pi_0'' \in \mathrm{Mor}\ \underline{S}^G$ gives exactness of sets in $\underline{S}^G$ at $H^1(M')$. Exactness at $H^1(M)$ follows since $\beta_1(x \cdot \alpha_1(g)) = \beta_1(x)$ for all $g \in \underline{Q}_\ell(j_0')(U)$ and $x \in Z^1(M)_S$. If $\beta_*^1(\{x\}) = \beta_*^1(\{y\})$, then we can choose representatives $x, y \in Z^1(M)_S$ for $\{x\}$, $\{y\}$ respectively such that

$\beta_1(x) = \beta_1(y)$ since $\overline{\beta}$, $\beta_1$, $\pi_0$, $\pi''_0 \in \mathrm{Mor}\ \underline{S}^G$ and $\overline{\beta}$ is onto. Then there is $g \in \underline{Q}_\ell(j'_0)(U)$ with $x = y \cdot \alpha_1(g)$. Hence $\{x\} = \{y\} \cdot \alpha^1_*(\{g\})$.

Define $\delta^1 : H^1(M'') \longrightarrow H^2(M')$ in the usual way. Thus if $x \in Z^1(M'')_S$, choose $y \in C\underline{Q}_\ell(j_0)_S(U)$ with $\overline{\beta}_1(y) = j''_1(x)$. Since the fifth row is exact in $\underline{S}^\cdot$ and $\pi''_1 j''_1$ is trivial, there is a unique $z \in Z^2(M')_S$ with $\alpha_2(z) = \pi_1(y)$. Let $\delta^1(\{x\}) = \{z\}$. Altering the choice of $y$ does not alter the cohomology class of $z$ since $\pi_1\overline{\alpha}_1 \in \mathrm{Mor}\ {}^G\underline{S}^G$ and the action of $C\underline{Q}_\ell(j'_0)(U)$ on $C\underline{Q}_\ell(j_0)_S(U)$ is independent of the side it acts on. Altering the cohomology class of $x$ alters $y$ by an element in $j_1\pi_0(CM_G(U))$ and so doesn't change $\{z\}$. The exactness of $\cdot \xrightarrow{j_1} \cdot \xrightarrow{\pi_1} \cdot$ and the above remark about the action of $C\underline{Q}_\ell(j'_0)(U)$ on $C\underline{Q}_\ell(j_0)_S(U)$ immediately gives exactness in $\underline{S}^\cdot$ at $H^1(M'')$. 3) follows immediately by using the exactness of $\cdot \xrightarrow{j''_1} \cdot \xrightarrow{\pi''_1} \cdot$ in $\underline{S}^\cdot$.

Since $\beta_2$ factors through $\underline{Q}_r(\alpha_2)$ the first part of 4) is trivial. Suppose $x \in Z^2(M)_S = \mathcal{O}_e({}_G\underline{Q}_r(j_1))(U)$ represents $\{x\} \in H^2(M)_S$ with $\beta^2_*(\{x\}) = \{\beta_2(x)\} \in H^2(M'')'_S$. Then, since $\overline{\beta}_1$ is onto in $\underline{S}^\cdot$, there is $y' \in C\underline{Q}_\ell(j_0)(U)$ with $\beta_2\pi_1(y') = \beta_2(x)$. Let $y = \pi_1(y')$. Since $\beta_2(y) = \beta_2(x)$, there is $g \in \underline{Q}_r(j'_1)(U)$ with $x \cdot \alpha_2(g) = y$. Thus $\{x\} \cdot \alpha^*_2(\{g\}) = \{y\} \in H^2(M)'_S$ as desired. ∎

Finally we must relate this cohomology theory to

the usual ones. If $M \in P(\underline{G})$ and $\{U_i \longrightarrow U\} \in \text{Cov}\,\underline{T}$, let

$$\check{Z}^1(\{U_i \longrightarrow U\};M) = \{g \in \prod_{I\times I} M(U_{i_1} \underset{U}{\times} U_{i_2}) \mid p_2^*(g)$$

$$= p_1^*(g) \cdot p_3^*(g) \in \prod_{I^3} M(U_{i_1} \underset{U}{\times} U_{i_2} \underset{U}{\times} U_{i_3})\},$$

and define $\check{H}^1(\{U_i \longrightarrow U\};M) = \check{Z}^1(\{U_i \longrightarrow U\};M)/\sim$ where $g_1 \sim g_2$ if there is $h \in \prod_I M(U_i)$ with $g_1 = p_1^*(h) \cdot g_2 \cdot p_2^*(h^{-1})$. As usual $p_i^*$ comes from the projection map onto all but the $i^{\underline{th}}$ factor. Given $\Phi\colon \{V_j \longrightarrow U\} \longrightarrow \{U_i \longrightarrow U\} \in \text{Mor}\, J_U$, there is an obvious induced map $\Phi_*^1\colon \check{H}^1(\{U_i \longrightarrow U\};M) \longrightarrow \check{H}^1(\{V_j \longrightarrow U\};M)$ which depends only on the domain and range [10, Proposition 1.2]. Let $\check{H}^1(U;M) = \varinjlim_{J_U^0} \check{H}^1(\{U_i \longrightarrow U\};M)$. Then it defines a functor $\check{H}^1(U;\ )\colon P(\underline{G}) \longrightarrow \underline{S}^{\cdot}$ which together with $\check{H}^0(U;M)$ give a different cohomology theory. In particular for $M \in S(\underline{G})$, $\check{H}^1(U;M)$ is the set of "locally trivial principal homogeneous spaces for $M$ in the topology over $U$."

Theorem 3.3

1) Let $M \in S(\underline{Ab})$. Then $H^n(M)$ is the $n^{\underline{th}}$ derived functor of $\Gamma_U\colon S(\underline{Ab}) \longrightarrow \underline{Ab}$ evaluated at $M$ where $\Gamma_U(A) = A(U)$ for all $A \in S(\underline{Ab})$.

2) $\check{H}^1(U;M)$ is naturally isomorphic to $H^1(M)_S$.

Proof: 1) Since the canonical resolution is indeed a resolution in the usual sense for $M \in S(\underline{Ab})$, it suffices to show that $CM \in S(\underline{Ab})$ is flask

[1, II, Corollary 4.4]. Given $\{U_i \xrightarrow{\varphi_i} U\} \in \text{Cov}\ \underline{T}$, we must show that the complex $S^*(CM)(\{U_i \xrightarrow{\varphi_i} U\})$ defined just before Theorem 2.7 where the boundary map $d^n$ is $\sum_{i=1}^{n+1} (-1)^i p_i^*$ has trivial cohomology. If $x \in \text{Ker}\ (d^n) \subseteq \Pi_{I^{n+1}} CM(U_{i_1} \underset{U}{\times} \cdots \underset{U}{\times} U_{i_{n+1}})$, then

$p_1^*(x) = \sum_{i=2}^{n+1} (-1)^i p_i^*(x)$. Theorem 2.7 shows that

$p_{1*}p_i^* = p_{i-1}^* p_{1*}$ for $i > 1$ where in the right hand term

$p_{i-1}^*: \Pi_{I^n} CM(U_{i_1} \underset{U}{\times} \cdots \underset{U}{\times} U_{i_n}) \longrightarrow \Pi_{I^{n+1}} CM(U_{i_1} \underset{U}{\times} \cdots \underset{U}{\times} U_{i_{n+1}})$

and $p_{1*}$ goes in the other direction. Moreover, $p_{1*}p_1^*(x) = x$. Hence $x = p_{1*}p_1^*(x) = \sum_{i=2}^{n+1} (-1)^i p_{1*}p_i^*(x)$

$= \sum_{i=2}^{n+1} (-1)^i p_{i-1}^*(x) p_{1*}(x) = d^{n-1}(-p_{1*}(x))$ as desired.

2) For $M \in S(\underline{G})$, consider $M \overset{j}{\rightarrowtail} CM \xrightarrow{\pi} \underline{Q}_r(j)$. Since $M$ and $CM$ are groups $Q_r^p(j)$ satisfies (+) and so $\check{H}^0(U;Q_r^p(j)) = \check{H}^0(U;\underline{Q}_r(j))$ by definition of $\#$. It is an easy argument to prove the results of Proposition 3.1 for a short exact sequence of presheaves in $P(\underline{S}^G)$ using $\check{H}^0(U;M)$ and $\check{H}^1(U;M)$ instead of $H^0(M)$ and $H^1(M)$ (see [10] for the definition of $\delta^0$ and most of the argument or [5, Chapter I, Theorem 3.1]). Thus if we can show that $\check{H}^1(U;CM)$ and $H^1(CM)$ are trivial for $M \in S(\underline{G})$, then the above sequence shows that

$\check{H}^1(U;M) \cong {}_GCM(U)\backslash \check{H}^0(U;Q_r^p(j))_S \cong {}_GCM(U)\backslash \underline{Q}_r(j)_S(U) \cong H^1(M)_S$

as desired.

But for any $\{U_i \longrightarrow U\} \in \mathrm{Cov}\,\underline{T}$ , $\check{H}^1(\{U_i \longrightarrow U\};CM) = e$ , for given $x \in \prod_{I\times I} CM(U_{i_1} \times_U U_{i_2})$ such that $p_2^*(x) = p_1^*(x) \cdot p_3^*(x)$, Theorem 2.7 shows that

$$x = p_{1*}p_1^*(x) = p_{1*}p_2^*(x) \cdot p_{1*}p_3^*(x)^{-1}$$

$$= p_1^*(p_{1*}(x)) \cdot p_2^*(p_{1*}(x))^{-1} .$$ Thus $x \sim e$ since $p_{1*}(x) \in \prod_I CM(U_i)$. Moreover $H^1(CM)$ is computed from $CM(U) \overset{j}{\rightarrowtail} C(CM)(U) \overset{\pi}{\longrightarrow} \underline{Q}_\ell(j)(U)$. But $j \in \mathrm{Mor}\, S(\underline{G})$ has a left inverse $\bar{j}: C(CM) \longrightarrow CM$ by GR4. Hence $C(CM)(U)$ is a semi-direct product of $(\mathrm{Ker}\,\bar{j})(U)$ and $(CM)(U)$, and a straightforward argument shows that the composite $(\mathrm{Ker}\,\bar{j})(U) \longrightarrow C(CM)(U) \longrightarrow Q_\ell^p(j)(U)$ is a set isomorphism. Since $\mathrm{Ker}\,\bar{j}$ is a sheaf, this shows that $\pi$ is onto. Thus $H^1(CM)_S = e$ as desired. ∎

## REFERENCES

[1] Artin, M., "Grothendieck Topologies", mimeographed notes, Harvard University, Cambridge, Mass. (1962).

[2] Artin, H. and Grothendieck, A., "Cohomologie étale des Schemas", Séminaire de Géometrie Algébrique, Institut des Hautes Etudes Scientifique, (1963-64).

[3] Dedecker, P., *C. R. Acad. Sci. Paris*, V. 258; 1117-1120, (1964), V. 259; 2054-2057, (1964), V. 260; 4137-4139, (1965).

[4] Eilenberg, S. and Moore, J. C., "Adjoint Functors and Triples", *Illinois J. Math.*, V. 9; 381-398, (1965).

[5] Garfinkel, G., "Amitsur Cohomology and an Exact Sequence Involving Pic and the Brauer Group", Dissertation, Cornell University, Ithaca, New York, (1968).

[6] Giraud, J., "Cohomologie Non Abelienne", mimeographed notes, Columbia University, New York, New York, (1965).

[7] Godement, R., *Théorie des Faisceaux*, Hermann, Paris (1958).

[8] Grothendieck, A., "Le Groupe de Brauer I; Algèbres d'Azumaya et interprétations diverses", Séminaire Bourbaki, No. 290, (May 1965).

[9] Grothendieck, A., "Sur Quelques Points d'Algebre Homologique", *Tohoku Math. J.*, (Series II) V. 9; 119-221, (1957).

[10] Hoobler, R., "A Generalization of the Brauer Group and Amitsur Cohomology", Dissertation, University of California, Berkeley, California, (1966).

[11] Serre, J. P., "Cohomologie Galoisienne", (Lecture Notes in Mathematics, No. 5) Springer-Verlag, (1965).

[12] Springer, T. A., "Nonabelian $H^2$ in Galois Cohomology", *Proc. Sympos. Pure Math.*, (Boulder, Col., 1965) V. 9; 164-182, AMS, (1966).

# FONCTEURS DERIVES ET K-THEORIE

par

Max Karoubi

On expose dans cet article certains résultats obtenus en appliquant des techniques connues d'algèbre homologique à la K-théorie. Des résultats plus complets accompagnés de leurs démonstrations paraitront prochainement [4].

Pour ne citer que cet example, on sait que la cohomologie à valeurs dans un faisceau $H^n(X;F)$, $n \geq 0$, $F$ variable, est caractérisée par les axioms suivants:

$$H^0(X;F) = \Gamma(X;F), \tag{1}$$

groupe des sections globales du faisceau $F$.

$$H^n(X;F) = 0, \forall n > 0 \tag{2}$$

si le faisceau $F$ est flasque. A toute suite exacte de faisceaux

$$0 \longrightarrow F' \longrightarrow F \longrightarrow F'' \longrightarrow 0 \tag{3}$$

est associée une suite exacte de cohomologie

$$\cdots \longrightarrow H^{n-1}(X;F) \longrightarrow H^{n-1}(X;F'') \xrightarrow{\delta^{n-1}} H^n(X;F') \longrightarrow H^n(X;F) \longrightarrow \cdots$$

Nous avons essayé d'adapter ce formalisme à la K-théorie des catégories additives et, plus généralement, à celle des catégories en groupes de Banach (Définition 1.2). On a pu ainsi aboutir à une définition axiomatique des foncteurs $K^n$, $n \geq 0$, de telles catégories. Dans certains cas, la périodicité de ces foncteurs a pu être demontrée (Théorème 2.3).

Cet article est divisé en deux parties. Dans la première nous développons la notion de "suite exacte de catégories" en nous inspirant de la théorie des opérateurs complètement continus dans les espaces de Hilbert. Dans la seconde partie nous définissons les "catégories flasques" et donnons une caractérisation axiomatique de la K-théorie semblable à celle de la cohomologie à valeurs dans un faisceau.

## I CATEGORIES EN GROUPES DE BANACH SUITES EXACTES

Definition 1.1

Soit $M$ un groupe abélien. Une quasi-norme sur $M$ est une application de $M$ dans $\mathbb{R}^+$ notée $x \longmapsto \|x\|$ jouissant des propriétés suivantes:

$$\|x\| = 0 \Longleftrightarrow x = 0 \quad (1)$$

$$\|x + y\| \leq \|x\| + \|y\| \quad (2)$$

$$\|-x\| = \|x\| \quad (3)$$

On appelle groupe quasi-normé un groupe abélien $M$ muni d'une quasi-norme. Le groupe $M$ est alors de manière naturelle un espace métrique pour la distance invariante par translation $d(x,y) = \|x-y\|$. Réciproquement tout groupe abélien muni d'une distance invariante par translation est quasi-normé si on pose $\|x\| = d(x,0)$. Un groupe de Banach est un groupe quasi-normé complet pour la distance definie par la quasi-norme.

Exemples. Un espace de Banach est évidemment un groupe de Banach. Il en est de même d'un groupe abélien quelconque muni de la quasi-norme suivante (dite "discrète"):

$$\|x\| = 0 \quad \text{si} \quad x = 0$$

$$\|x\| = 1 \quad \text{si} \quad x \neq 0$$

Un espace de Fréchet dont la topologie est définie par une famille dénombrable de semi-normes $p_i$ est aussi un groupe de Banach pour la quasi-norme

$$\|x\| = \sum 2^{-i} \operatorname{Inf}(1, p_i(x))$$

Tous les sorites développés pour les espaces de Banach se démontrent aussi bien pour les groupes de Banach. On pourra par exemple faire le quotient d'un groupe de Banach par un sous-groupe fermé. Si $M$ et $N$ sont deux groupes de Banach, les applications bornées de $M$ dans $N$ forment un groupe de Banach pour la quasi-norme

$$\|f\| = \sup_{x \neq 0} \frac{\|f(x)\|}{\|x\|}$$

Definition 1.2

Une catégorie en groupes de Banach est une catégorie additive $\tau$ où $\mathrm{Hom}_\tau(M,N)$ est muni d'une structure de groupe de Banach de telle sorte que, quels que soient les objets $M$, $N$ et $P$ de $\tau$ et les morphismes $u: M \longrightarrow N$, $v: N \longrightarrow P$, on ait l'inégalité $\|v \circ u\| \leq C\|v\| \times \|u\|$, $C$ étant une constante ne dépendant que de $M$, $N$ et $P$.

Exemples. L'exemple le plus important en K-théorie est sans doute celui de la catégorie $\tau = \xi(X)$ des fibrés vectoriels (réels ou complexes) de rang fini sur un espace compact X. En effet, si E et F sont deux fibrés vectoriels, $\mathrm{Hom}_\tau(E,F)$ s'identifie à l'espace de Banach des sections du fibré en homomorphismes HOM(E,F). Plus généralement, toute catégorie prébanachique dans le sens de [3] est une catégorie en groupes de Banach. Enfin une simple catégorie additive en est aussi un exemple, $\mathrm{Hom}_\tau(M,N)$ étant muni de la quasi-norme discrète.

Remarque 1. Dans une catégorie en groupes de Banach, l'application définie par la composition des morphismes

$$\mathrm{Hom}_\tau(M,N) \times \mathrm{Hom}_\tau(N,P) \longrightarrow \mathrm{Hom}_\tau(M,P)$$

est continue.

Remarque 2. Comme pour les espaces de Banach, on convient d'identifier deux quasi-normes sur un groupe abélien lorsque celles-ci sont équivalentes. La même remarque s'applique aux catégories en groupes de Banach.

Si $\mathcal{D}$ est une catégorie quelconque et si E et F sont deux objets de $\mathcal{D}$ on appelle morphisme direct de E dans F la donnée de deux flèches $s: E \longrightarrow F$ et $p: F \longrightarrow E$ telles que $p \cdot s = \mathrm{Id}_E$. On voit aisément que s (resp. p) est un monomorphisme (resp. un épimorphisme) et que les morphismes directs sont les flèches d'une catégorie dont les objets sont les objets de $\mathcal{D}$. On notera $(s,p): E \longrightarrow F$

une telle flèche. Supposons maintenant que $\mathcal{D}$ soit une catégorie en groupes de Banach et considérons une sous-catégorie additive pleine $\tau$ de $\mathcal{D}$. Si $E$ est un objet de $\mathcal{D}$ on appelle $\tau$-filtration de $E$ la donnée d'objets $E_i$ de $\tau$, $i \in I$ ensemble d'indices quelconque, et de $\mathcal{D}$-morphismes directs $f_i = (s_i, p_i): E_i \longrightarrow E$ satisfaisant à l'axiome suivant:

F 1. Si $E_i$ et $E_j$ sont deux objets de la filtration de $E$, il existe un troisième objet $E_k$ de la filtration qui rend commutatif le diagramme

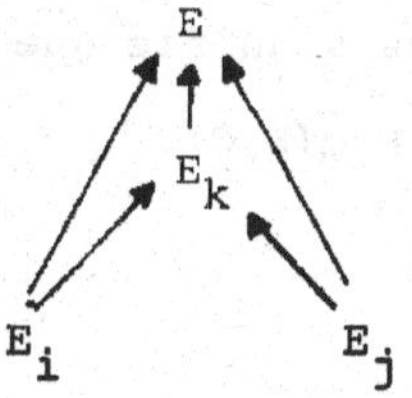


Soient $\{E_i\}$ et $\{E'_{i'}\}$ deux filtrations de $E$. On dira que la filtration $\{E_i\}$ est moins fine que la filtration $\{E'_{i'}\}$ si, pour tout indice $i$, on peut trouver un indice $i'$ et des morphismes directs $h_{i'i}$ qui rendent commutatif le diagramme

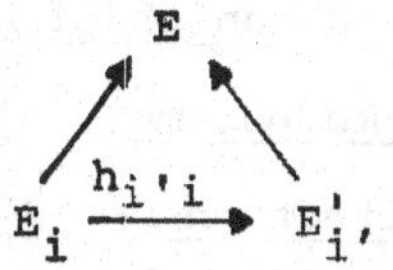


On dira que les deux filtrations sont équivalentes si l'une est plus fine que l'autre et réciproquement.

*Diagramme Commutatif à $\varepsilon$-près.* Soit de nouveau $\mathcal{D}$ une catégorie en groupes de Banach et soit $\Delta$ un diagramme quelconque dans $\mathcal{D}$. Nous dirons que le diagramme $\Delta$ est commutatif à $\varepsilon$-près si, pour tout couple d'objets $(E,F)$ de ce diagramme et pour tout couple de morphismes $(f,g)$ joignant $E$ à $F$ dans ce diagramme on a $\|f - g\| < \varepsilon$ dans $\mathrm{Hom}(E,F)$. Par exemple, dire que le diagramme

$$\begin{array}{ccc} E & \xrightarrow{f} & H \\ {\scriptstyle h}\downarrow & & \downarrow{\scriptstyle g} \\ G & \xrightarrow{k} & F \end{array}$$

est commutatif à $\varepsilon$-près signifie que l'on a l'inégalité $\|g\circ f - k\circ h\| < \varepsilon$ dans $\mathrm{Hom}(E,F)$.

*Definition 1.3*

*Soit $\mathcal{D}$ une catégorie en groupes de Banach et soit $\tau$ une sous-catégorie additive pleine de $\mathcal{D}$. Une $\tau$-filtration sur la catégorie $\mathcal{D}$* (1) *est la donnée, pour tout objet $E$ de $\mathcal{D}$, d'une $\tau$-filtration $E_i$, $i \in I$, sur $E$ ($I$ ne dépendant que de $E$) vérifiant les axiomes suivants:*

F 2. *Soient $E$ un objet de $\tau$, $F$ un objet de $\mathcal{D}$ et $f: E \longrightarrow F$ un $\mathcal{D}$-morphisme. Alors, $\forall \varepsilon > 0$, il existe un objet $F_j$ de la filtration de $F$ et un $\tau$-morphisme $g_j: E \longrightarrow F_j$ tels que le diagramme*

---

(1) On dit aussi que $\tau$ est sous-catégorie idéale de $\mathcal{D}$.

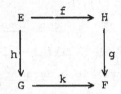

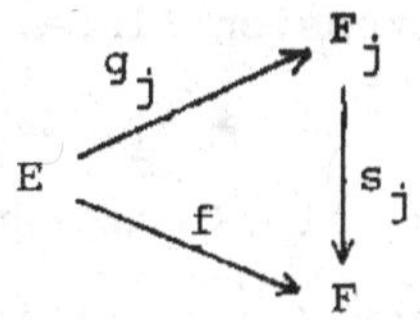


soit commutatif à $\varepsilon$-près.

F 3. Soient E un objet de $\mathcal{D}$, F un objet de $\tau$ et f: E $\longrightarrow$ F un $\mathcal{D}$-morphisme. Alors, $\forall\varepsilon > 0$, il existe un objet $E_i$ de la filtration de E et un $\tau$-morphisme $g_i$: $E_i \longrightarrow$ F tel que le diagramme

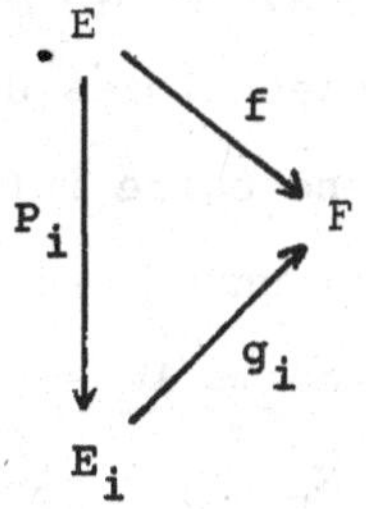


soit commutatif à $\varepsilon$-près.

F 4. Si E et F sont des objets de $\mathcal{D}$, la filtration $E_i \oplus F_j$ de $E \oplus F$ est équivalente à la filtration $(E \oplus F)_k$.

Exemples

1. Soit $\mathcal{H}$ la catégorie des espaces de Hilbert et soit $\xi$ la catégorie des espaces de dimension finie. On peut alors considérer $\mathcal{H}$ comme $\xi$-filtrée de la manière suivante: pour tout espace de Hilbert E, $E_i$ sera la collection

de ses sous-espaces de dimension finie, $s_i: E_i \longrightarrow E$ étant l'injection canonique, $p_i: E \longrightarrow E_i$ la projection orthogonale. Plus généralement, si $X$ est un espace compact on voit aisément (en utilisant une partition de l'unité) que $\xi_T(X)$ est une sous-catégorie idéale de $\mathcal{H}_T(X)$, $\xi_T(X)$ (resp. $\mathcal{H}_T(X)$) désignant la catégorie des fibrés vectoriels triviaux de dimension finie (resp. hilbertiens).

2. Soit $A$ un anneau de Banach (i.e., un groupe de Banach muni d'une multiplication telle que $\|xy\| \leq C\|x\|\times\|y\|$ ; exemples: une algèbre de Banach ou un anneau discret) et soit $\mathcal{L}(A)$ la catégorie des modules libres de type fini sur $A$. Soit $(M_1, \ldots, M_n, \ldots)$ une suite infinie d'objets de $\mathcal{L}(A)$ les $M_i$ étant choisis parmi un nombre fini d'objets de $\mathcal{L}(A)$. On définit leur $L^1$-somme comme le sous-ensemble du produit $M_1 \times \ldots \times M_n \times \ldots$ formé des suites $x = (x_1, \ldots, x_n, \ldots)$ telles que $\Sigma\|x_n\| < +\infty$ . Ce sous-ensemble est en fait un groupe de Banach pour la "quasi-norme" $L^1$, à savoir $\|x\| = \|x_1\| + \ldots + \|x_n\| + \ldots$ . On désigne par $\mathcal{L}_1(A)$ la catégorie dont les objets sont de telles $L^1$-somme, les morphismes étant les homomorphismes bornés de A-modules. Ceci dit, soit $\tau$ une catégorie en gronpes de Banach quelconque. Nous allons définir une catégorie en groupes de Banach $\mathcal{D}$ qui sera $\tau'$-filtrée, $\tau'$ étant une catégorie équivalente a $\mathcal{D}$. Les objets de $\mathcal{D}$ sont les suites $(E_1, \ldots, E_n, \ldots)$ où $E_i \in \mathrm{Ob}\tau$, les $E_i$ étant (pour chaque suite) choisis parmi un nombre fini $H, K, \ldots, L$ d'objets de

$\tau$. En particulier, chaque objet $E_i$ de la suite est facteur direct de $H \oplus K \oplus \dots \oplus L$. De plus, pour tout couple (i, j), $Hom(E_i, E_j)$ s'identifie à un sous-groupe fermé de l'anneau de Banach $A = End(H \oplus K \oplus \dots \oplus L)$. Soit maintenant $F = (F_1, \dots, F_n, \dots)$ un deuxième objet de $\mathcal{D}$. Quitte à changer les objets H, K, ..., L, on peut supposer que les objets $F_1, \dots, F_n, \dots$ sont aussi choisis parmi H, K, ..., L. Donc $Hom(E_i, F_j)$ est pour les mêmes raisons un facteur direct de A. Considérons à présent une matrice infinie

$$f = (f_{ji})$$

où $f_{ji} \in Hom(E_i, F_j)$. Cette matrice peut être interprétée d'après la discussion précédente comme une application de $A \oplus \dots \oplus A \oplus \dots$ dans $A \times A \dots \times A \times \dots$. On dira que f est "$L^1$-bornée" si f se prolonge en une application de la somme $L^1$ de $\aleph_0$-exemplaires de A dans elle-même. D'autre part, la matrice f est dite "permutante" si elle est $L^1$-bornée et si, sur chaque ligne et sur chaque colonne, il y a au plus un élément non nul: en d'autres termes il existe une bijection $k: \mathbb{N} \longrightarrow \mathbb{N}$ telle que $f_{ji} = \delta_{k(i)i} f_{ji}$ où $\delta$ désigne le symbole de Kronecker. La matrice f est dite "$\Sigma$-permutante" si elle est somme finie de matrices permutantes. Enfin la matrice est dite $\Sigma^1$-permutante s'il existe une suite $f_r$ de matrices $\Sigma$-permutantes qui converge vers f pour la quasi-norme $L^1$ définie précédemment. Les morphismes de $\mathcal{D}$ sont alors les matrices $\Sigma^1$-permutantes qu'on vient de décrire. On vérifie aisément que cette définition des morphismes est indépendante du choix des objets H, K, ..., L qui ont servi à définir la quasi-norme $L^1$ et que l'on obtient ainsi une catégorie (pour le produit des matrices). On définit également la quasi-norme d'une flèche de $\mathcal{D}$ comme la quasi-norme de la flèche $L_1(A)$ qu'elle définit. Celle-ci dépend évidemment du choix de A mais sa "classe d'équivalence" n'en dépend pas. Pour cette quasi-norme, la catégorie $\mathcal{D}$ est bien une catégorie en groupes de Banach. Soit maintenant $\tau'$ la sous-catégorie pleine de $\mathcal{D}$ dont les objets sont les suites $(E_1,\dots,E_n,\dots)$ nulles à partir d'un certain rang. Alors un calcul facile montre que $\tau'$ est équivalente à $\tau$ et que la catégorie $\mathcal{D}$ est $\tau'$-filtrée. Pour des raisons qui apparaîtront plus loin, on notera $\tau_1$ la catégorie $\mathcal{D}$.

Proposition et Definition 1.4

Soit $\mathcal{D}$ une catégorie $\tau$-filtrée et soit $f: E \longrightarrow F$ un $\mathcal{D}$-morphisme. Les deux assertions suivantes sont alors équivalentes:

(i) $\forall\varepsilon > 0$, il existe un objet $F_j$ de la filtration de F et un morphisme $f_j: E \longrightarrow F_j$ tel que le diagramme

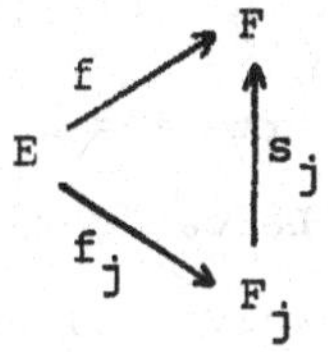


soit commutatif à $\varepsilon$-près.

(ii) $\forall \varepsilon > 0$, il existe un objet $E_i$ de la filtration de $E$ et un morphisme $f_i : E_i \longrightarrow F$ tel que le diagramme

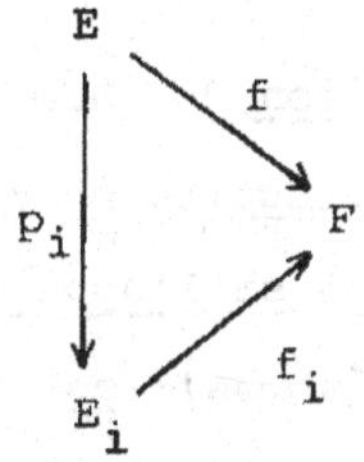


soit de même commutatif à $\varepsilon$-près.

Un morphisme $f$ vérifiant l'une des conditions équivalentes (i) ou (ii) est dit complètement continu.

Remarque. Cette définition est évidemment inspirée de celle des opérateurs complètement continus (ou compacts) dans les espaces de Hilbert (cf. exemple 1 des catégories filtrées).

Soit $\tau$ une sous-catégorie idéale de $\mathcal{D}$. On peut alors définir une catégorie quotient $\mathcal{D}/\tau$ de la manière suivante: les objets de $\mathcal{D}/\tau$ sont les objets de $\mathcal{D}$, les morphismes sont ceux de $\mathcal{D}$ modulo les morphismes complètement continus. En d'autres termes on a

$$\mathrm{Hom}_{\mathcal{D}/\tau}(E,F) = \mathrm{Hom}_{\mathcal{D}}(E,F)/\mathcal{K}(E,F)\,,\ \mathcal{K}(E,F)$$

désignant le sous-groupe fermé des morphismes complètement continus de $E$ dans $F$. La catégorie $\mathcal{D}/\tau$ est évidemment la solution d'un problème universel dont nous laissons la formulation au lecteur.

Definition 1.5

Soient $\tau$ et $\tau'$ deux catégories en groupes de Banach. Un foncteur additif $\varphi: \tau \longrightarrow \tau'$ est dit "de Serre"[2] si l'application de $\mathrm{Hom}_{\tau}(M,N)/\mathrm{Ker}\varphi_*$ dans $\mathrm{Hom}_{\tau'}(\varphi M,\varphi N)$ est une bijection bornée ainsi que son inverse. Le foncteur est dit borné si l'application $\varphi_*: \mathrm{Hom}_{\tau}(M,N) \longrightarrow \mathrm{Hom}_{\tau'}(\varphi M,\varphi N)$ est bornée.

Pour pouvoir parler maintenant de suites exactes de catégories il nous faut introduire la "catégorie" $\mathcal{B}$ suivante: les objets de $\mathcal{B}$ sont les catégories en groupes de Banach; un morphime $\varphi: \tau \longrightarrow \tau'$ de $\mathcal{B}$ est un foncteur de Serre.

Definition 1.6

Soit

$$\tau' \xrightarrow{\ \theta\ } \tau \xrightarrow{\ \chi\ } \tau''$$

(2) Cette terminologie est justifiée par le fait que l'application naturelle $\mathrm{Iso}_{\tau}(M,N) \longrightarrow \mathrm{Iso}_{\tau'}(\varphi M,\varphi N)$ est une fibration de Serre dans le cas des catégories de Banach classiques.

<u>une suite d'objets et de morphismes de $\mathcal{B}$. Cette suite est dite exacte si le noyau de l'application</u>

$$\mathrm{Hom}_{\tau}(E,F) \longrightarrow \mathrm{Hom}_{\tau''}(\mathcal{X}E,\mathcal{X}F)$$

<u>est l'ensemble des $\tau$-morphismes de E dans F complètement continus pour une filtration de $\tau$ par la "catégorie image" $\theta(\tau')$</u> (3).

Soit $\tau$ une catégorie en groupes de Banach quelconque. On peut munir la catégorie $\tau$ de deux filtrations évidentes. La première (dite grossière) consiste à prendre comme filtration d'un objet E de $\tau$ l'objet E seulement. Dans la seconde (dite discrète) la filtration de E se réduit à l'objet nul 0 de la catégorie $\tau$. Dans le premier cas (resp. le second) la sous-catégorie idéale est égale à $\tau$ (resp. à 0).

<u>Application</u>. Avec cette définition, la suite

$$0 \longrightarrow \tau \longrightarrow \mathcal{D} \longrightarrow \mathcal{D}/\tau \longrightarrow 0$$

est bien exacte, la catégorie $\tau$ (resp. $\mathcal{D}/\tau$) étant munie de la filtration discrète (resp. grossière). Réciproquement, si on a une suite exacte

$$0 \longrightarrow \tau' \xrightarrow{\theta} \tau \longrightarrow \tau'' \longrightarrow 0$$

la catégorie $\tau''$ s'identifie à la catégorie quotient $\tau/\theta(\tau')$.

(3) La catégorie image $\theta(\tau')$ est la sous-catégorie pleine de $\tau$ dont les objets sont isomorphes aux images des objets de $\tau'$ par $\theta$. On démontre en fait que la filtration de $\tau$ par $\theta(\tau')$ est unique (à équivalence près).

## II CARACTERISATION AXIOMATIQUE DE LA K-THEORIE

Definition 2.1

Soit $\mathcal{D}$ une catégorie en groupes de Banach. La catégorie $\mathcal{D}$ est dite flasque s'il existe un foncteur borné $\tau: \mathcal{D} \longrightarrow \mathcal{D}$ tel que les foncteurs $\tau$ et $\tau \oplus Id_{\mathcal{D}}$ soient isomorphes.

Exemples. La catégorie des espaces de Hilbert est flasque. En effet, il suffit de poser $\tau(E) = E \oplus \dots \oplus E \oplus \dots$ (somme hilbertienne de $\aleph_0$-exemplaires de E). On démontre de même que la catégorie $\tau_1$ (exemple 2 des catégories filtrées au § 1) est une catégorie flasque.

Théorème 2.2

Soit $\tau$ une catégorie en groupes de Banach. Il existe alors une suite exacte (dépendant canoniquement de $\tau$)

$$0 \longrightarrow \tau \xrightarrow{\alpha_0} \mathcal{D} \longrightarrow \tau' \longrightarrow 0$$

où $\mathcal{D}$ est une catégorie flasque.

En effet, on choisit $\mathcal{D} = \tau_1$ (cf. exemple 2 des catégories filtrées au § 1) et $\alpha_o$ = le foncteur défini par $\alpha_o(E) = (E,0,0,\dots)$.

Application. Le théorème 2.2 nous permet ainsi d'affirmer, par des raisonnements standard, l'existence d'une "résolution flasque canonique"

$$0 \longrightarrow \tau \xrightarrow{\alpha_0} \tau_1 \xrightarrow{\alpha_1} \tau_2 \xrightarrow{\alpha_2} \cdots \xrightarrow{\alpha_{n-1}} \tau_n \longrightarrow \cdots$$

de toute catégorie en groupes de Banach $\tau$. Il est à noter que si $\tau$ est une catégorie prébanachique dans le sens de [3], il en est de même des catégories $\tau_i$. Dans le cas général on définit la suspension $n^{\text{ième}}$ de $\tau$ comme la catégorie quotient $\tau_n/\alpha_{n-1}(\tau_{n-1})$. Rappelons (cf. [3] §1.2) que si $\tau$ est une catégorie additive on a désigné par $\tilde{\tau}$ la catégorie pseudo-abélienne associée à $\tilde{\tau}$.

Définition et Théorème 2.3

*Soit $\tau$ une catégorie en groupes de Banach. On désigne par $K^n(\tau)$, $n \geq 0$, le groupe de Grothendieck de $\widetilde{S^n(\tau)}$. Dans le cas où $\tau$ est une catégorie prébanachique, ces groupes sont périodiques de période 8 dans le cas réel et 2 dans le cas complexe.*

La démonstration de ce théorème est délicate et nécessite l'introduction des algèbres de Clifford (cf. [3],[4]).

Définition 2.4

*Une théorie de la cohomologie sur $\mathcal{B}$ est par définition la donnée de foncteurs $F^n$ de la catégorie $\mathcal{B}$ dans la catégorie des groupes abéliens et d'homomorphismes naturels*

$$\delta^{n-1}: F^{n-1}(\tau'') \longrightarrow F^n(\tau')$$

*définis pour toute suite exacte*

$$0 \longrightarrow \tau' \longrightarrow \tau \longrightarrow \tau'' \longrightarrow 0 \quad .$$

On suppose que $F^n(\tau) = 0$ si $\tau$ est flasque et que la suite suivante est exacte

$$F^{n-1}(\tau) \longrightarrow F^{n-1}(\tau'') \xrightarrow{\delta^{n-1}} F^n(\tau') \longrightarrow F^n(\tau) \longrightarrow F^n(\tau'') .$$

Théorème 2.5

Il existe une théorie de la cohomologie et une seule à isomorphisme près sur $\mathcal{B}$ telle que $F^0(\tau) = K(\tilde{\tau})$. De plus, les groupes $F^n(\tau)$ coincident avec les groupes $K^n(\tau)$ définis plus haut.

Le démonstration de ce théorème n'est pas non plus très évidente. On doit se servir du groupe $K_1$ introduit par Bass dans [1] et remarquer que le foncteur $K$ est le "premier foncteur dérivé" du foncteur $K_1$.

Considérons maintenant un foncteur de Serre essentiellement surjectif $\varphi: \tau \longrightarrow \tau'$ entre deux catégories en groupes de Banach. Pour simplifier les raisonnements, on supposera que $\varphi_*: \mathrm{Ob}\tau \longrightarrow \mathrm{Ob}\tau'$ est bijectif. On associe alors à $\varphi$ la catégorie $\mathcal{D}(\varphi)$ suivante: les objets de $\mathcal{D}(\varphi)$ sont les objets de $\tau_1$ (catégorie flasque associée à $\tau$), donc aussi de $\tau_1'$. Les flèches de $\mathcal{D}(\varphi)$ sont les classes de flèches de $\tau_1$ pour la relation d'équivalence suivante: $\alpha \sim 0 \Longleftrightarrow \alpha$ est complètement continu (pour la $\tau$-filtration) et $\varphi_1(\alpha) = 0$. On a alors un foncteur évident

$$\tau' \xrightarrow{\ \theta\ } \mathcal{D}(\varphi)$$

défini par $\theta(E) = (E,0,0, \ldots)$ sur les objets et par

$$\theta(f) = \begin{pmatrix} \bar{f} & 0 & 0 & \cdots \\ 0 & 0 & 0 & \cdots \\ 0 & 0 & 0 & \cdots \\ \vdots & & & \end{pmatrix} ,$$

avec $\varphi(\bar{f}) = f$, sur les morphismes. Le foncteur $\theta$ est d'ailleurs filtrant pour la filtration induite par celle de de telle sorte qu'on a la suite exacte

$$0 \longrightarrow \tau' \longrightarrow \mathcal{D}(\varphi) \longrightarrow S^1\tau \longrightarrow 0$$

<u>Définition et Théorème 2.6</u>

<u>Pour tout foncteur de Serre essentiellement surjectif</u> $\varphi: \tau \longrightarrow \tau'$ , <u>on pose</u>

$$K^{n+1}(\varphi) = K^n(\mathcal{D}(\varphi))$$

si $n \geq 0$ (4). <u>On a alors la suite exacte</u>

$$K^{n-1}(\tau) \longrightarrow K^{n-1}(\tau') \xrightarrow{\delta^{n-1}} K^n(\varphi) \longrightarrow K^n(\tau) \longrightarrow K^n(\tau'), \; n \geq 1,$$

<u>où tous les homomorphismes sont naturels à l'exception de</u> $\delta^{n-1}$ <u>qui est induit par le foncteur</u> $\tau' \longrightarrow \mathcal{D}(\varphi)$.

<u>Remarque</u>. Ce théorème s'applique en particulier au foncteur "restriction des fibrés" $\xi_T(X) \longrightarrow \xi_T(Y)$ où $X$ est un espace compact et $Y$ un sous-espace fermé. On en déduit la suite exacte de cohomologie en K-théorie topologique

---

(4) Si $\tau' = 0$ on retrouve bien (à isomorphisme près) le groupe $K^{n+1}(\tau) = K^n(S^1\tau)$.

$$K^{n-1}(X) \longrightarrow K^{n-1}(Y) \longrightarrow K^n(X,Y) \longrightarrow K^n(X) \longrightarrow K^n(Y),\ n \geq 1\ ,$$

sans évoquer les algèbres de Clifford ou la périodicité de Bott. Il résulte une construction relativement élémentaire de la K-théorie en tant que théorie cohomologique sur les espaces compacts. Bien entendu la périodicité des groupes $K^n$ (théorème 2.3) n'est pas évidente avec ce point de vue et doit être démontrée séparément.

## BIBLIOGRAPHIE

[1] Bass, H. "K-theory and Stable Algebra", Publications Mathématiques de l'I.H.E.S. No. 22; 5-60. (1964).

[2] Godement, R. Théorie des Faisceaux.

[3] Karoubi, M. "Algèbres de Clifford et K-théorie", Annales Scientifiques de l'Ecole Normale Supérieure. p.161-270. (1968).

[4] Karoubi, M. Séminaire sur la K-théorie (à paraître).

# RELATIVE FUNCTORIAL SEMANTICS: ADJOINTNESS RESULTS*

by

F. E. J. Linton

## INTRODUCTION

Central to any exposition of the theory of triples are the Godement construction [4] of the triple arising from an adjoint functor situation, the Eilenberg-Moore construction [3] of the category of algebras over a triple, with the associated adjoint pair of "free" and "underlying" functors, and the Kleisli construction [6] of the clone of tripleary tuples of operations [8], along with the various adjointness relations available among these constructions and their cotriple analogues. If the usual base category $S$ of sets and functions is replaced by an arbitrary closed or monoidal category $V$, one expects to develop a satisfactorily parallel exposition at the level of $V$-categories [2]. Bunge [1] and Kock [7] have, indeed, gone far in this direction; unfortunately, however, both these treatments add the assumption of a compatible symmetric monoidal structure on $V$ whenever they

* Research supported by N.S.F. grant NSF-GP 6325, Wesleyan Faculty Research Grant 5427-143, and Battelle Memorial Institute, Seattle Research Center.

deal with the case that the base category $V$ is closed.

In what follows, therefore, we shall present the rudiments of two separate but parallel (and, when $V$ is closed monoidal, equivalent) $V$-theories of triples, one for monoidal $V$, the other for $V$ closed — no symmetry assumptions are needed in either case. Just as in the work [1] of Bunge, however, one must assume, in either case, that $V$ has difference kernels (equalizers) in order to get any theory at all; moreover, when $V$ is closed, one must know that each left represented endofunctor $L^A = V(A,-): V \longrightarrow V$ of $V$ preserves difference kernels (as it automatically will in the closed monoidal case). These assumptions shall therefore be in force throughout the paper, except in the first paragraph of §1 and in Lemma 2.

An awkward consequence of our refusal to use symmetry is the minor nuisance that the cotriple theory cannot be obtained simply by carrying out the triple theory on the duals of all the $V$-categories involved — there are no such duals, in general. This is no real difficulty, however: one merely reverses enough arrows in the ensuing triple theory exposition to obtain a valid exposition of the cotriple theory. This exercise in judicious arrow reversal will be left to the reader; a good time for him to engage in it is immediately before §3.

Throughout this work, we assume familiarity with the elementary notions of closed and monoidal categories, and categories, functors, and natural transformations over them: see [2] for full information. The indispensable adjointness notions relevant to this setting are presented in [1] and [5].

The definition of $V$-triple is found in §1 of the present work, along with the Eilenberg-Moore construction of the $V$-category of algebras over a $V$-triple and the $V$-adjointness between the "free" and "underlying" $V$-functors. These matters also appear in Bunge's work [1], at least in the case of monoidal $V$; however, the argument presented here centers around a split difference kernel phenomenon (Lemma 1) which seems not to have been publicly recorded even in the case $V = S$.

The $V$-triple analogue of the structure-semantics adjointness, which is the concern of §2, goes through without a hitch. The main tool is the fact that, unlike most ordinary functors between $V$-categories, a functor $P: X \longrightarrow A^{\mathbf{T}}$ satisfying

$$U^{\mathbf{T}} \circ P = U \qquad (*)$$

at the level of ordinary functors, where $U$ is a fixed $V$-functor $X \longrightarrow A$, admits at most one enrichment to a $V$-functor compatible with the validity of equation (*) at the level of $V$-functors. A general form of this fact, accompanied by a manageable criterion for such an enrichment to exist, is recorded as Lemma 2, and is used all through the remainder of the paper.

Assuming familiarity with the $\mathcal{V}$-cotriple analogues of the results of the preceding sections, §3 exposes the $\mathcal{V}$-category generalization of Lawvere's as yet unpublished triple-cotriple, (Kleisli category)-(Eilenberg-Moore construction) adjointness theorem presented in his lectures at Battelle.

A description, as in [8], of algebras over a $\mathcal{V}$-triple $\mathbf{T}$ in terms of $\mathcal{V}$-valued $\mathcal{V}$-functors from the Kleisli category of $\mathbf{T}$, which was originally intended for inclusion here, has been omitted, for the reason that neither an adequate background in contravariant $\mathcal{V}$-functors nor any $\mathcal{V}$-valued Yoneda lemmas are yet available without use of symmetry (cf. [2]) or smallness of the domain (cf. [1], Theorem 4.7), respectively. It is hoped to remedy these omissions elsewhere.

## §1 ALGEBRAS OVER A $\mathcal{V}$-TRIPLE

A $\mathcal{V}$-triple $\mathbf{T} = (T,\eta,\mu)$ on a $\mathcal{V}$-category $A$ consists of a $\mathcal{V}$-functor $T: A \longrightarrow A$ and $\mathcal{V}$-natural transformations $\eta: id_A \longrightarrow T$, $\mu: TT \longrightarrow T$ satisfying the familiar triple identities

$$\mu \circ \eta T = \mu \circ T\eta = id_T ,$$

$$\mu \circ \mu T = \mu \circ T\mu .$$

A $\mathbb{T}$-algebra structure $\alpha$ on $A \in \operatorname{obj} \mathcal{A}$ is, as usual, an $\mathcal{A}$-morphism $\alpha\colon TA \longrightarrow A$ for which

$$\alpha \circ \eta_A = id_A \ ,$$

$$\alpha \circ \mu_A = \alpha \circ T\alpha \ .$$

The usual set of $\mathbb{T}$-algebra maps $f\colon (A,\alpha) \longrightarrow (B,\beta)$, that is, the set of all those $\mathcal{A}$-morphisms $f\colon A \longrightarrow B$ making the diagram

$$\begin{array}{ccc} TA & \xrightarrow{Tf} & TB \\ {\scriptstyle\alpha}\downarrow & & \downarrow{\scriptstyle\beta} \\ A & \xrightarrow[f]{} & B \end{array}$$

commute, can be viewed as the difference kernel (or equalizer) of the pair of functions

$$\begin{array}{ccccc} \mathcal{A}_0(A,B) & & \xrightarrow{\quad \mathcal{A}_0(\alpha,B) \quad} & & \mathcal{A}_0(TA,B) \\ & {\scriptstyle T_0}\searrow & & \nearrow{\scriptstyle \mathcal{A}_0(TA,\beta)} & \\ & & \mathcal{A}_0(TA,TB) & & \end{array} \ .$$

When $T$ is a $\mathcal{V}$-functor, this diagram can be lifted to $\mathcal{V}$, and its difference kernel, if any, makes a lovely candidate for a $\mathcal{V}$-object of $\mathbb{T}$-algebra maps from $(A,\alpha)$ to $(B,\beta)$. We therefore <u>henceforth</u> <u>assume</u> $\mathcal{V}$ <u>has</u> <u>difference</u> <u>kernels</u> and define the $\mathcal{V}$-object $\mathcal{A}^{\mathbb{T}}((A,\alpha),\ (B,\beta))$, frequently abbreviated as $\mathcal{A}^{\mathbb{T}}(\alpha,\beta)$, to be the difference kernel of the lifted diagram

$$\begin{array}{ccccc} \mathcal{A}(A,B) & & \xrightarrow{\quad \mathcal{A}(\alpha,B) \quad} & & \mathcal{A}(TA,B) \\ & {\scriptstyle T_{A,B}}\searrow & & \nearrow{\scriptstyle \mathcal{A}(TA,\beta)} & \\ & & \mathcal{A}(TA,TB) & & \end{array} \ .$$

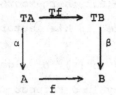

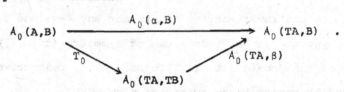

We shall write $U^{\mathbb{T}}_{(A,\alpha),(B,\beta)}$ , or simply $U^{\mathbb{T}}$, for the (monic) $V$-morphism

$$U^{\mathbb{T}}\colon A^{\mathbb{T}}((A,\alpha),(B,\beta)) = A^{\mathbb{T}}(\alpha,\beta) \longrightarrow A(A,B)$$

canonically associated with this difference kernel.

Proposition 1

*Assume that* $V$ *is a closed category with difference kernels that are preserved by each left-represented functor* $V(X,-)\colon V \longrightarrow V$ (*resp., that* $V$ *is a monoidal category with difference kernels*). *Let* $\mathbb{T}$ *be a* $V$*-triple on a* $V$*-category* $A$.

(a) *There is precisely one* $V$*-morphism*

$$j_{(A,\alpha)}\colon I \longrightarrow A^{\mathbb{T}}((A,\alpha),(A,\alpha))$$

*making commutative the triangle*

$$\begin{array}{ccc} A^{\mathbb{T}}((A,\alpha),(A,\alpha)) & \xrightarrow{\;U^{\mathbb{T}}\;} & (A,A) \\ {\scriptstyle j_{(A,\alpha),(A,\alpha)}}\nwarrow & & \nearrow{\scriptstyle j_A} \\ & I & \end{array}$$

(b) *There is precisely one map*

$$L^{(A,\alpha)}_{(B,\beta),(C,\gamma)}\colon A^{\mathbb{T}}(\beta,\gamma) \longrightarrow V(A^{\mathbb{T}}(\alpha,\beta),A^{\mathbb{T}}(\alpha,\gamma))$$

(*resp.* $M^{(B,\beta)}_{(A,\alpha),(C,\gamma)}\colon A^{\mathbb{T}}(\beta,\gamma) \otimes A^{\mathbb{T}}(\alpha,\beta) \longrightarrow A^{\mathbb{T}}(\alpha,\gamma))$

*making commutative the diagram*

$$\begin{array}{ccc}
A^{\mathbb{T}}(\beta,\gamma) & \xrightarrow{\;L^{(A,\alpha)}_{(B,\beta),(C,\gamma)}\;} & \mathcal{V}(A^{\mathbb{T}}(\alpha,\beta),A^{\mathbb{T}}(\alpha,\gamma)) \\
{\scriptstyle U^{\mathbb{T}}}\big\downarrow & & \\
A(B,C) & & \big\downarrow{\scriptstyle \mathcal{V}(id,U^{\mathbb{T}})} \\
{\scriptstyle L^{A}_{B,C}}\big\downarrow & & \\
\mathcal{V}(A(A,B),A(A,C)) & \xrightarrow[\;\mathcal{V}(U^{\mathbb{T}},id)\;]{} & \mathcal{V}(A^{\mathbb{T}}(\alpha,\beta),A(A,C))
\end{array}$$

(resp.
$$\begin{array}{ccc}
A^{\mathbb{T}}(\beta,\gamma)\otimes A^{\mathbb{T}}(\alpha,\beta) & \xrightarrow{\;M^{(B,\beta)}_{(A,\alpha),(C,\gamma)}\;} & A^{\mathbb{T}}(\alpha,\gamma) \\
{\scriptstyle U^{\mathbb{T}}\otimes U^{\mathbb{T}}}\big\downarrow & & \big\downarrow{\scriptstyle U^{\mathbb{T}}} \\
A(B,C)\otimes A(A,B) & \xrightarrow[\;M^{B}_{A,C}\;]{} & A(A,C))
\end{array} .$$

(c) *With the structure provided by* (a) *and* (b), *$\mathbb{T}$-algebras and the $\mathcal{V}$-objects* $A^{\mathbb{T}}((A,\alpha),(B,\beta))$ *of $\mathbb{T}$-morphisms between them form a $\mathcal{V}$-category* $A^{\mathbb{T}}$, *while the $\mathcal{V}$-morphisms* $U^{\mathbb{T}}$ *make the passage* $(A,\alpha) \longmapsto A$ *a* *$\mathcal{V}$-functor* $U^{\mathbb{T}}: A^{\mathbb{T}} \longrightarrow A$ .

*Proof.* (a) It suffices to show that the diagram

$$\begin{array}{ccccc}
 & & A(A,A) & \xrightarrow{\;T_{AA}\;} & A(TA,TA) \\
 & {\scriptstyle j_A}\nearrow & & & \big\downarrow{\scriptstyle A(TA,\alpha)} \\
I & & & & \\
 & {\scriptstyle j_A}\searrow & & & \\
 & & A(A,A) & \xrightarrow[\;A(\alpha,A)\;]{} & A(TA,A)
\end{array}$$

commutes. Since $T_{AA} \circ j_A = j_{TA,TA}$, however, it's merely a matter of knowing that $A(TA,\alpha) \circ j_{TA,TA} = A(\alpha,A) \circ j_{A,A}$, which is elementary (see [2], Chapter I, diagram (9.10) and Chapter II, diagram (8.10)).

(b) In the case that $V$ is monoidal, it suffices to prove that the perimeter of the diagram

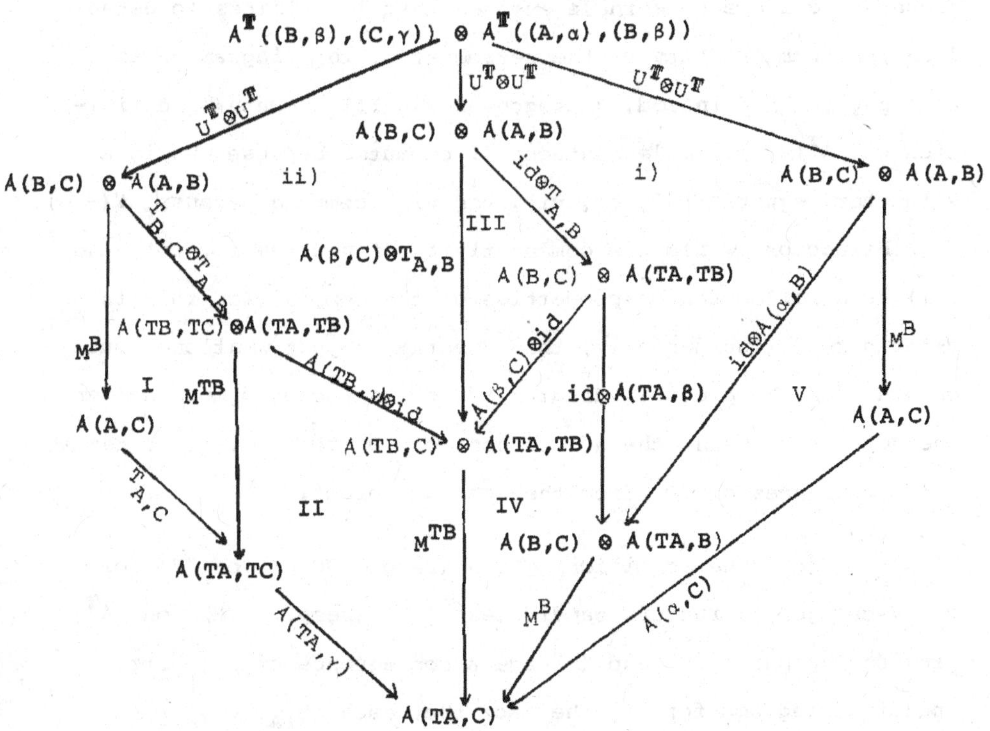


commutes. Indeed, square I commutes because T is a $V$-functor. Squares and triangles II, III, IV, and V commute because $A$ is a $V$-category. Region i) is obtained from the difference kernel definition of $A^{\mathbb{T}}((A,\alpha),(B,\beta))$ by tensoring with $U^{\mathbb{T}}\colon A^{\mathbb{T}}((B,\beta),(C,\gamma)) \longrightarrow A(B,C)$, and therefore commutes; similarly, region ii) commutes, because it is obtained from the difference kernel definition of $A^{\mathbb{T}}((B,\beta),(C,\gamma))$ by tensoring with $T_{A,B} \circ U^{\mathbb{T}}\colon A^{\mathbb{T}}((A,\alpha),(B,\beta)) \longrightarrow A(TA,TB)$.

In the case $V$ closed, the assumption that $V(X,-)$ preserves difference kernels ensures that it suffices to establish the commutativity of the perimeter of the diagram on the next page. And indeed, pentagons I and III commute by definition of $A^{\mathbb{T}}(-,-)$, while pentagon II commutes because $T$ is a $V$-functor; squares iv), v), vi), and vii) commute because $V(-,-)$ is a bifunctor, while the commutativity of squares i), ii), and iii) is a reflection of properties of the composition rule in a $V$-category. (Incidentally, this diagram is just what one would obtain from the previous diagram if $V$ were closed monoidal and one set about, using the adjointness of $\otimes$ to $V(-,-)$, to remove all occurrences of $\otimes$ from the monoidal proof.

(c) The conditions VC n (resp. VC n') of [2] for the $V$-category structure candidates $j,L$ (resp. $j,M$) on $A^{\mathbb{T}}$ provided by parts (a) and (b) are a consequence of the corresponding diagrams for $A$, the fact that each $U^{\mathbb{T}}_{(A,\alpha),(B,\beta)}$ is a monomorphism, and the commutativity of the diagrams established in (a) and (b) (these commutativities are essentially the conditions VF 1 and VF 2 (or VF 2') of [2], so that the

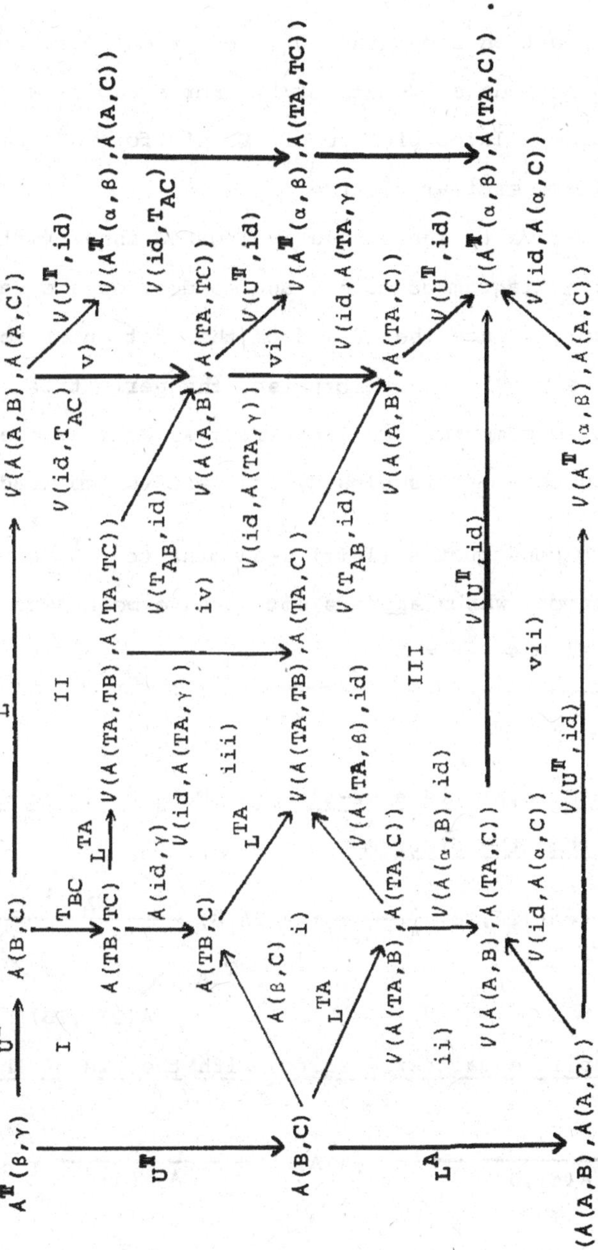

A^T(β,γ)
U^T
A(B,C)
L^A
V(A(A,B),A(A,C))
V(U^T,id)
V(A^T(α,β),A(A,C))
A(TB,TC)
T_BC
L^TA
V(A(TA,TB),A(TA,TC))
V(id,T_AC)
V(A(A,B),A(TA,TC))
V(A^T(α,β),A(TA,TC))
A(TB,C)
A(id,γ)
V(id,A(TA,γ))
V(T_AB,id)
V(A(TA,TB),A(TA,C))
V(id,A(TA,γ))
V(A(A,B),A(TA,C))
V(A^T(α,β),A(TA,C))
A(β,C)
V(A(TA,β),id)
V(A(TA,B),A(TA,C))
V(A(α,B),id)
V(A(A,B),A(TA,C))
V(id,A(α,C))
V(A(A,B),A(A,C))
V(U^T,id)
V(A^T(α,β),A(A,C))
V(id,A(α,C))
I
II
III
i)
ii)
iii)
iv)
v)
vi)
vii)

$V$-functor assertion regarding $U^{\mathbb{T}}$ is automatic as soon as $A^{\mathbb{T}}$ is known to be made a $V$-category). For example, when $V$ is monoidal, draw a large picture of VC n' for $A^{\mathbb{T}}$; within it, draw a smaller, similar picture of VC n' for $A$. Now $U^{\mathbb{T}}$ maps each outer vertex to the similarly located inner vertex, the inner VC n' diagram commutes, and so does each square along the perimeter. Since the $U^{\mathbb{T}}$ from the last outer vertex to the last inner vertex is a monomorphism, the perimetric VC n' diagram must also commute. Full details, as well as the analogous arguments in case $V$ is closed, are left to the reader.

To construct a (left) $V$-adjoint to $U^{\mathbb{T}}$, we use the following lemma, which appears not to have been recorded even in the classical case $V = S$.

Lemma 1

*If* $(B,\beta)$ *is a* $\mathbb{T}$*-algebra* ($\mathbb{T}$ *a* $V$*-triple on* $A$) *and* $A \in \operatorname{obj} A$, *then the diagram*

$$A(A,B) \xrightarrow{T_{A,B}} A(TA,TB) \xrightarrow{A(TA,\beta)} A(TA,B) \underset{A(TTA,\beta)\,\circ\, T_{TA,B}}{\overset{A(\mu_A,B)}{\rightrightarrows}} A(TTA,B)$$

$$T_{TA,B} : A(TA,B) \to A(TTA,TB), \qquad A(TTA,\beta) : A(TTA,TB) \to A(TTA,B)$$

*becomes a split equalizer diagram, with the aid of the splitting maps*

$$A(A,B) \xleftarrow{A(\eta_A,B)} A(TA,B) \xleftarrow{A(T\eta_A,B)} A(TTA,B) \; .$$

<u>Proof.</u> Four identities must be established:

a) $\mathcal{A}(\eta_A,B)\circ(\mathcal{A}(TA,\beta)\circ T_{A,B}) = id_{\mathcal{A}(A,B)}$ ;

b) $\mathcal{A}(\mu_A,B)\circ(\mathcal{A}(TA,\beta)\circ T_{A,B}) = (\mathcal{A}(TTA,\beta)\circ T_{TA,B})\circ(\mathcal{A}(TA,\beta)\circ T_{A,B})$ ;

c) $\mathcal{A}(T\eta_A,B)\circ\mathcal{A}(\mu_A,B) = id_{\mathcal{A}(TA,B)}$ ;

d) $\mathcal{A}(T\eta_A,B)\circ(\mathcal{A}(TTA,\beta)\circ T_{TA,B}) = (\mathcal{A}(TA,\beta)\circ T_{A,B})\circ\mathcal{A}(\eta_A,B)$ .

The proof of a) resides in the diagram

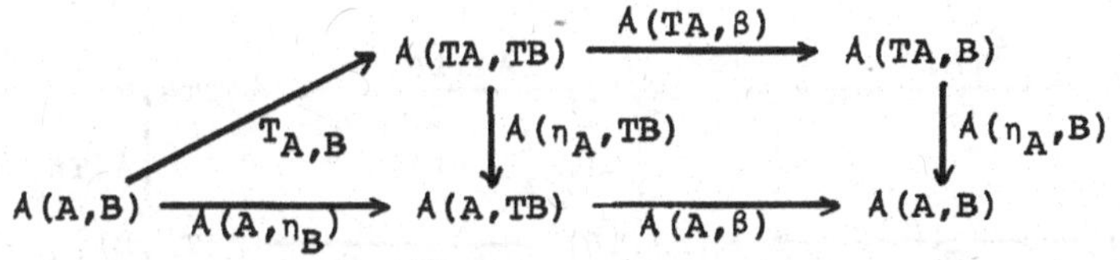


.

The triangle commutes (condition VN of [2]) because $\eta$ is $\mathcal{V}$-natural; the square commutes because $\mathcal{A}(-,-)$ is a bifunctor; the base is $\mathcal{A}(A,\beta\eta_B) = \mathcal{A}(A,id_B) = id_{\mathcal{A}(A,B)}$ .

The proof of b) resides in the diagram

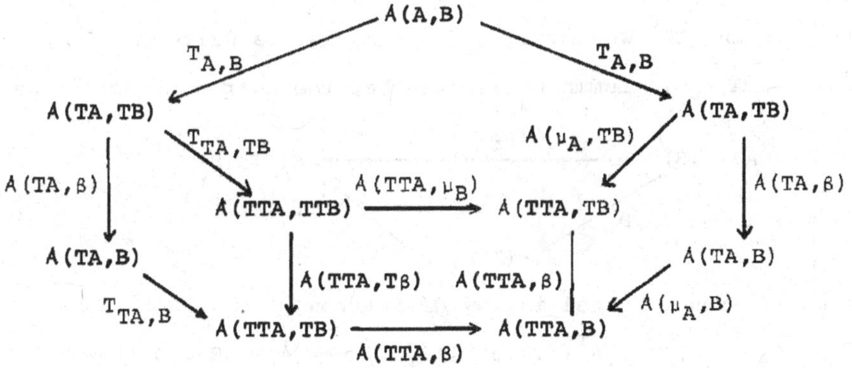


.

The pentagon commutes because $\mu$ is $\mathcal{V}$-natural. The left hand diamond commutes because $T$ is a $\mathcal{V}$-functor, the central square, because $\beta$ is a $\mathbb{T}$-algebra structure on $B$ ($\beta \circ T\beta = \beta \circ \mu_B$), and the commutativity of the right hand diamond follows from the bifunctor character of $\mathcal{A}(-,-)$.

Relation c) follows easily from the triple identity $\mu_A \circ T\eta_A = id_{TA}$, while relation d) results from the commutativity of both squares in the diagram

$$\begin{array}{ccccc}
\mathcal{A}(TA,B) & \xrightarrow{T_{TA,B}} & \mathcal{A}(TTA,TB) & \xrightarrow{\mathcal{A}(TTA,\beta)} & \mathcal{A}(TTA,B) \\
\downarrow{\scriptstyle \mathcal{A}(\eta_A,B)} & & \downarrow{\scriptstyle \mathcal{A}(T\eta_A,TB)} & & \downarrow{\scriptstyle \mathcal{A}(T\eta_A,B)} \\
\mathcal{A}(A,B) & \xrightarrow[T_{A,B}]{} & \mathcal{A}(TA,TB) & \xrightarrow[\mathcal{A}(TA,\beta)]{} & \mathcal{A}(TA,B)
\end{array} ,$$

the first commuting because $T$ is a $\mathcal{V}$-functor, the second because $\mathcal{A}(-,-)$ is a bifunctor. The proof of Lemma 1 is thus complete.

Recalling now that $\mu_A : TTA \longrightarrow TA$ is a $\mathbb{T}$-algebra structure on $TA$, whatever $A \in \text{obj}\,\mathcal{A}$, let us write $F^{\mathbb{T}}(A) = (TA,\mu_A)$. Lemma 1 asserts that the pair of $\mathcal{V}$-morphisms

$$\begin{array}{ccc}
\mathcal{A}(TA,B) & \xrightarrow{\mathcal{A}(\mu_A,B)} & \mathcal{A}(TTA,B) \\
{\scriptstyle T_{TA,B}} \searrow & & \nearrow {\scriptstyle \mathcal{A}(TTA,\beta)} \\
 & \mathcal{A}(TTA,TB) &
\end{array} ,$$

whose difference kernel is, by definition,

$$U^{\mathbb{T}}_{F^{\mathbb{T}}(A),(B,\beta)} : \mathcal{A}^{\mathbb{T}}((TA,\mu_A),(B,\beta)) \longrightarrow \mathcal{A}(TA,B) ,$$

fits, in fact, in a split equalizer diagram with difference kernel $\mathcal{A}(A,B)$. There result isomorphisms

$$\mathcal{A}(A,U^{\mathbb{T}}(B,\beta)) = \mathcal{A}(A,B) \xrightarrow{\cong} \mathcal{A}^{\mathbb{T}}((TA,\mu_A),(B,\beta)) = \mathcal{A}^{\mathbb{T}}(F^{\mathbb{T}}A,(B,\beta))$$

which can be used (see [1] or [5]) to endow $F^{\mathbb{T}}$ with the structure of a $\mathcal{V}$-functor for which these isomorphisms constitute a $\mathcal{V}$-adjointness relation between $U^{\mathbb{T}}$ and $F^{\mathbb{T}}$. It is, however, equally simple to describe this $\mathcal{V}$-functor structure on $F^{\mathbb{T}}$ directly. Referring back to the diagram used in the proof of Lemma 1 to establish identity b), commutativity of the upper pentagon indicates that the map

$$T_{A,B}\colon \mathcal{A}(A,B) \longrightarrow \mathcal{A}(TA,TB)$$

uniquely factors, by a map we shall call $F^{\mathbb{T}}_{A,B}$, through $U^{\mathbb{T}}\colon \mathcal{A}^{\mathbb{T}}(F^{\mathbb{T}}A,F^{\mathbb{T}}B) \longrightarrow \mathcal{A}(TA,TB)$ :

$$\begin{array}{ccc} & & \mathcal{A}^{\mathbb{T}}(F^{\mathbb{T}}A,F^{\mathbb{T}}B) \\ & \nearrow^{F^{\mathbb{T}}_{A,B}} & \downarrow{\scriptstyle U^{\mathbb{T}}} \\ \mathcal{A}(A,B) & \xrightarrow[T_{A,B}]{} & \mathcal{A}(TA,TB) \end{array} .$$

By the use of Lemma 2 below (see §2), the proof that $F^{\mathbb{T}}$ is a $\mathcal{V}$-functor is negligible; the identity that $T = U^{\mathbb{T}}F^{\mathbb{T}}$ and the naturality of the above isomorphisms

$$\mathcal{A}(A,U^{\mathbb{T}}(B,\beta)) \xrightarrow{\cong} \mathcal{A}(F^{\mathbb{T}}A,(B,\beta))$$

being easy (associated front adjunction is the $\mathcal{V}$-natural transformation $\eta\colon \mathrm{id} \longrightarrow T = U^{\mathbb{T}}F^{\mathbb{T}}$, while the back adjunction $\varepsilon^{\mathbb{T}}$ satisfies $U^{\mathbb{T}}\varepsilon^{\mathbb{T}}_{(A,\alpha)} = \alpha$), we have essentially completed the proof of

<u>Theorem 1</u>

<u>The</u> $\mathcal{V}$-<u>functor</u> $U^{\mathbb{T}}: \mathcal{A}^{\mathbb{T}} \longrightarrow \mathcal{A}$ <u>is</u> (<u>right</u>) $\mathcal{V}$-<u>adjoint</u> <u>to the</u> $\mathcal{V}$-<u>functor</u> $F^{\mathbb{T}}: \mathcal{A} \longrightarrow \mathcal{A}^{\mathbb{T}}$ <u>determined as follows</u>:

$$F^{\mathbb{T}}A = (TA, \mu_A)$$

$$U^{\mathbb{T}}_{F^{\mathbb{T}}A, F^{\mathbb{T}}B} \circ F^{\mathbb{T}}_{A,B} = T_{A,B}: \mathcal{A}(A,B) \longrightarrow \mathcal{A}(TA,TB) \quad .$$

<u>The front adjunction is</u> $\eta: \mathrm{id} \longrightarrow T = U^{\mathbb{T}}F^{\mathbb{T}}$; <u>the back adjunction</u> $\varepsilon^{\mathbb{T}}$ <u>has the effect</u> $\varepsilon^{\mathbb{T}}_{(A,\alpha)} = \alpha: F^{\mathbb{T}}A \longrightarrow (A,\alpha)$; <u>and the adjunction isomorphism is that induced by the fact registered in Lemma</u> 1 <u>that</u> $\mathcal{A}(A,B)$ <u>is difference kernel of the pair of maps whose difference kernel</u> $\mathcal{A}^{\mathbb{T}}(F^{\mathbb{T}}A, (B,\beta))$ <u>is defined to be</u>.

<u>Remarks on Lemma 1</u>. If $\mathbb{T} = (T,\eta,\mu)$ is a $\mathcal{V}$-triple on $\mathcal{A}$, a cotriple $\check{\mathbb{T}} = (\check{T},\check{\eta},\check{\mu})$ is obtained on the functor category $\mathcal{V}^{\mathcal{A}^{op}}$ by the following means:

$$\left.\begin{aligned} \check{T}(X) &= X \circ T^{op} \\ ((\check{\eta})_X)_A &= X(\eta_A) \\ ((\check{\mu})_X)_A &= X(\mu_A) \end{aligned}\right\} \quad (X: \mathcal{A}^{op} \longrightarrow \mathcal{V},\ A \in \mathrm{obj}\ \mathcal{A}) \quad .$$

Lemma 1 may be interpreted as asserting that, for each $\mathbb{T}$-algebra $(B,\beta)$, the objects $\mathcal{A}(A,B)$ are the values of the $\check{\mathbb{T}}$-coalgebra obtained by associating with the functor $\mathcal{R}^B = \mathcal{A}(-,B): \mathcal{A}^{op} \longrightarrow \mathcal{V}$ the natural transformation $\check{\beta}: \mathcal{R}^B \longrightarrow \check{T}\mathcal{R}^B$ whose components are $\check{\beta}_A = \mathcal{A}(TA,\beta) \circ T_{A,B}$: $\mathcal{R}^B(A) = \mathcal{A}(A,B) \longrightarrow \mathcal{A}(TA,TB) \longrightarrow \mathcal{A}(TA,B) = \mathcal{R}^B TA = (\check{T}(\mathcal{R}^B))(A)$.

It must, of course, be verified that $\check{\beta}$ really is a natural transformation; the coalgebra structure, however, is then fully guaranteed by Lemma 1.

This remark can be extended, and the converse to the extension proved, as soon as the necessary Yoneda Lemma machinery for contravariant $V$-valued $V$-functors (needed elsewhere as well, as noted at the end of the introduction) has been constructed.

## §2. V-STRUCTURE AND V-SEMANTICS

Although an ordinary functor $P: X \longrightarrow Y$ between $V$-categories may, in general, carry several enrichments to a $V$-functor, if any, we have already met (in $F^{\mathbf{T}}$), and shall in what follows continue to meet instances in which, subject to minimal side conditions, the $V$-functor structure, if any, of $P$ is unique. To isolate the ideas, we state the following lemma, for which the usual standing hypotheses on $V$ are relaxed.

Lemma 2

*Let $X$, $Y$, $A$ be $V$-categories, let $U: X \longrightarrow A$ and $V: Y \longrightarrow A$ be $V$-functors. Assume that each $V$-morphism $V_{Y,Z}: Y(Y,Z) \longrightarrow A(VY,VZ)$ is a monomorphism (and, if $V$ is not assumed monoidal, that each $V(X,V_{Y,Z})$ is a monomorphism, as well). Then a functor $P: X \longrightarrow Y$, satisfying*

$$V \circ P = U \tag{2.1}$$

*at the level of ordinary functors, admits at most one enrichment to a V-functor satisfying* (2.1) *at the level of V-functors; moreover, it admits such an enrichment if (and, of course, only if) the dotted arrow in each diagram*

$$\begin{array}{ccc} \mathcal{X}(W,X) & \dashrightarrow & \mathcal{Y}(PW,PX) \\ {\scriptstyle U_{W,X}}\downarrow & & \downarrow{\scriptstyle V_{PW,PX}} \\ \mathcal{A}(UW,UX) & = & \mathcal{A}(VPW,VPX) \end{array}$$

*can be filled in by a V-morphism* $P_{W,X}$ *rendering the square commutative.*

*Proof.* An enrichment of $P$ to a $V$-functor satisfying (2.1) involves precisely such fillings in, subject to the side conditions named VF n (resp. VF n') (n = 1,2) in [2]. Since each $V_{PW,PX}$ is monic, there can be at most one such system of fillings in, hence at most one such enrichment. So much for the uniqueness (and the "only if" assertion). Now assume such maps $P_{W,X}$ are indeed available. Then we have the diagrams

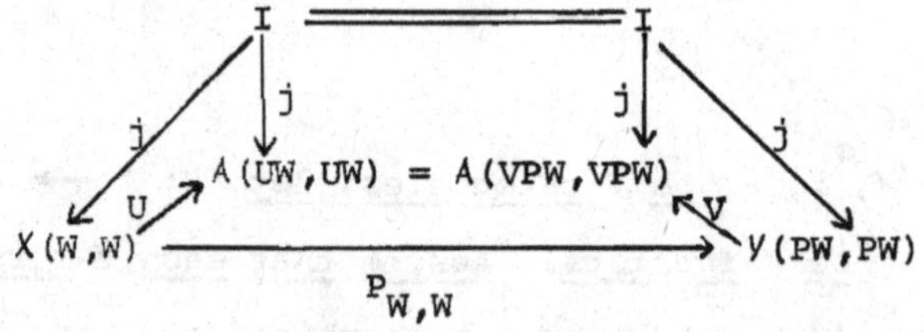


,

and, if $V$ is monoidal,

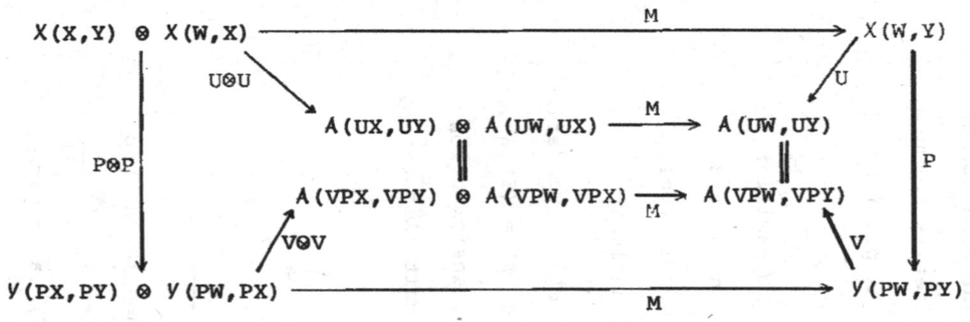


or, if $V$ is closed,

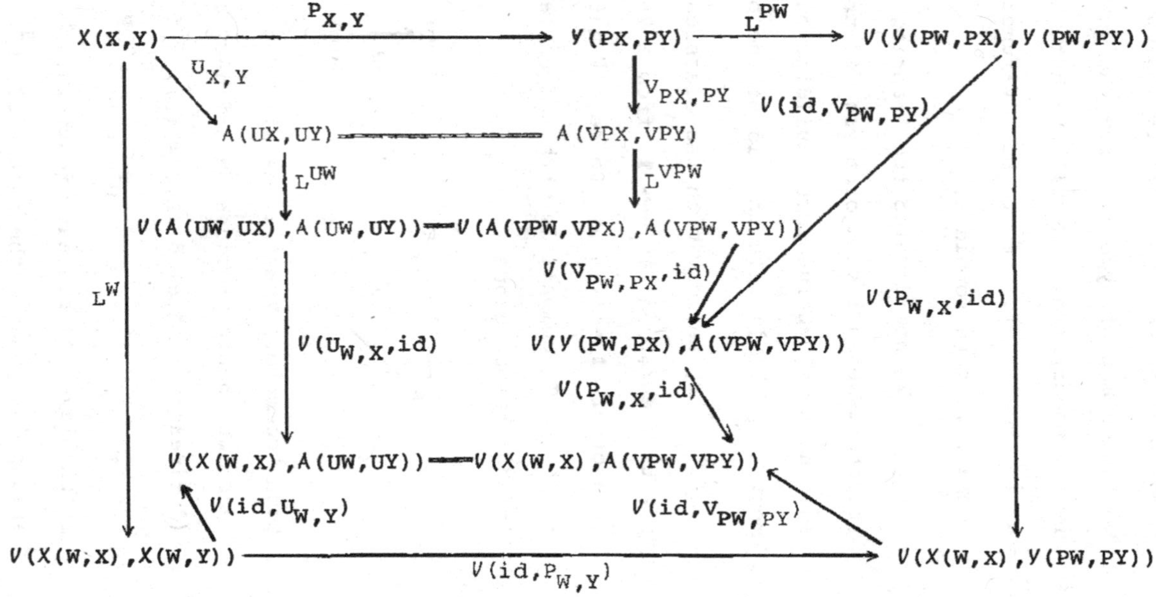

in each of which each small region commutes and the north-westwards map of type V from the lower right hand corner is a monomorphism. It follows that the perimeters commute, as is required for P to be a $\mathcal{V}$-functor. The proof of Lemma 2 is thus complete.

We now develop the $\mathcal{V}$-analogue of the familiar structure-semantics adjointness for triples on $\mathcal{A}$ and adjoint $\mathcal{A}$-valued functors, building on the usual development (e.g., [8]) with the aid of the preceding lemma.

By analogy with the classical situation, we define a $\mathcal{V}$-triple map $\tau\colon \mathbb{T}' \longrightarrow \mathbb{T}$ from a $\mathcal{V}$-triple $\mathbb{T}' = (T', \eta', \mu')$ to another $\mathbb{T} = (T, \eta, \mu)$ to be a $\mathcal{V}$-natural transformation $\tau\colon T' \longrightarrow T$ compatible with the units and multiplications, in that the diagrams

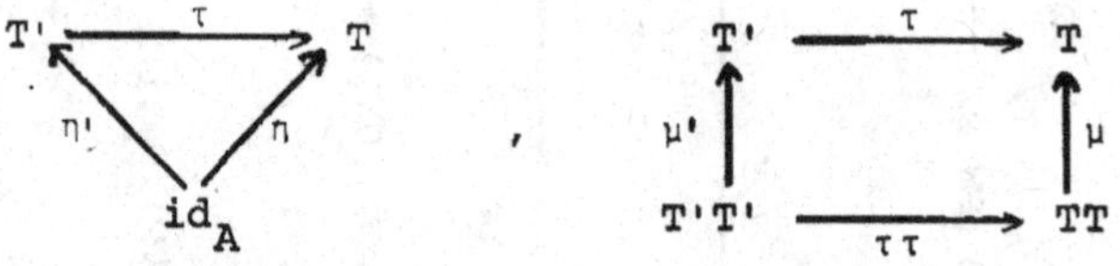


both commute. Thus, $\tau$ is also a triple map in the usual sense and hence provides an ordinary functor $\mathcal{A}^{\tau}\colon \mathcal{A}^{\mathbb{T}} \longrightarrow \mathcal{A}^{\mathbb{T}'}$ (defined by $\mathcal{A}^{\tau}(A,\alpha) = (A, \alpha \circ \tau_A)$, $\mathcal{A}^{\tau}(f) = f$) satisfying

$$U^{\mathbb{T}'} \circ \mathcal{A}^{\tau} = U^{\mathbb{T}} \tag{2.2}$$

at the level of ordinary functors. Using the standing assumptions on $\mathcal{V}$, the nature of $U^{\mathbb{T}'}$ and $\mathcal{A}^{\tau}$ and the definition of the $\mathcal{V}$-structure on $\mathcal{A}^{\mathbb{T}'}$, the preceding lemma guarantees a unique $V$-functor structure for $A^{\tau}$ such that (2.2) holds at the level of $V$-functors, provided only the perimeter of the diagram

$$\begin{array}{ccccc}
A^{\mathbb{T}}((A,\alpha),(B,\beta)) & \xrightarrow{U^{\mathbb{T}}} & A(A,B) & \xrightarrow{T'_{A,B}} & A(T'A,T'B) \\
 & & \downarrow{\scriptstyle T_{A,B}} & & \downarrow{\scriptstyle A(id,\tau_B)} \\
\downarrow{\scriptstyle U^{\mathbb{T}}} & & A(TA,TB) & \xrightarrow[A(\tau_A,id)]{} & A(T'A,TB) \\
 & & {\scriptstyle A(id,\beta)}\downarrow & & \downarrow{\scriptstyle A(id,\beta)} \\
A(A,B) & \xrightarrow[A(\alpha,B)]{} & A(TA,B) & \xrightarrow[A(\tau_A,B)]{} & A(T'A,B)
\end{array}$$

commutes. Now the left hand pentagon commutes by definition of $A^{\mathbb{T}}$, $V$-naturality of $\tau$ guarantees that the upper right hand square commutes, and either leg of the lower right hand square is $A(\tau_A,\beta)$. So the perimeter does commute, and we have proved the first part of

Proposition 2

Each functor $A^{\tau}: A^{\mathbb{T}} \longrightarrow A^{\mathbb{T}'}$ ($\tau: \mathbb{T}' \longrightarrow \mathbb{T}$ a $V$-triple map) is in a unique way a $V$-functor satisfying $U^{\mathbb{T}'}A^{\tau} = U^{\mathbb{T}}$ at the level of $V$-functors. Moreover, the familiar relations

$$A^{\tau\tau'} = A^{\tau'} \circ A^{\tau}, \quad A^{id_{\mathbb{T}}} = id_{A^{\mathbb{T}}}$$

expressing the functionality of triple semantics, are valid even at the level of $V$-functors when $\tau$ and $\tau'$ are $V$-triple maps and $\mathbb{T}$ is a $V$-triple.

<u>Proof</u>. The validity of the displayed relations at the level of $\mathcal{V}$-functors is a simple consequence of the uniqueness assertion of Lemma 2.

Let us then gather together, on the one hand, all $\mathcal{V}$-triples on the $\mathcal{V}$-category $\mathcal{A}$, along with all $\mathcal{V}$-triple maps between them – forming a category $\mathcal{V}\text{-Trip}(\mathcal{A})$ – and, on the other hand, all $\mathcal{V}$-adjoint $\mathcal{A}$-valued $\mathcal{V}$-functors (that is, $\mathcal{V}$-functors $U\colon \mathcal{X} \longrightarrow \mathcal{A}$ equipped with a $\mathcal{V}$-functor $F\colon \mathcal{A} \longrightarrow \mathcal{X}$ and $\mathcal{V}$-natural front and back adjunctions making $U$ $\mathcal{V}$-adjoint (on the right) to $F$), with all $\mathcal{V}$-functors between their domains making commutative triangles

$$\begin{array}{ccc} \mathcal{X} & \longrightarrow & \mathcal{X}' \\ & {\scriptstyle U}\searrow \quad \swarrow{\scriptstyle U'} & \\ & \mathcal{A} & \end{array}$$

as morphisms – forming a category $\mathcal{V}\text{-Adj}(\mathcal{V}\text{-Cat}, \mathcal{A})$. Proposition 2 then asserts that the passages

$$[(T,\eta,\mu) = \mathbb{T}] \longmapsto [U^{\mathbb{T}}\colon \mathcal{A}^{\mathbb{T}} \longrightarrow \mathcal{A}; F^{\mathbb{T}},\eta,\varepsilon^{\mathbb{T}}],$$

$$[\tau\colon \mathbb{T}' \longrightarrow \mathbb{T}] \longmapsto [\mathcal{A}^{\tau}\colon \mathcal{A}^{\mathbb{T}} \longrightarrow \mathcal{A}^{\mathbb{T}'}],$$

constitute a contravariant functor – $\mathcal{V}$-triple semantics – from $\mathcal{V}\text{-Trip}(\mathcal{A})$ to $\mathcal{V}\text{-Adj}(\mathcal{V}\text{-Cat}, \mathcal{A})$.

The $\mathcal{V}$-triple structure functor (also contravariant) in the other direction is easier: if $U\colon \mathcal{X} \longrightarrow \mathcal{A}$ and $F\colon \mathcal{A} \longrightarrow \mathcal{X}$ are $\mathcal{V}$-adjoint $\mathcal{V}$-functors with ($\mathcal{V}$-natural) front and back adjunctions $\eta\colon id_{\mathcal{A}} \longrightarrow UF$, $\varepsilon\colon FU \longrightarrow id_{\mathcal{X}}$, there is no trouble to see that $(UF,\eta,U\varepsilon F)$ is a $V$-triple. If $U'\colon X' \longrightarrow A$ and $F'\colon A \longrightarrow X'$ is another pair of $V$-adjoint $V$-functors, with front and back adjunctions $\eta'$ and $\varepsilon'$, respectively, and if $P\colon X \longrightarrow X'$ is a $V$-functor for which $U' \circ P = U$, it is well known how to construct an ordinary natural transformation $\tau_P\colon U'F' \longrightarrow UF$ that actually is a triple map from $(U'F',\eta',U'\varepsilon'F')$ to $(UF,\eta,U\varepsilon F)$. In fact, however, $\tau_P$ is a $V$-natural, and hence a $V$-triple map, being given explicitly as the composition

$$U'F' \xrightarrow[U'F'\eta]{} U'F'UF = U'F'U'PF \xrightarrow[U'\varepsilon'PF]{} U'PF = UF$$

of $V$-natural transformations. The functoriality of the ordinary contravariant triple structure functor then guarantees that the passages

$$(U;F,\eta,\varepsilon) \longrightarrow (UF,\eta,U\varepsilon F) \;,$$
$$P \longrightarrow \tau_P \;,$$

constitute a contravariant functor – $V$-triple structure – from $V\text{-Adj}(V\text{-Cat},A)$ to $V\text{-Trip}(A)$.

In Theorem 2 below, we shall prove that $V$-triple semantics and $V$-triple structure are adjoint on the right. To this end, we observe first that the $V$-triple structure of the $V$-adjoint $V$-functor $(U^{\mathbb{T}};F^{\mathbb{T}},\eta,\varepsilon^{\mathbb{T}})$ arising from the $V$-triple $\mathbb{T} = (T,\eta,\mu)$ is quite obviously nothing other than $\mathbb{T}$ itself again. For, $U^{\mathbb{T}}F^{\mathbb{T}} = T$, $\eta = \eta$, and $U^{\mathbb{T}}\varepsilon^{\mathbb{T}}F^{\mathbb{T}} = \mu$ (for this recall $(U^{\mathbb{T}}\varepsilon^{\mathbb{T}})_{F^{\mathbb{T}}A} = (U^{\mathbb{T}}\varepsilon^{\mathbb{T}})_{(TA,\mu_A)} = \mu_A$). This identification is one of the adjunction maps for the advertised adjointness between $V$-structure and $V$-semantics. The other is described as follows.

Let $(U;F,\eta,\varepsilon)$ be as before, and let $\mathbb{T} = (UF,\eta,U\varepsilon F)$. The well known "semantical comparison functor" $\Phi\colon X \longrightarrow A^{\mathbb{T}}$, defined by $\Phi X = (UX, U\varepsilon_X)$, $\Phi f = Uf$, satisfies $U^{\mathbb{T}} \circ \Phi = U$ at the level of ordinary functors. We use Lemma 2 to prove

Proposition 3

The usual semantical comparison functor $\Phi\colon X \longrightarrow A^{\mathbb{T}}$ arising from the $V$-adjoint $V$-functor situation $(U\colon X \longrightarrow A;\ F,\eta,\varepsilon)$ with $\mathbb{T} = (UF,\eta,U\varepsilon F)$, admits a unique enrichment to a $V$-functor $\Phi\colon X \longrightarrow A^{\mathbb{T}}$ satisfying

$$U^{\mathbb{T}} \circ \Phi = U \tag{2.3}$$

at the level of $V$-functors; moreover, $\Phi$ is the only $V$-functor satisfying (2.3), and in addition, we have

$$\Phi \circ F = F^{\mathbb{T}} \tag{2.4}$$

and

$$\tau_{\Phi} = \mathrm{id}_{\mathbb{T}}\ . \tag{2.5}$$

Proof. The first part of Proposition 3 is proved, using Lemma 2, by showing that the exterior of the diagram

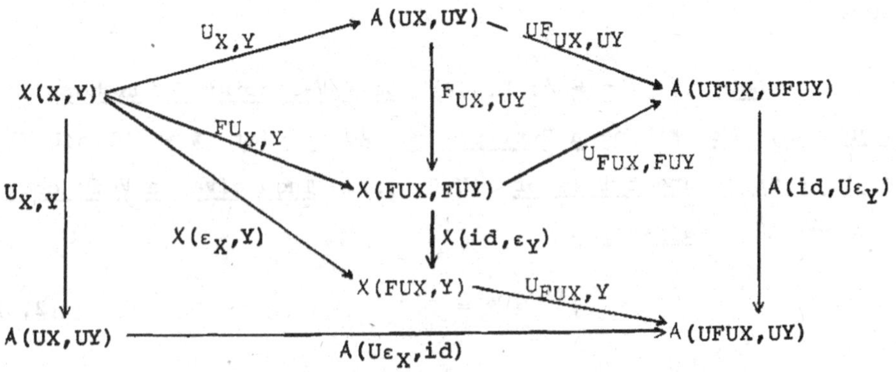


commutes. But the two upper triangles commute by the definition of the $V$-functor structure of a composition of $V$-functors; the remaining triangle commutes because $\varepsilon$ is $V$-natural; and the other regions commute for even easier reasons.

Since the ordinary semantical comparison functor $\Phi$ is the only functor for which (2.3) holds at the lowest level, the second uniqueness statement now follows from the first.

Since (2.4) is known at the level of ordinary functors and since both $\Phi F$ and $F^{\mathbb{T}}$ are $V$-functors to $A^{\mathbb{T}}$, the uniqueness portion of Lemma 2 establishes the validity of (2.4) at the level of $V$-functors. Finally, since relation (2.5) is well known for ordinary triple-structure-semantics, it is true here as well.

To give the reader two strategy options for proving Theorem 2, and to make our exposition more complete, we state

Lemma 3

Let $(U: X \longrightarrow A; F,\eta,\varepsilon)$ be a $V$-adjoint $V$-functor situation, let $\mathbb{T}'$ be a $V$-triple on $A$, and let $\mathbb{T} = (UF,\eta,U\varepsilon F)$ be the $V$-structure triple of $(U;F,\eta,\varepsilon)$. Then given a $V$-functor $P: X \longrightarrow A^{\mathbb{T}'}$ satisfying

$$U^{\mathbb{T}'} \circ P = U \tag{2.6}$$

at the level of $V$-functors, the unique ordinary triple map $\tau: \mathbb{T}' \longrightarrow \mathbb{T}$, solving the equation

$$P = A^{\tau} \circ \Phi \tag{2.7}$$

of ordinary functors (available by ordinary triple-structure-semantics adjointness), is in fact a $V$-triple map, and (2.7) is valid at the level of $V$-functors as well.

Proof. $\tau$ is, of course, just $\tau_P$ which we have already observed is a $V$-triple map. The validity of (2.7) at the level of $V$-functors is a by now familiar consequence of Lemma 2.

With this lemma behind us, nothing can prevent two proofs of

Theorem 2 (Adjointness of $V$-structure and $V$-semantics)

The contravariant functors $V$-triple-structure and $V$-triple-semantics are adjoint on the right, with one adjunction transformation given by the system of $V$-semantical comparison

<u>$\mathcal{V}$-functors</u> $\Phi$ , <u>and the other given by the identifications</u>

$$\mathbb{T} = (U^{\mathbb{T}}F^{\mathbb{T}}, \eta, U^{\mathbb{T}}\varepsilon^{\mathbb{T}}F^{\mathbb{T}}) \quad . \tag{2.8}$$

<u>First Proof in Outline</u>. First verify that the systems of $\mathcal{V}$-semantical comparison functors and of identifications (2.8) are natural; then relations (2.5) and (2.8) deliver the adjointness.

<u>Second Proof in Outline</u>. Lemma 3 provides one-one correspondence between $\mathcal{V}$-Trip($\mathcal{A}$)($\mathbb{T}'$,$\mathbb{T}$) and the class of all $\mathcal{V}$-functors $P: \mathcal{X} \longrightarrow \mathcal{A}^{\mathbb{T}'}$ satisfying (2.6). That delivers the adjointness, and the information that the family of semantical comparison functors constitutes one of the adjunctions; there remains only the easy verification that the identifications (2.8) constitute the other.

## §3. LAWVERE'S KLEISLI-EILENBERG-MOORE ADJOINTNESS

Since we never suppose the base category $\mathcal{V}$ to be symmetric, we cannot blindly apply duality to obtain counterpart results for cotriples. Nevertheless, diligent mimicry of the first sections of this paper will produce all that we shall assume known about $\mathcal{V}$-cotriples $\mathbb{G}$ on a $\mathcal{V}$-category $\mathcal{A}$, about the $\mathcal{V}$-category $\mathcal{A}_{\mathbb{G}}$ of $\mathbb{G}$-coalgebras, and about the structure-semantics adjointness relations appropriate to

$\mathcal{V}$-coadjoint $\mathcal{V}$-functors versus $\mathcal{V}$-cotriples.

Bearing $\mathcal{V}$-cotriples in mind, let us reconsider the $\mathcal{V}$-triple semantics functor. Given the $\mathcal{V}$-triple $\mathbb{T} = (T,\eta,\mu)$ on a $\mathcal{V}$-category $A$, we have produced the $\mathcal{V}$-category $A^{\mathbb{T}}$ with a new $\mathcal{V}$-adjointness situation $(U^{\mathbb{T}}: A^{\mathbb{T}} \longrightarrow A;\ F^{\mathbb{T}},\eta,\varepsilon^{\mathbb{T}})$. Clearly $(F^{\mathbb{T}}U^{\mathbb{T}},\varepsilon^{\mathbb{T}},F^{\mathbb{T}}\eta U^{\mathbb{T}})$ is a $\mathcal{V}$-cotriple on $A^{\mathbb{T}}$, which we shall designate briefly as $\mathbb{G}^{\mathbb{T}}$. This passage from a $\mathcal{V}$-triple on one category to a $\mathcal{V}$-cotriple on another is, as we shall see, actually part of a functor from the category $\mathcal{V}$-Trip (whose objects are all pairs $(A,\mathbb{T})$ with $A$ a $\mathcal{V}$-category and $\mathbb{T}$ a $\mathcal{V}$-triple on $A$, where as morphisms $(A,\mathbb{T}) \longrightarrow (A',\mathbb{T}')$ we allow all $\mathcal{V}$-functors $X: A \longrightarrow A'$ such that

$$\left.\begin{aligned} & T'X = XT, \\ & \eta'X = X\eta: X \longrightarrow T'X = XT, \\ \text{and}\quad & \mu'X = X\mu: T'T'X = XTT \longrightarrow XT = T'X) \end{aligned}\right\} \qquad (3.1)$$

to the analogously defined category $\mathcal{V}$-Cotrip of all $\mathcal{V}$-cotriples on various $\mathcal{V}$-categories and strictly $\mathcal{V}$-cotriple-preserving $\mathcal{V}$-functors.

To describe the effect on a morphism $X: A \longrightarrow A'$ from $(A,\mathbb{T})$ to $(A',\mathbb{T}')$, we shall once again have recourse to Lemma 2. To begin with, we define an ordinary functor $\tilde{X}: A^{\mathbb{T}} \longrightarrow A'^{\mathbb{T}'}$ satisfying

$$\left.\begin{aligned} XU^{\mathbb{T}} &= U^{\mathbb{T}'}\tilde{X} \,, \\ \tilde{X}F^{\mathbb{T}} &= F^{\mathbb{T}'}X \,, \\ \tilde{X}\varepsilon^{\mathbb{T}} &= \varepsilon^{\mathbb{T}'}\tilde{X} \,. \end{aligned}\right\} \qquad (3.2)$$

Namely, if $(A,\alpha\colon TA \longrightarrow A)$ is a $\mathbb{T}$-algebra in $\mathcal{A}$, let $\tilde{X}(A,\alpha) = (XA, X\alpha\colon T'XA = XTA \longrightarrow XA)$. It is left to the reader to decide that $\tilde{X}(A,\alpha)$ is in fact a $\mathbb{T}'$-algebra – all he needs is relations (3.1) and the defining properties of algebra structure maps. In much the same way, the reader can easily verify that, whatever the $\mathbb{T}$-algebra map $f\colon (A,\alpha) \longrightarrow (B,\beta)$, the $\mathcal{A}'$-morphism $Xf\colon XA \longrightarrow XB$ is actually a $\mathbb{T}'$-algebra map, call it $\tilde{X}f$, from $\tilde{X}(A,\alpha)$ to $\tilde{X}(B,\beta)$.

It should then be clear that $\tilde{X}$ is a functor and that relations (3.2) are at least satisfied at the level of ordinary functors. In fact, $\tilde{X}$ is easily the only functor having these properties. Now an application of Lemma 2 enriches $\tilde{X}$ uniquely to a $\mathcal{V}$-functor satisfying the first of the equations (3.2) at the level of $\mathcal{V}$-functors, and another application of Lemma 2 guarantees the validity of the second of these equations at the level of $\mathcal{V}$-functors as well. The validity of the third equation not being interfered with, it now follows that $\tilde{X}$ is in fact a $\mathcal{V}$-Cotrip morphism from $(\mathcal{A}^{\mathbb{T}},\mathbb{G}^{\mathbb{T}})$ to $(\mathcal{A}'^{\mathbb{T}'},\mathbb{G}^{\mathbb{T}'})$. Since there is no problem in seeing that $\widetilde{YX} = \tilde{Y}\tilde{X}$ and that $\widetilde{id_{(\mathcal{A},\mathbb{T})}} = id_{\mathcal{A}^{\mathbb{T}}}$, the passages

$$(\mathcal{A},\mathbb{T}) \longmapsto (\mathcal{A}^{\mathbb{T}},\mathbb{G}^{\mathbb{T}}) \,,$$
$$X \longmapsto \tilde{X}$$

constitute a functor $EM\colon \mathcal{V}\text{-Trip} \longrightarrow \mathcal{V}\text{-Cotrip}$.

The purpose of this section is to describe the left adjoint to EM.

Consider, therefore, a $\mathcal{V}$-triple $\mathbb{T} = (T, \eta, \mu)$ on a $\mathcal{V}$-category $\mathcal{A}$. Let $K\ell(\mathbb{T})$ be the usual Kleisli category for $\mathbb{T}$, with objects those of $\mathcal{A}$ and $K\ell(\mathbb{T})$-morphisms from A to B all $\mathcal{A}$-morphisms from A to TB. We make $K\ell(\mathbb{T})$ into a $\mathcal{V}$-category by setting $K\ell(\mathbb{T})(A,B) = \mathcal{A}(A,TB)$, and using the adjointness isomorphisms

$$K\ell(\mathbb{T})(A,B) = \mathcal{A}(A,TB) \xrightarrow{\cong} \mathcal{A}^{\mathbb{T}}(F^{\mathbb{T}}A, F^{\mathbb{T}}B) \qquad (3.3)$$

to find (uniquely) units and composition rules for $K\ell(\mathbb{T})$ in such a way that the (iso-) morphisms (3.3) make the passage

$$A \longrightarrow F^{\mathbb{T}}A$$

a ($\mathcal{V}$-fully faithful) $\mathcal{V}$-functor $I^{\mathbb{T}}\colon K\ell(\mathbb{T}) \longrightarrow \mathcal{A}^{\mathbb{T}}$ . Explicitly, the units are

$$I \xrightarrow[j_A]{} \mathcal{A}(A,A) \xrightarrow[\mathcal{A}(A,\eta_A)]{} \mathcal{A}(A,TA) = K\ell(\mathbb{T})(A,A) \ ,$$

and the composition rules are

$$\begin{array}{ccccc}
K\ell(\mathbb{T})(B,C) \otimes K\ell(\mathbb{T})(A,B) & & \dashrightarrow & & K\ell(\mathbb{T})(A,C) \\
\| & & & & \| \\
\mathcal{A}(B,TC) \otimes \mathcal{A}(A,TB) & & & & \| \\
\big\downarrow{\scriptstyle T_{B,TC} \otimes \mathrm{id}} & & & & \| \\
\mathcal{A}(TB,TTC) \otimes \mathcal{A}(A,TB) & \xrightarrow[M]{} & \mathcal{A}(A,TTC) & \xrightarrow[\mathcal{A}(A,\mu_C)]{} & \mathcal{A}(A,TC)
\end{array} \ ,$$

or, if $\mathcal{V}$ is closed,

$$\begin{array}{ccc} K\ell(\mathbb{T})(B,C) & \dashrightarrow & V(K\ell(\mathbb{T})(A,B),K\ell(\mathbb{T})(A,C)) \\ \| & & \| \\ A(B,TC) & & V(A(A,TB),A(A,TC)) \\ \downarrow T_{B,TC} & & \uparrow V(id,A(A,\mu_C)) \\ A(TB,TTC) & \xrightarrow[L^A]{} & V(A(A,TB),A(A,TTC)) \end{array} .$$

By Lemma 2, the passages

$$A \longmapsto A$$

$$A(A,B) \xrightarrow[A(A,\eta_B)]{} A(A,TB) = K\ell(\mathbb{T})(A,B)$$

constitute a $V$-functor $f^{\mathbb{T}}: A \longrightarrow K\ell(\mathbb{T})$ satisfying $I^{\mathbb{T}}f^{\mathbb{T}} = F^{\mathbb{T}}$. The $V$-fully faithfulness of $I^{\mathbb{T}}$ shows that $U^{\mathbb{T}}I^{\mathbb{T}} =_{def} u^{\mathbb{T}}$ serves as (right) $V$-adjoint to $f^{\mathbb{T}}$, with front adjunction $\eta$ and back adjunction $(I^{\mathbb{T}})^{-1}(\varepsilon^{\mathbb{T}}F^{\mathbb{T}})$. It follows that the $V$-triple structure of $(u^{\mathbb{T}};f^{\mathbb{T}},\eta,(I^{\mathbb{T}})^{-1}(\varepsilon^{\mathbb{T}}F^{\mathbb{T}}))$ is just $\mathbb{T}$ again.

On the other hand, if $(U: X \longrightarrow A;\ F,\eta,\varepsilon)$ is a $V$-adjoint $V$-functor situation, with $\mathbb{T} = (UF,\eta,U\varepsilon F)$ the associated $V$-triple, there is a canonical $V$-functor $\Psi: K\ell(\mathbb{T}) \longrightarrow X$ satisfying $U\Psi = u^{\mathbb{T}}$ – indeed, $K\ell(\mathbb{T})$ is clearly $V$-isomorphic with the full image $V$-category of $F: A \longrightarrow X$. Moreover, $\Psi f^{\mathbb{T}} = F$ and the $V$-triple map induced by $\Psi$ is obviously $id_{\mathbb{T}}$. This shows that the system of $\Psi$'s is one of the adjunction transformations making the Kleisli $V$-category construction and $V$-triple structure adjoint on the left. Alternatively, viewing $V$-triple structure as a covariant functor

$$(V\text{-}Adj(V\text{-}Cat,\ A))^{op} \longrightarrow Trip(A) ,$$

it has $V$-triple-semantics – the Eilenberg-Moore construction of $A^{\mathbb{T}}$ et al – as left adjoint, and the above described Kleisli $V$-category construction as right adjoint.

As earlier in this section, we wish to let $A$ vary. Clearly, each Trip-morphism $X: (A,\mathbb{T}) \longrightarrow (A',\mathbb{T}')$, inducing a $V$-functor $\tilde{X}: A^{\mathbb{T}} \longrightarrow A'^{\mathbb{T}'}$ satisfying (3.2), actually winds up inducing a unique $V$-functor $\hat{X}: K\ell(\mathbb{T}) \longrightarrow K\ell(\mathbb{T}')$ satisfying

$$Xu^{\mathbb{T}} = u^{\mathbb{T}'}\hat{X} ,$$
$$\hat{X}f^{\mathbb{T}} = f^{\mathbb{T}'}X ,$$
$$\tilde{X}I^{\mathbb{T}} = I^{\mathbb{T}'}\hat{X} .$$

Furthermore, $\hat{X}$ actually turns out to be a Cotrip-morphism from $(K\ell(\mathbb{T}),(f^{\mathbb{T}}u^{\mathbb{T}},(I^{\mathbb{T}})^{-1}(\varepsilon^{\mathbb{T}}F^{\mathbb{T}}),f^{\mathbb{T}}\eta u^{\mathbb{T}}))$ to $(K\ell(\mathbb{T}'),(f^{\mathbb{T}'}u^{\mathbb{T}'},(I^{\mathbb{T}'})^{-1}(\varepsilon^{\mathbb{T}'}F^{\mathbb{T}'}),f^{\mathbb{T}'}\eta' u^{\mathbb{T}'}))$ .

Thus, passage to the Kleisli $V$-category along with the $V$-adjointness cotriple for $u^{\mathbb{T}} \vdash f^{\mathbb{T}}$ becomes a functor $\overline{\mathrm{KL}}$: $V$-Trip $\longrightarrow$ $V$-Cotrip.

Actually, we must dualize the above considerations, by the mimicry procedure mentioned at the head of this section, to obtain another Kleisli-type functor KL: $V$-Cotrip $\longrightarrow$ $V$-Trip.

<u>Theorem 3</u>

KL: $V$-Cotrip $\longrightarrow$ $V$-Trip <u>has as (right) adjoint</u> EM: $V$-Trip $\longrightarrow$ $V$-Cotrip.

<u>Proof Sketch</u>. We content ourselves with presenting the front and back adjunctions. Let $\mathbb{G}$ be a $V$-cotriple on the $V$-category $X$. Form the Kleisli $V$-category $K\ell(\mathbb{G})$ for $\mathbb{G}$. We have the $V$-triple $\mathbb{T}_{\mathbb{G}}$ on $K\ell(\mathbb{G})$ coming from the $V$-adjoint pair of $V$-functors at the left in the inset diagram. There is then the $V$-semantical comparison $V$-functor $\Phi$ from $X$ to the $V$-category of $\mathbb{T}_{\mathbb{G}}$-algebras on $K\ell(\mathbb{G})$. $\Phi$ makes both triangles commute, as was seen before. It can then easily be seen to be a $V$-Cotrip morphism from $(X,\mathbb{G})$ to $EM(KL(X,\mathbb{G})) = ((K\ell(\mathbb{G}))^{\mathbb{T}_{\mathbb{G}}}, \mathbb{G}^{\mathbb{T}_{\mathbb{G}}})$ .

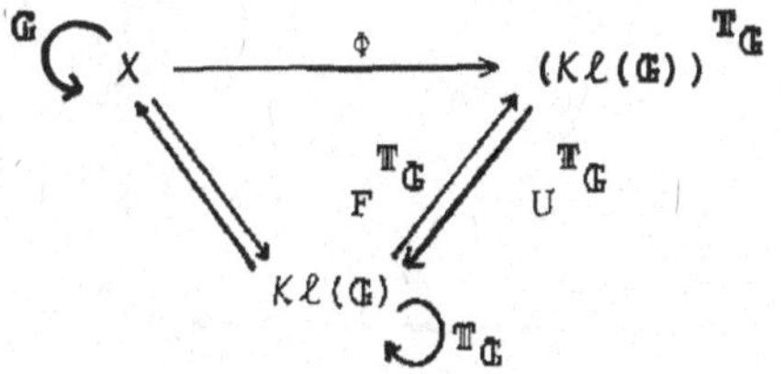


For the back adjunction, let $\mathbb{T}$ be a $V$-triple on the $V$-category $A$. Form the $V$-category $A^{\mathbb{T}}$ of $\mathbb{T}$-algebras in $A$. This bears the cotriple $\mathbb{G}^{\mathbb{T}}$, whose Kleisli $V$-category we now form. There is then the $V$-functor $\Psi: K\ell(\mathbb{G}^{\mathbb{T}}) \longrightarrow A$ making both triangles commute, indeed, giving a $V$-Trip morphism from (next page) $KL(EM(A,\mathbb{T})) = (K\ell(\mathbb{G}^{\mathbb{T}}), \mathbb{T}_{\mathbb{G}^{\mathbb{T}}})$ to $(A,\mathbb{T})$. We leave to the reader the verifications that, using these adjunctions, KL is left adjoint to EM.

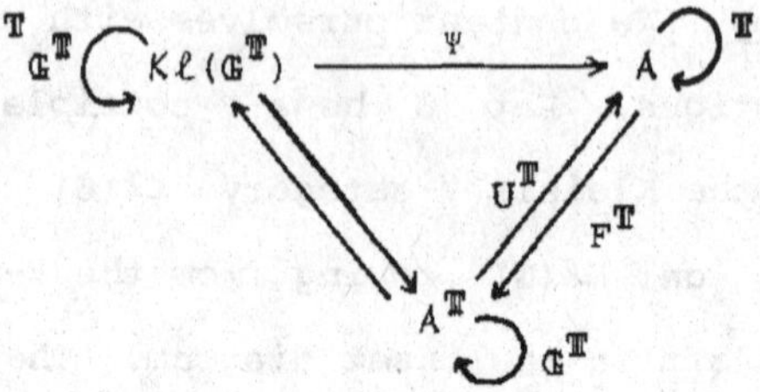


Incidentally, the same sort of thing can be done replacing EM by $\overline{\text{EM}}$: $V$-Cotrip $\longrightarrow$ $V$-Trip, assigning to $(X,\mathbb{G})$ the canonical $V$-triple on the $V$-category $X_{\mathbb{G}}$ of $\mathbb{G}$-coalgebras. Then $\overline{\text{EM}}$ has $\overline{\text{KL}}$ as left adjoint. Once again, purely formal mimicry delivers the proof.

## REFERENCES

[1] Bunge, Marta, "Relative Functor Categories and Categories of Algebras", *J. Alg.* (to appear).

[2] Eilenberg, Samuel, and Kelly, G. Max, "Closed Categories", pp. 421-562 in *Proc. Conf. Categ. Alg.* (La Jolla, 1965), Springer, Berlin, (1966).

[3] Eilenberg, Samuel, and Moore, J. C., "Adjoint Functors and Triples", *Ill. J. Math.*, 9; 381-398 (1965).

[4] Godement, Roger, *Théorie des Faisceaux*. Hermann, Paris, (1958).

[5] Kelly, G. Max, "Tensor Products in Categories", *J. Alg.*, 2; 15-37 (1965).

[6] Kleisli, Heinrich, "Every Standard Construction is Induced by a Pair of Adjoint Functors", *Proc. Amer. Math. Soc.*, 16; 544-546 (1965). Review by P. J. Huber, *Math. Reviews*, 31; #1289, (1966).

[7] Kock, Anders, "On Monads in Symmetric Monoidal Closed Categories", *Preprint Series* 1967/68 No. 14 (April, 1968), Mat. Inst., Aarhus Univ.

[8] Linton, F. E. J., "An Outline of Functorial Semantics", in *Seminar on Triples and Categorical Homology Theory*, (Springer Lecture Notes in Math., Vol. 80) (1968).

# MINIMAL SUBALGEBRAS FOR DYNAMIC TRIPLES[1]

by

Ernest Manes

This paper is a preliminary report on a larger project dedicated to the proposition that universal algebra and compact topological dynamics have a lot to learn from each other. The author has tried so hard to make this paper accessible to topological dynamicists (as opposed to categorists) that the work "adjoint" doesn't seem to come up. The reader is referred to [Eb] and the references there for the dynamical origin of the algebra we study here. The prerequisite for reading the paper is a knowledge of universal algebra in the language of triples in sets such as may be found in [Ma] or [Mb]. It is hoped that the meaning of the main theorems is clear without knowledge of triple-theory; in such a case think of a "$\mathbb{T}$-algebra" as just a "universal algebra", which is entirely accurate though sufficiently non-classical to include exotic infinitary examples such as compact transformation groups; the enveloping semigroup of the algebra (as defined in 2.1) is the set of all derived unary operations; "$(1T, 1\mu)$" is the free algebra on one generator.

---

[1]Research supported by Harvey Mudd College.

## 1. DYNAMIC MONOIDS

Let $E$ be an abstract monoid with unit $e$. For $p \in E$, $\{L_p\}$ $\{R_p\}$ denotes the {left} {right} multiplication function induced by $p$.

### 1.1 Definition

$E$ is a *dynamic* monoid if $E$ possesses a minimal right ideal $I$ such that $I$, qua semigroup, is left cancellative (i.e. all left multiplications of $I$ are injective).

### 1.2 Definition

$\Delta \subset E$ is a *division set in* $E$ if $\Delta$ is non-empty and if for all $p,q \in E$ there exists $x \in E$ such that $\delta px = \delta q$ for all $\delta \in \Delta$.

### 1.3 Theorem

Assume that $E$ possesses a minimal right ideal $I$ and a maximal division set $\Delta$. Then the following statements are valid.

a. $\Delta$ is a left ideal in $E$.

b. There exists $u \in I \cap \Delta$ such that $\delta u = \delta$ $(\delta \in \Delta)$ and $up = p$ $(p \in I)$; (in particular, $uu = u$).

c. $I \cap \Delta = I\Delta$ and $I \cap \Delta$ is a group.

d. $I$ is a left cancellative semigroup (and hence $E$ is a dynamic monoid).

Proof. a. This is clear from the maximality of $\Delta$.

b. Let $a \in I$. There exists $x \in E$ such that $\delta ax = \delta e = \delta$ $(\delta \in \Delta)$. Define $u = ax$. Then $u \in I$ and $\delta u = \delta$ $(\delta \in \Delta)$. Let $p,q \in E$. There exists $x \in E$ with $\delta px = \delta q$ $(\delta \in \Delta)$. As $upxI = I$ there exists $y \in I$ with $upxy = uq$. For $\delta \in \Delta$, $\delta pxy = \delta upxy = \delta uq = \delta q$. By the maximality of $\Delta$, $u \in \Delta$. In particular $uu = u$. That $up = p$ $(p \in I)$ is a general fact about idempotents in a minimal right ideal: as $uI = I$ there exists $q \in I$ with $uq = p$ and then $up = uuq = uq = p$.

c. $I\Delta \subset I \cap \Delta$ since $I$ is a right ideal and $\Delta$ is a left ideal. If $p \in I \cap \Delta$ then $p = up \in I\Delta$. Clearly $I \cap \Delta$ is a subsemigroup with $u$ as two-sided unit. Let $p \in I$, $\delta \in \Delta$ so that $p\delta$ is a typical element of $I \cap \Delta$. There exists $x \in I$ with $p\delta x = u$ and then there exists $y \in I$ with $xp\delta y = x$. Since $x = xp\delta y = xup\delta y$ $= xp\delta xp\delta y = xp\delta x = xu$, $x \in \Delta$. To see $x$ is also a left inverse let $z \in I$ with $xp\delta z = u$; then $u = xp\delta z = xp\delta xp\delta z$ $= xp\delta u = xp\delta$.

d. Let $a,p,q \in I$ with $ap = aq$. As $u, au \in I \cap \Delta$ there exists $\delta \in I \cap \Delta$ with $\delta u = au$. Hence $p = uup = \delta^{-1}\delta up = \delta^{-1}aup = \delta^{-1}ap = \delta^{-1}aq = q$. The proof is complete.

1.4 Definition

$E$ is compactible if there exists a compact Hausdorff

topology on $E$ with respect to which $L_p$ is continuous for all $p \in E$.

## 1.5 Definitions

For sets $X, \Gamma$, the fine-power topology on $X^\Gamma$ is the cartesian power topology induced by the discrete topology on $X$. For $\Gamma \xrightarrow{\gamma} E$ any $E$-valued function define $E^\gamma$ to be the subset $\{\Gamma \xrightarrow{\gamma} E \xrightarrow{R_p} E : p \in E\}$ of $E^\Gamma$. $E$ is a quasicompactible monoid if $E$ has a minimal right ideal and if for every $\Gamma \xrightarrow{\gamma} E$, $E^\gamma$ is closed in the fine-power topology on $E^\Gamma$.

## 1.6 Hierarchy Theorem

Each of the following four conditions on $E$ implies those beneath it.

a. $E$ is compactible.

b. $E$ is quasicompactible.

c. $E$ has a minimal right ideal and a maximal division set.

d. $E$ is dynamic.

Proof. a implies b. Let $\mathcal{T}$ be a compact Hausdorff topology on $E$ making each $L_p$ continuous. By Zorn's Lemma and compactness, $E$ has a minimal closed right ideal $I$. For $p \in I$, $pE$ is closed since $L_p$ is closed, so $pE = I$. Hence $I$ is a minimal right ideal. Let $\Gamma \xrightarrow{\gamma} E$. Consider the continuous map $f$ defined by

$$\begin{array}{ccc} (E,T) & \overset{f}{\dashrightarrow} & (E,T)^{\Gamma} \\ & \searrow_{L_{x\gamma}} & \downarrow pr_x \qquad (x \in \Gamma) \quad . \\ & & (E,T) \end{array}$$

As $E^{\gamma}$ is the image of $f$, $E^{\gamma}$ is closed in $(E,T)^{\Gamma}$ and hence is closed in the fine-power topology on $E^{\Gamma}$.

<u>b implies c</u>. Let $I$ be a minimal right ideal and let $a \in I$. If $p,q \in E$ then $aEE = I = aE$ so that $apx = aq$ for some $x \in E$. Hence $\{a\}$ is a division set. Let $(\Delta_{\alpha})$ be a chain of division sets and set $\Delta = \cup \Delta_{\alpha}$. Let $\Delta \xrightarrow{\gamma} E$ be the inclusion function. Let $F$ be a finite subset of $\Delta$. There exists $\alpha$ with $F \subset \Delta_{\alpha}$. There exists $x \in E$ with $\delta px = \delta q$ $(\delta \in \Delta_{\alpha})$. We have shown that for every finite subset of $\Delta$ there exists some $\psi \in E^{\gamma R_p}$ (i.e., $\psi = \gamma R_p R_x$) such that $\psi$ agrees with $\gamma R_q$ on $F$. This says that $\gamma R_q$ is in the fine-power closure of $E^{\gamma R_p}$. By the hypothesis, $\gamma R_q \in E^{\gamma R_p}$ so that $\gamma R_q = \gamma R_p R_y$ for some $y \in E$. Hence $\Delta$ is a division set. By Zorn's Lemma, there exists a maximal division set.

<u>c implies d</u>. This is 1.3d. The proof is complete.

Examples 1.8, 1.9 and 1.10 below show that none of the implications in the hierarchy theorem 1.6 have true converses.

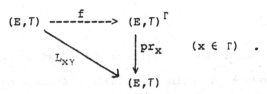

1.7 Theorem

Let $E$ be any lattice which has all suprema (including 0), and provide $E$ with its binary infimum monoid multiplication. Then $E$ is quasicompactible.

Proof. $\{0\}$ is a minimal right ideal. Let $\Gamma \xrightarrow{\gamma} E$, and let $\psi$ be in the fine-power closure of $E^{\gamma}$. Define $x = \sup\{\lambda\psi: \lambda \in \Gamma\}$. If $\lambda \in \Gamma$ then there exists $p \in E$ with $\lambda\psi = \lambda\gamma R_p = \inf\{\lambda\gamma, p\}$ so that $\lambda\psi = \inf\{\lambda\gamma, \lambda\psi\}$ $\leq \inf\{\lambda\gamma, x\} = \lambda\gamma R_x$. Let $\lambda' \in \Gamma$. There exists $q \in E$ with $\lambda\psi = \lambda\gamma R_q$ and $\lambda'\psi = \lambda'\gamma R_q$. Therefore $\inf\{\lambda\gamma, \lambda'\psi\}$ $= \inf\{\lambda\gamma, \lambda'\gamma, q\} \leq \inf\{\lambda\gamma, q\} = \lambda\psi$. As $\lambda'$ is arbitrary, $\inf\{\lambda\gamma, x\} \leq \lambda\psi$. Hence $\psi = \gamma R_x$. The proof is complete.

1.8 Example

A quasicompactible monoid that is not compactible. Let $E$ be the disjoint union of the real intervals $A = [0,1]$ and $B = [0,1]$. Defining $p \leq q$ if and only if ($p,q \in A$ or $p,q \in B$ or $p \in A$, $q \in B$) and $p \leq q$ as numbers, $E$ is a complete lattice. As in 1.7, $E$ is a quasicompactible monoid. Since any compact topology on $A$ with respect to which closed rays are closed sets must contain- hence be equal to- the usual topology on $[0,1]$, $A$ is not compactible. Since $A = \{p \in E: pL_x = 0\}$, where $x = \inf(B)$, it follows that $E$ is not compactible.

## 1.9 Example

A monoid with a minimal right ideal and a maximal division set that is not quasicompactible. Let $X$ be an infinite set and set $E = \{f \in X^X\colon f$ is the identity function or $f$ is not injective$\}$. $E$ is a submonoid of $X^X$. For $x \in X$ let $\tilde{x}$ denote the corresponding constant function. $I = \{\tilde{x} : x \in X\}$ is a minimal right ideal. Let $\delta \in X$. Then $\Delta = \{\tilde{\delta}\}$ is a division set. Suppose $\psi \in E$ were such that $\{\tilde{\delta}, \psi\}$ is a division set with $\psi \neq \tilde{\delta}$. Let $x \in X$ with $x\psi \neq \delta$. There exist $f,g,h \in E$ such that $x\psi f = \delta f$, $x\psi g \neq \delta g$, $\tilde{\delta} fh = \tilde{\delta} g$, $\psi fh = \psi g$. But then $\delta g = \delta fh = x\psi fh = x\psi g$, a contradiction. To prove that $E$ is not quasicompactible, let $X \xrightarrow{\gamma} E$ be the function $x\gamma = \tilde{x}$ and let $f \in X^X$ be any function not in $E$. It is easy to check that $f\gamma$ is in the fine-power closure of $E^\gamma$ but that $f\gamma \notin E^\gamma$.

## 1.10 Example

A dynamic monoid that has no maximal division set. Let $X$ be an infinite set with $x_0, x_1$ distant elements of $X$. Set $E = \{f \in X^X\colon f$ induces a bijection of $X - \{x_0, x_1\}$ onto $X - \{x_0, x_1\}$ and $x_0 f = x_0 = x_1 f\} \cup \{\tilde{x}_0, \tilde{x}_1, 1_X\}$. $E$ is a submonoid of $X^X$. $I = \{\tilde{x}_0, \tilde{x}_1\}$ is a minimal right ideal which qua semigroup is left cancellative, so $E$ is dynamic. Suppose $\Delta$ were a maximal division set. By 1.3b, $\Delta = \{\tilde{x}_0\}$ or $\Delta = \{\tilde{x}_1\}$. But $\{\tilde{x}_0, \tilde{x}_1\}$ is a division set. For let $f,g \in E$. If $f \notin I$, $ff^{-1}g = g$ (what we mean by "$f^{-1}$" being clear). If $g \in I$,

$fg = g$. Finally, if $f \in I$, $g \notin I$ then $fg = g$ on $\{x_0, x_1\}$.

1.11 Theorem

Let $E$ be a left cancellative monoid. Then $E^\gamma$ is closed in the fine-power topology for all $\Gamma \xrightarrow{\gamma} E$

Proof. Let $\Gamma \xrightarrow{\gamma} E$ and let $\psi$ be in the fine-power closure of $E^\gamma$. Let $a \in E$. If $F$ is a finite subset of $\Gamma$ then there exists $p \in E$ with $\lambda\psi = \lambda\gamma R_p$ $(\lambda \in F)$. Hence $\lambda\psi L_a = \lambda\gamma R_p L_a = \lambda\gamma L_a R_p$ $(\lambda \in F)$. This shows that $\psi L_a$ is in the fine-power closure of $E^{\gamma L_a}$. Let $\lambda_0 \in \Gamma$. For each $\lambda \in \Gamma$ there exists $p(\lambda) \in E$ with $\lambda_0\psi L_a = \lambda_0\gamma L_a R_{p(\lambda)}$ and $\lambda\psi L_a = \lambda\gamma L_a R_{p(\lambda)}$. For all $\lambda \in \Gamma$ we have $a(\lambda_0\gamma)p(\lambda)$ $= a(\lambda_0\psi) = a(\lambda_0\gamma)p(\lambda_0)$; It follows from the hypothesis on $E$ that $p(\lambda) = p(\lambda_0)$ for all $\lambda \in \Gamma$. Therefore $\psi L_a = \gamma L_a R_{p(\lambda_0)}$ $= \gamma R_{p(\lambda_0)} L_a$. Therefore $\psi = \gamma R_{p(\lambda_0)} \in E^\gamma$. The proof is complete.

1.12 Corollary

Every group is quasicompactible.

We remark that $E$ is a group if and only if $E$ is left cancellative and possesses a minimal right ideal (so that to prove $E$ is quasicompactible using 1.11, 1.12 must apply). For let $E$ be left cancellative and let $I$ be a minimal right ideal in $E$. There exists $p \in I$. Since $pI = I$, $pu = p$ for some $u \in I$. By left cancellativity, $u$ is the unit of $E$, so that $I = E$. But then $pE = E$ for all $p \in E$. It follows from [L, II.2.18] that $E$ is a group.

The following theorem provides many examples of quasicompactible, non-compactible monoids.

1.13 Theorem

A countably infinite abelian group is not compactible.

Proof. To begin with, suppose $X$ is a countable set and that $\mathcal{T}$ is a compact Hausdorff topology on $X$. Letting $\mathcal{S}$ be the topology generated by choosing a pair of separating open sets for each two-element subset of $X$, $\mathcal{S} = \mathcal{T}$ (as $\mathcal{T}$ is compact and $\mathcal{S}$ is Hausdorff). It follows that $\mathcal{T}$ must be second countable.

Now, let $A$ be a countably infinite abelian group and suppose $A$ were compactible via the compact Hausdorff topology $\mathcal{T}$. Then $\mathcal{T}$ is second countable and addition is separately continuous. By the theorem of [W], $A$ is a compact Hausdorff topological group. Let $h$ be the unique Haar measure on $A$ with $h(A) = 1$. By countable additivity and invariance we have the absurd equation $h(\{0\}) \cdot \infty \doteq 1$. The proof is complete.

1.14 Example

Let $X$ be a set. Then $E = X^X$ is compactible. For there exists a compact Hausdorff topology on $X$ (e.g., remove a point, discretify, and restore the point with the one-point compactification topology). Let $E$ have the induced cartesian power topology. If $p \in E$ then $L_p pr_x = pr_{xp}$ for all $x \in X$; this proves $L_p$ is continuous.

1.15 Theorem

Let $E$ be a dynamic monoid and let $I$ be a minimal right ideal in $E$ which qua semigroup is left cancellative. Then the following statements are valid.

a. For all $p \in I$ the unique $u \in I$ with $pu = p$ is an idempotent.

b. $\{Iu : u \in I$ and $uu = u\}$ partitions $I$ into groups.

c. If $u,v \in I$ are idempotents then if $Iu \cap Iv$ is nonempty, $u = v$.

Proof. If $pu = p$ let $q \in I$ with $uq = u$; then $pq = puq = pu$ so that $q = u$. This proves (a). If $u,v \in I$ are idempotents and $p \in Iu \cap Iv$ then $pu = p = pv$ and so $u = v$. That $Iu$ is a group doesn't require cancellability, since any semigroup $S$ with a right unit such that $pS = S$ $(p \in S)$ is a group (e.g., see [L, II.2.18]). The proof is complete.

1.16 Example

A monoid with a minimal right ideal that has no idempotents (and so which is not dynamic). Let $X$ be an infinite set. Define $E = \{f \in X^X : f = 1_X$ or $f$ is injective and $X - \text{im } f$ is countably infinite$\}$. $E$ is a submonoid of $X^X$. Define $I = \{f \in E : f \neq 1_X\}$. It is trivial to check that $I$ is a minimal right ideal with no idempotents.

1.17 Definition

Let $E$ be a monoid. The left compactification of $E$ is the set, $E\beta$, of all ultrafilters on the set $E$ with the binary multiplication

$$\mathcal{U} \cdot \mathcal{V} = \{A \subset E : \exists V \in \mathcal{V}\ \forall v \in V\ \exists U \in \mathcal{U} .\ Uv \subset A\} \quad .$$

1.18 Theorem

Let $E$ be a monoid with left compactification $E\beta$. Then the following statements are valid.

a. $E\beta$ is a monoid with unit $\dot{e}$ (where $e \in E$ is the unit of $E$ and for $p \in E$, $\dot{p}$ denotes the principal ultrafilter induced by $p$).

b. The map $E \xrightarrow{E\eta} E\beta$ sending $p$ to $\dot{p}$ is a monoid homomorphism.

c. $E\beta$, with its usual compact Hausdorff topology (making it the free compact space on $E$ generators) becomes a compactible monoid.

d. For $\mathcal{U} \in E\beta$ and $p \in E$, $\mathcal{U} \cdot \dot{p} = \mathcal{U}R_p$ and $\dot{p} \cdot \mathcal{U} = \mathcal{U}L_p$.

<u>Proof</u>. <u>a</u>. Let $\mathcal{U},\mathcal{V},\mathcal{W} \in E\beta$. To begin with we must show $\mathcal{U} \cdot \mathcal{V}$ is an ultrafilter. That $A \cap B \in \mathcal{U} \cdot \mathcal{V}$ when $A,B \in \mathcal{U} \cdot \mathcal{V}$ is trivial. Suppose $A \notin \mathcal{U} \cdot \mathcal{V}$. Define $V = \{p \in E : \forall U \in \mathcal{U} \,.\, Up \not\subset A\}$; clearly $V \in \mathcal{V}$. Let $v \in V$. Define $U = \{p \in E : pv \notin A\}$; in view of the definition of $V$, $U \in \mathcal{U}$. This shows that $E - A \in \mathcal{U} \cdot \mathcal{V}$. For the associative law, let $A \in (\mathcal{U} \cdot \mathcal{V}) \cdot \mathcal{W}$. $\exists W \in \mathcal{W}\ \forall w \in W\ \exists B_w \in \mathcal{U} \cdot \mathcal{V}$ with $B_w w \subset A$. For each $w \in W$ $\exists V_w \in \mathcal{V}\ \forall v \in V_w\ \exists U \in \mathcal{U}.$ $Uv \subset B_w$. Define $\mathbb{C} = \bigcup_{w \in W} V_w w \in \mathcal{V} \cdot \mathcal{W}$. Let $c \in \mathbb{C}$. For some $w \in W$, $v \in V_w$, $c = vw$. There exists $U \in \mathcal{U}$ with $Uv \subset B_w$. Then $Uc = Uvw \subset B_w w \subset A$. Hence $A \in \mathcal{U} \cdot (\mathcal{V} \cdot \mathcal{W})$.

The proofs of (b) and (d) are trivial. To prove (c) we must recall that $\{\mathring{A} : A \subset E\}$ is a base for the topology of $E\beta$, where $\mathring{A} = \{\mathcal{V} \in E\beta : A \in \mathcal{V}\}$. Let $\mathcal{U},\mathcal{V} \in E\beta$, $A \subset E$ with $\mathcal{U} \cdot \mathcal{V} \in \mathring{A}$. $\exists V \in \mathcal{V}\ \forall v \in V\ \exists U \in \mathcal{U}.$ $Uv \subset A$. It is obvious that $\mathcal{U} \cdot \mathcal{W} \in \mathring{A}$ whenever $\mathcal{W} \in \mathring{V}$. This proves that $L_{\mathcal{U}}$ is continuous at $\mathcal{V}$. The proof is complete.

<u>1.19 Theorem</u>

Let $E$ be a compactible monoid via the compact Hausdorff topology $\mathcal{T}$. Let $D$ be any submonoid of $\{p \in E : R_p$ is continuous$\}$. Let $D\beta \xrightarrow{\xi} E$ be the unique continuous extension of the inclusion map of D into E (i.e., $\mathcal{U}\xi$ is the unique point of E to which $\mathcal{U}$ converges in $\mathcal{T}$). Then $\xi$ is a monoid homomorphism.

<u>Proof</u>. Let $\mathcal{U} \in D\beta$ and set $\mathcal{F} = \{\mathcal{V} \in D : (\mathcal{U} \cdot \mathcal{V})\xi = \mathcal{U}\xi \cdot \mathcal{V}\xi\}$. Let $p \in D$. Since $R_p$ is continuous, the diagram

$$\begin{array}{ccc} D\beta & \xrightarrow{R_p\beta} & D\beta \\ \downarrow \xi & & \downarrow \xi \\ E & \xrightarrow{R_p} & E \end{array}$$

commutes. By 1.18d, $\dot{p} \in \mathcal{F}$. Since $\mathcal{F}$ is the equalizer of the continuous maps $L_{\mathcal{U}}\xi$, $\xi L_{\mathcal{U}\xi}$, $\mathcal{F}$ is closed. Hence $\mathcal{F} = D\beta$, which completes the proof.

<u>1.20 Theorem</u>

$\beta$ is a functor from the category of monoids into the category of compact Hausdorff monoids with continuous left multiplications.

<u>Proof</u>. Let $E_1 \xrightarrow{f} E_2$ be a monoid homomorphism. We must show that $E_1\beta \xrightarrow{f\beta} E_2\beta$ is a monoid homomorphism. That $f\beta$ preserves the unit is clear. Let $\mathcal{U}, \mathcal{V} \in E_1\beta$ and let $A \in (\mathcal{U} \cdot \mathcal{V})f$. There exists $A_1 \in \mathcal{U} \cdot \mathcal{V}$ with $A_1 f \subset A$. $\exists V \in \mathcal{V}, \forall v \in V, \exists U_v \in \mathcal{U}.\ U_v v \subset A_1$. Then $Vf \in \mathcal{V}f$ and for all $v \in V$, $(U_v f)vf = (U_v v)f \subset A_1 f \subset A$. Hence $A \in \mathcal{U}f \cdot \mathcal{V}f$. The proof is complete.

## 2. DYNAMIC TRIPLES

Let $\mathbb{T} = (T,\eta,\mu)$ be a triple in the category of sets.

### 2.1 Definition

The structure monoid of $\mathbb{T}$, which we denote $E_{\mathbb{T}}$, is the set of natural transformations from the identity functor to $T$ equipped with the binary multiplication $g \cdot h = 1 \xrightarrow{gh} TT \xrightarrow{\mu} T$.

For $(X,\xi)$ a $\mathbb{T}$-algebra and $g \in E_{\mathbb{T}}$ define $\xi^g = X \xrightarrow{Xg} XT \xrightarrow{\xi} X$. The enveloping semigroup of $(X,\xi)$ is the set $\{\xi^g : g \in E_{\mathbb{T}}\}$.

### 2.2 Theorem

The following statements are valid.

a. $E_{\mathbb{T}}$ is a monoid with unit $\eta$.

b. For every algebra $(X,\xi)$, $E_{(X,\xi)}$ is both a submonoid of $X^X$ and the subalgebra of $(X,\xi)^X$ generated by $\{1_X\}$.

c. For every algebra $(X,\xi)$, $g \longmapsto \xi^g$ is a monoid epimorphism of $E_{\mathbb{T}}$ onto $E_{(X,\xi)}$ and an algebra homomorphism of $(1T,1\mu)$ onto $E_{(X,\xi)}$ (where $E_{\mathbb{T}}$ is identified with $1T$ by the Yoneda correspondence).

d. If $(X,\xi) = (1T,1\mu)$, the homomorphisms of (c) are isomorphisms.

<u>Proof</u>. Let $(X,\xi)$ be a $\mathbb{T}$-algebra. Reviewing some generalities about $\mathbb{T}$-algebras, the structure map $\xi^{(X)}$ of $(X,\xi)^X$ is defined by $\xi^{(X)}.pr_x = pr_xT.\xi$ $(x \in X)$ and the unique homomorphism $1T \xrightarrow{\psi} (X,\xi)^X$ sending 1 to the constant function $\tilde{1}_X$ is $\tilde{1}_XT.\xi^{(X)}$. The diagram

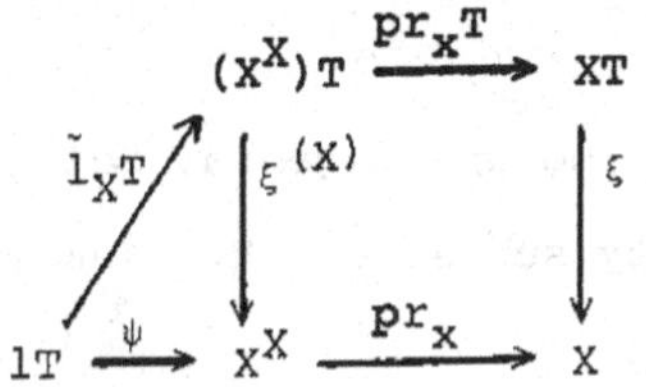


and the Yoneda lemma shows that $g\psi = \xi^g$ for all $g \in 1T$. Hence $\langle 1_X \rangle = \mathrm{Im}\ \psi = E_{(X,\xi)}$. The remaining details follow easily from the diagrams

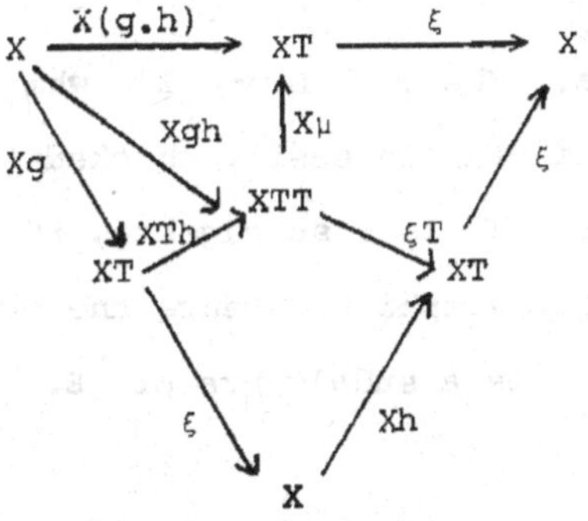


and

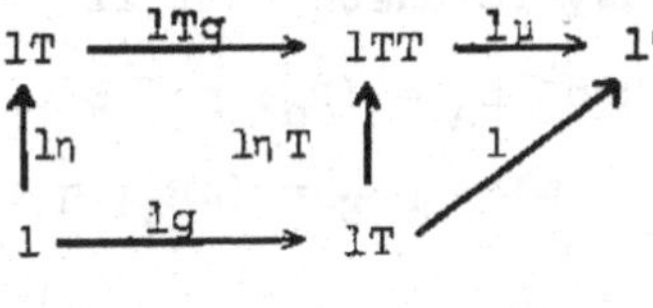


the second of which shows how to recover $g$ from $(1\mu)^g$.

2.3 Remark

Let $(X,\xi)$ be a $\mathbb{T}$-algebra, $E = E_{(X,\xi)}$. Then for all $x \in X$, $\langle x \rangle = xE$. For $\langle x \rangle = \langle 1_X pr_x \rangle = \langle 1_X \rangle pr_x = E\, pr_x = xE$.

2.4 Theorem

Let $(X,\xi)$ be a $\mathbb{T}$-algebra, set $E = E_{(X,\xi)}$ and let $I$ be a non-empty subset of $E$. The following statements are valid.

a. If $I$ is a subalgebra of $E$, $I$ is a right ideal of $E$.

b. If $I$ is a right ideal in $E$ then for all $p \in I$, $\langle p \rangle \subset I$.

Proof. a. The map $E \longrightarrow E^I$ which sends $p$ to the I-restriction of $R_p$ is easily checked to be a $\mathbb{T}$-homomorphism. Since $I$ is a subalgebra, the inclusion $I^I \longrightarrow E^I$ is a $\mathbb{T}$-homomorphism. Hence the pullback $P = \{p \in E\colon Ip \subset I\}$ is a subalgebra of $E$. Since $1_X \in P$, $P = E$.

b. It is easy to check that if $A$ is any subalgebra of $(X,\xi)^X$ then $E_A = \{R_p\colon p \in E\}$. Using 2.3 we have for $p \in I$ that $\langle p \rangle = pE_E = \{pR_q\colon q \in E\} \subset I$.

2.5 Corollary

Let $(X,\xi)$ be a $\mathbb{T}$-algebra, set $E = E_{(X,\xi)}$ and let

$I \subset E$. Then $I$ is a minimal (i.e., minimal non-empty) subalgebra of $E$ if and only if $I$ is a minimal right ideal of $E$.

2.6 Theorem

The following statements are equivalent.

a. Every non-empty $\mathbb{T}$-algebra has a minimal subalgebra.

b. $E_{\mathbb{T}}$ has a minimal right ideal.

Proof. a implies b. By 2.5, the enveloping semigroup $E$ of $(1T, 1\mu)$ has a minimal right ideal. By 2.2d, $E_{\mathbb{T}}$ is isomorphic to $E$.

b implies a. By reversing the argument of "a implies b", $(1T, 1\mu)$ has a minimal subalgebra $M$. If $(X, \xi)$ is a non-empty $\mathbb{T}$-algebra, there exists a homomorphism $\psi$ from $(1T, 1\mu)$ to $(X, \xi)$. $M\psi$ is a minimal subalgebra of $(X, \xi)$. The proof is complete.

2.7 Definition

A universal minimal $\mathbb{T}$-algebra is a $\mathbb{T}$-algebra $M$ satisfying the following three properties.

UM1. $M$ is a minimal $\mathbb{T}$-algebra.

UM2. If $N$ is a minimal $\mathbb{T}$-algebra then there exists a $\mathbb{T}$-homomorphism of $M$ onto $N$.

UM3. Every $\mathbb{T}$-endomorphism of $M$ is an automorphism.

It is clear that if a universal minimal $M$ exists,

it is isomorphic to any algebra satisfying UM1 and UM2. Hence we should speak of the universal minimal algebra. Any minimal subalgebra of the product of a representative set of all minimal algebras will satisfy UM1 and UM2 and hence will be the universal minimal algebra when it exists.

2.8 Definition

$\mathbb{T}$ is a dynamic triple if $E_{\mathbb{T}}$ is a dynamic monoid.

2.9 Theorem

The following conditions on $\mathbb{T}$ are equivalent.

a. $\mathbb{T}$ is dynamic.

b. Every non-empty $\mathbb{T}$-algebra contains a minimal $\mathbb{T}$-algebra and there exists a universal minimal $\mathbb{T}$-algebra.

Proof. a implies b. 2.6 is half the work. Let $E$ be the enveloping semigroup of $(1T, 1\mu)$, and let $M$ be a minimal right ideal of $E$ which qua semigroup is left cancellative. $M$ is a minimal algebra by 2.5. Since $E$ is isomorphic to $(1T, 1\mu)$, $M$ admits a homomorphism onto every minimal algebra. Suppose $M \xrightarrow{f} M$ is a $\mathbb{T}$-homomorphism. Of course $f$ is onto. As remarked in the proof of 2.4b, $E_M = \{R_p : p \in E\}$. Let $p \in E$. If $MT \xrightarrow{\xi} M$ is the structure map of $M$, there exists $g \in E_{\mathbb{T}}$ with $\xi^g = R_p$. In the diagram

$$\begin{array}{ccccc} M & \xrightarrow{Mg} & MT & \xrightarrow{\xi} & M \\ \downarrow f & & \downarrow fT & & \downarrow f \\ M & \xrightarrow{Mg} & MT & \xrightarrow{\xi} & M \end{array}$$

both squares commute. Hence $fR_p = R_p f$ for all $p \in E$. By 1.13a $M$ has an idempotent $u$. For $x \in M$ we have $xf = (ux)f = uR_x f = ufR_x = (uf)x$, so $f = L_{uf}$ which proves that $f$ is injective.

<u>b implies a</u>. Let $E$ be the enveloping semigroup of $(1T, 1\mu)$. Let $M$ be a minimal subalgebra of $E$. Since $M$ satisfies UM1, UM2, $M$ is the universal minimal algebra. By 2.5, $M$ is a minimal right ideal. Let $p \in M$. $L_p$ is a $\mathbb{T}$-endomorphism of $(1T, 1\mu)^{1T}$ since $L_p.pr_x = pr_{xp}$ for all $x \in 1T$. In particular, $L_p$ is a $\mathbb{T}$-endomorphism of $M$, hence a $\mathbb{T}$-isomorphism. As $E_{\mathbb{T}}$ is isomorphic to $E$, $E_{\mathbb{T}}$ is dynamic. The proof is complete.

## <u>2.10 Definition</u>

$\mathbb{T}$ is <u>distal</u> if $E_{\mathbb{T}}$ is a group.

## <u>2.11 Theorem</u>

The following conditions on $\mathbb{T}$ are equivalent.

a. $\mathbb{T}$ is distal.

b. $(1T, 1\mu)$ is the universal minimal algebra.

c. $(1T, 1\mu)$ is a minimal algebra.

<u>Proof</u>. Let $E$ be the enveloping semigroup of $(1T, 1\mu)$.

a implies b. Since E is already a minimal right ideal, E is the universal minimal algebra in view of the proof of 2.9.

c implies a. Since E is a minimal right ideal with a right unit, E is a group. The proof is complete.

2.12 Definition

A $\mathbb{T}$-algebra $(X,\xi)$ is homogeneous if whenever $x,y \in X$ there exists a $\mathbb{T}$-endomorphism $(X,\xi) \xrightarrow{f} (X,\xi)$ with $xf = y$.

2.13 Definition

A $\mathbb{T}$-algebra $(X,\xi)$ is partially free on one generator if there exists a triple $\$ = (S,\eta',\mu')$ and an algebraic functor $\Phi$

$$\begin{array}{ccc} S^{\$} & \xrightarrow{\Phi} & S^{\mathbb{T}} \\ & {\scriptstyle U^{\$}}\searrow \quad \swarrow{\scriptstyle U^{\mathbb{T}}} & \\ & S & \end{array}$$

such that $(1S, 1\mu')\Phi = (X,\xi)$.

2.14 Theorem

Let $(X,\xi)$ be a minimal $\mathbb{T}$-algebra. Then the following statements are equivalent.

a. $(X,\xi)$ is partially free on one generator.

b. $(X,\xi)$ is homogeneous.

Proof. a implies b. Let $\Phi$ be the algebraic functor of 2.13. Let $x,y \in X$ and let $1 \in X$ be the free generator of $(1S,1\mu')$. There exist \$-endomorphisms - and hence $\mathbb{T}$-endomorphisms - $f,g$ with $1f = x$, $1g = y$. Since $(X,\xi)$ is minimal, $f$ is onto so $Zf = 1$ for some $Z \in X$. Using partial freeness again, there exists a $\mathbb{T}$-endomorphism $h$ with $1h = Z$. By minimality, the equalizer of $fh$, $1_X$ is all of $X$ so $fh = 1_X$. Hence $f$ is injective. $f^{-1}g$ is the desired endomorphism.

b implies a. Since this result is not crucial for the rest of the paper, we lapse into language recognizable only to dyed-in-the-wool categorists. Let $A$ be the full subcategory $\{(X,\xi)\}$ of $S^{\mathbb{T}}$. The functor $A \longrightarrow S^{\mathbb{T}} \longrightarrow S$ is tractable because $A$ is small, and so has Linton structure triple \$. Let $S^{\$} \xrightarrow{\Phi} S^{\mathbb{T}}$ be the unique algebraic functor induced by the reflectivity of the semantics comparison functor $A \longrightarrow S^{\$}$. Using the definition of \$ it is not hard to show that $(1S,1\mu')\Phi = \varprojlim[(1,i) \longrightarrow S^{\mathbb{T}}]$ where $i$ is the inclusion functor $A \longrightarrow S^{\mathbb{T}}$. Using the sort of reasoning that appeared in the proof of "a implies b" we know that for all $x,y \in X$ there exists a unique $\mathbb{T}$-automorphism sending $x$ to $y$. That $\varprojlim[(1,i) \longrightarrow S^{\mathbb{T}}] = (X,\xi)$ is then clear. The proof is complete.

2.15 Remark

The proof of 2.14 "b implies a" shows that if

$(X,\xi)$ is minimal then $(X,\xi)^n$ is partially free on one generator where $n$ is the set of homogeneous components of $(X,\xi)$.

2.16 Definition

$\mathbb{T}$ is homogeneous if $E_{\mathbb{T}}$ possesses a minimal right ideal which is a group. Notice that $\mathbb{T}$ is dynamic if $\mathbb{T}$ is homogeneous.

2.17 Theorem

Let $\mathbb{T}$ be dynamic. The following conditions on $\mathbb{T}$ are equivalent.

a. $\mathbb{T}$ is homogeneous.

b. The universal minimal set is homogeneous.

c. The universal minimal set is partially free on one generator.

Proof. a implies b. Let $M$ be a minimal right ideal in the enveloping semigroup of $(1T,1\mu)$ which is a group. Then $M$ is the universal minimal $\mathbb{T}$-algebra. If $x,y \in M$ there exists $p \in M$ with $px = y$. But $L_p$ is a $\mathbb{T}$-endomorphism of $M$.

b implies a. Let $M$ be as above. Let $x,y \in M$. Then there exists a $\mathbb{T}$-endomorphism $f$ with $xf = y$. By the proof of 2.9 there exists $p \in M$ with $f = L_p$. Hence $px = y$. Hence $M$ is a group. In view of 2.14, the proof is complete.

## 3. EXAMPLES

### 3.1 Definition

$\mathbb{T}$ is compactible if $E_{\mathbb{T}}$ is compactible.

### 3.2 Theorem

Let $\mathbb{T}$ be any triple in sets and let $\$$ be the triple corresponding to any Birkhoff subcategory $\mathcal{B}$ of the category $S^{\mathbb{T} \otimes \mathbb{B}}$ of compact $\mathbb{T}$-algebras. Let $X = (1S, 1\mu)$ and let $E$ be the enveloping semigroup of $X$. Then $\$$ is compactible and there exists a continuous monoid epimorphism of $E_{\mathbb{T}}\beta$ onto $E$.

Proof. $E = \langle 1_X \rangle_{\mathbb{T} \otimes \mathbb{B}}$ (noting that there is no ambiguity with respect to "subalgebra generated by" since $\mathcal{B}$ is closed under subalgebras and products). Hence $E_1 = \langle 1_X \rangle_{\mathbb{T}}$ is a dense submonoid of $E$. As in the proof of 2.4b, $E \xrightarrow{R_p} E$ is in the $\mathbb{T}$-enveloping semigroup of $E$ for all $p \in E_1$. By the definition of $\mathbb{T} \otimes \mathbb{B}$, $R_p$ is continuous for all $p \in E_1$. By 1.19, $E$ is a continuous monoid quotient of $E_1\beta$ (onto because $E_1$ is dense; also $E$ is compactible, since $L_p$ is even a $\mathbb{T} \otimes \mathbb{B}$-homomorphism for all $p \in E$). But $E_1$ is a monoid quotient of $E_{\mathbb{T}}$ by 2.2c so that $E_1\beta$ is a continuous quotient of $E_{\mathbb{T}}\beta$ by 1.20 ($\beta$ preserves outoness since it is also a functor from sets to compact spaces). The proof is complete.

3.2 is potentially a powerful structure theorem for a large class of universal minimal algebras. The associated algebraic problem is to determine the nature of the closed subsemigroups of $E\beta \times E\beta$ which are equivalence relations, for an arbitrary monoid $E$.

3.3 Theorem

Let $\mathbb{T}$ be a triple in sets such that there exists an algebraic functor $S^{\mathbb{T}} \xrightarrow{\Phi} S^{\beta}$ (this includes all the $\mathbb{S}$'s of 3.2, but conceivably there are others). Then $\mathbb{T}$ is compactible.

Proof. Let $X = (1T, 1\mu)$ and let $E$ be the enveloping semigroup of $X$. As $E = \langle 1_X \rangle_{\mathbb{T}}$, $E\Phi$ is a closed subset of $(X\Phi)^X$ and so becomes a compact Hausdorff space. For $p \in E$, $L_p$ is a $\mathbb{T}$-homomorphism and, hence, is a continuous endomorphism of $E\Phi$. The proof is complete.

3.4 Theorem

Let $E$ be any monoid and let $\mathbb{T}(E)$ be the associated triple (whose algebras are right $E$-sets). Then $E_{\mathbb{T}(E)}$ is isomorphic to $E$. Hence, any monoid can be the structure monoid to a triple.

Proof. The free $\mathbb{T}(E)$-algebra on one generator is just $E$. $\langle 1_E \rangle$ = the orbit of $1_X$ in $E^E = \{R_p : p \in E\} \simeq E$.

The proof is complete.

3.5 Example

Let E be the monoid of 1.16 Then $\mathbb{T}(E)$ is not dynamic, but every E-set has a minimal E-invariant subset (2.6). By 2.7, there is no universal minimal E-set.

3.6 Theorem

Let $\mathbb{T}$ be any triple in sets. Then there exists an algebraic functor $S^{\mathbb{T}} \xrightarrow{\Phi} E_{\mathbb{T}}$-sets which sends every minimal $\mathbb{T}$-algebra to a minimal $E_{\mathbb{T}}$-set. Hence, in some sense, all questions concerning minimal algebras reduce to questions about minimal monoid actions.

Proof. For each $\mathbb{T}$-algebra $(X,\xi)$ define $(X,\xi)\Phi$ to be the action

$$X \times E_{\mathbb{T}} \longrightarrow X$$
$$x, g \longmapsto x\xi^{g} \quad .$$

It is easy to check that $\Phi$ is well-defined on objects and sends $\mathbb{T}$-homomorphisms to equivariant maps. For $x \in X$ we have from 2.2c and 2.3 that $\langle x\rangle_{\mathbb{T}} = xE_{(X,\xi)} = xE_{\mathbb{T}} = \langle x\rangle_{E(\mathbb{T})}$ which completes the proof.

3.7 Theorem

Let $\mathbb{T}$ be any triple in sets. Then there exists an algebraic functor $S^{\mathbb{T}\otimes\beta} \xrightarrow{\Phi}$ compact $E_{\mathbb{T}}$-sets which sends every minimal compact $\mathbb{T}$-algebra to a minimal compact $E_{\mathbb{T}}$-set.

Proof. If $X$ is a compact $\mathbb{T}$-algebra, make $X$ an $E_{\mathbb{T}}$-set as in 3.6, and leave the topology alone. Since each $\xi^g$ is continuous, the action is continuous ($E_{\mathbb{T}}$ being considered discrete, of course). If $x \in X$, we have $\langle x \rangle_{\mathbb{T} \otimes \beta}$ $= \langle \langle x \rangle_{\mathbb{T}} \rangle_{\beta} = \langle \langle x \rangle_{E(\mathbb{T})} \rangle_{\beta} = \langle x \rangle_{E(\mathbb{T}) \otimes \beta}$ which completes the proof.

The following example arose in conversation with J.F. Kennison.

3.8 Example

Let $\mathbb{T}$ be the triple corresponding to the equational class of algebras $X$ equipped with binary operation $m$ and unary operations $u_1, u_2$ subject to the equations

$$x u_1 x u_2 m = x$$
$$x y m u_1 = x$$
$$x y m u_2 = y \quad .$$

A $\mathbb{T}$-algebra amounts to being a set $X$, together with a specified bijection $X \times X \overset{m}{\rightarrowtail\!\!\!\twoheadrightarrow} X$. A $\mathbb{T} \otimes \beta$-algebra, then, amounts to being a compact Hausdorff space $X$, together with a specified homeomorphism $X \times X \overset{m}{\rightarrowtail\!\!\!\twoheadrightarrow} X$. The cantor set becomes a minimal $\mathbb{T} \otimes \beta$-algebra as follows: let $2 = \{0,1\}$ and define $2 \xrightarrow{-} 2$ by $\bar{0} = 1$, $\bar{1} = 0$. Let $\mathbb{N}$ be the positive integers $\{1,2,\ldots\}$. Define a homeomorphism $2^{\mathbb{N}} \times 2^{\mathbb{N}} \xrightarrow{m} 2^{\mathbb{N}}$ by

$(x_i)(y_i)m = (\bar{x}_1\bar{y}_1\bar{x}_2\bar{y}_2\ldots)$. Let $(x_i) \in 2^{\mathbb{N}}$ and let $n \geq 2$ be an even integer. We show, by induction on $n$, that given any $(y_i) \in 2^{\mathbb{N}}$ some element of $\langle (x_i) \rangle_{\mathbb{T}}$ agrees with $(y_i)$ for $1 \leq i \leq n$; for then $(y_i) \in \langle\langle (x_i) \rangle_{\mathbb{T}} \rangle_{\beta} = \langle (x_i) \rangle_{\mathbb{T} \otimes \beta}$ for all $(y_i) \in 2^{\mathbb{N}}$ and $(2^{\mathbb{N}}, m)$ is minimal. The proof for $n = 2$ is clear:

$$(x_1\ldots)\quad (x_1\ldots)\ m = (\bar{x}_1\bar{x}_1\ldots)$$
$$(x_1\ldots)\quad (\bar{x}_1\ldots)\ m = (\bar{x}_1 x_1\ldots)$$
$$(\bar{x}_1\ldots)\quad (x_1\ldots)\ m = (x_1\bar{x}_1\ldots)$$
$$(\bar{x}_1\ldots)\quad (\bar{x}_1\ldots)\ m = (x_1 x_1\ldots)\ .$$

For $n > 2$, let $(y_i) \in 2^{\mathbb{N}}$. By the induction hypothesis, there exist $(a_i), (b_i) \in \langle (x_i) \rangle_{\mathbb{T}}$ with $a_i = \bar{y}_{2i-1}$, $b_i = \bar{y}_{2i}$ $(1 \leq i \leq \frac{n}{2})$. But then $(a_i)(b_i)m \in \langle (x_i) \rangle_{\mathbb{T}}$ agrees with $(y_i)$ for $1 \leq i \leq n$.

In particular, $E_{\mathbb{T}}$ acts minimally on the cantor set, by 3.7.

<u>3.9 Theorem</u>

Let $E$ be any monoid. Then $E\beta$, with action

$$E\beta \times E \longrightarrow E\beta\ ,$$
$$\mathcal{U}, p \longmapsto \mathcal{U} \cdot \dot{p}$$

is the free compact E-set on one generator.

<u>Proof</u>. Since $\mathcal{U} \cdot p = \mathcal{U}R_{\dot{p}}$, $R_{\dot{p}} = R_p\beta$ is indeed continuous for all $p \in E$. Let $X \times E \longrightarrow X$ be another

compact E-set and let $x \in X$. Let $E \xrightarrow{f} X$ be the unique equivariant map sending $e \in E$ to $x$ (i.e., $pf = xp$). Let $\tilde{f}$ be the unique continuous extension of $f$ to $E\beta$:

$$\begin{array}{ccc} E & \xrightarrow{E\eta} & E\beta \\ & {\scriptstyle f}\searrow \quad \swarrow{\scriptstyle \tilde{f}} & \\ & X & \end{array} .$$

Since the map $x \longmapsto xp$ is continuous for all $p \in E$, $\tilde{f}$ is equivariant, the proof being entirely similar to that of 1.19. The uniqueness of $\tilde{f}$ follows from the uniqueness of f. The proof is complete.

It follows from 3.6 and the proof of 2.9, that any minimal closed E-invariant subset of $E\beta$ is the universal minimal compact E-set.

## REFERENCES

[Ea] Ellis, Robert, "A Semigroup Associated With a Transformation Group", *Trans. Amer. Math. Soc.*, 94; 272-281, (1960).

[Eb] Ellis, Robert, "Universal Minimal Sets", *Proc., Amer. Math. Soc.*, 11; 540-543, (1960).

[L] Ljapin, E. S., "Semigroups", *Amer. Math. Soc. Trans. Math. Mon.*, (1960).

[Ma] Manes, Ernest, *A Triple Miscellany: Some Aspects of the Theory of Algebras Over a Triple*, dissertation, Wesleyan University, (1963).

[Mb] Manes, Ernest, *A Triple-Theoretic Construction of Compact Algebras*, to appear in the proceedings of the 1966-67 Zurich Conference on Categorical Algebra.

[W] Wu, Ta-sun, "Continuity in Topological Groups", *Proc. Amer. Math. Soc.*, 13; 452-453, (1962).

# CATEGORIES OF SPECTRA AND INFINITE LOOP SPACES

by

J. Peter May

At the Seattle conference, I presented a calculation of $H_*(F;Z_p)$ as an algebra, for odd primes $p$, where $F = \varinjlim F(n)$ and $F(n)$ is the topological monoid of homotopy equivalences of an n-sphere. This computation was meant as a preliminary step towards the computation of $H^*(BF;Z_p)$. Since then, I have calculated $H^*(BF;Z_p)$, for all primes $p$, as a Hopf algebra over the Steenrod and Dyer-Lashof algebras. The calculation, while not difficult, is somewhat lengthy, and I was not able to write up a coherent presentation in time for inclusion in these proceedings. The computation required a systematic study of homology operations on n-fold and infinite loop spaces. As a result of this study, I have also been able to compute $H_*(\Omega^n S^n X;Z_p)$, as a Hopf algebra over the Steenrod algebra, for all connected spaces $X$ and prime numbers $p$. This result, which generalizes those of Dyer and Lashof [3] and Milgram [8], yields explicit descriptions of both $H_*(\Omega^n S^n X;Z_p)$ and $H_*(QX;Z_p)$, $QX = \varinjlim \Omega^n S^n X$, as functors of $H_*(X;Z_p)$.

An essential first step towards these results was a systematic categorical analysis of the notions of n-fold and infinite loop spaces. The results of this analysis will be presented here. These include certain adjoint functor relationships that provide the conceptual reason that $H_*(\Omega^n S^n X; Z_p)$ and $H_*(QX; Z_p)$ are functors of $H_*(X; Z_p)$ and that precisely relate maps between spaces to maps between spectra. These categorical considerations motivate the introduction of certain non-standard categories, $\mathcal{I}$ and $\mathcal{L}$, of (bounded) spectra and $\Omega$-spectra, and the main purpose of this paper is to propagandize these categories. It is clear from their definitions that these categories are considerably easier to work with topologically than are the usual ones, but it is not clear that they are sufficiently large to be of interest. We shall remedy this by showing that, in a sense to be made precise, these categories are equivalent for the purposes of homotopy theory to the standard categories of (bounded) spectra and $\Omega$-spectra. We extend the theory to unbounded spectra in the last section.

The material here is quite simple, both as category theory and as topology, but it turns out nevertheless to have useful concrete applications. We shall indicate two of these at the end of the paper. In the first, we observe that there is a natural epimorphism, realized by a map of spaces, from the stable homotopy groups of an infinite loop space to its ordinary homotopy groups. In the second, by coupling our results with other information, we shall construct a collection of interesting topological spaces and maps; the other information by itself gives no hint of the possibility of performing this construction.

## 1 THE CATEGORIES $\mathcal{L}_n$ AND HOMOLOGY

In order to sensibly study the homology of iterated loop spaces, it is necessary to have a precise categorical framework in which to work. It is the purpose of this section to present such a framework.

We let $\mathcal{T}$ denote the category of topological spaces with base-point and base-point preserving maps, and we let

$$\mu\colon \mathrm{Hom}_{\mathcal{T}}(X,\Omega Y) \longrightarrow \mathrm{Hom}_{\mathcal{T}}(SX,Y) \tag{1.1}$$

denote the standard adjunction homeomorphism relating the loop and suspension functors.

We define the category of n-fold loop sequences, $\mathcal{L}_n$, to have objects $B = \{B_i \mid 0 \le i \le n\}$ such that $B_i = \Omega B_{i+1} \in \mathcal{T}$ and maps $g = \{g_i \mid 0 \le i \le n\}$ such that $g_i = \Omega g_{i+1} \in \mathcal{T}$; clearly $B_0 = \Omega^i B_i$ and $g_0 = \Omega^i g_i$ for $0 \le i \le n$. We define $\mathcal{L} = \mathcal{L}_\infty$ to be the category with objects $B = \{B_i \mid i \ge 0\}$ such that $B_i = \Omega B_{i+1} \in \mathcal{T}$ and maps $g = \{g_i \mid i \ge 0\}$ such that $g_i = \Omega g_{i+1} \in \mathcal{T}$; clearly $B_0 = \Omega^i B_i$ and $g_0 = \Omega^i g_i$ for all $i \ge 0$. We call $\mathcal{L}_\infty$ the category of perfect $\Omega$-spectra (or of infinite loop sequences). For all n, we define forgetful functors $U_n\colon \mathcal{L}_n \longrightarrow \mathcal{T}$ by $U_n B = B_0$ and $U_n g = g_0$. Of course, if $n < \infty$, $U_n B$ and $U_n g$ are n-fold loop spaces and maps. We say that a space $X \in \mathcal{T}$ is a perfect infinite loop space if $X = U_\infty B$ for some object $B \in \mathcal{L}_\infty$ and we say that a map $f \in \mathcal{T}$ is a perfect infinite loop map if $f = U_\infty g$ for some map $g \in \mathcal{L}_\infty$.

We seek adjoints $Q_n: \mathcal{T} \longrightarrow \mathcal{L}_n$, $1 \leq n \leq \infty$, to the functors $U_n$. For $n < \infty$, define $Q_n X = \{\Omega^{n-i} S^n X \mid 0 \leq i \leq n\}$ and $Q_n f = \{\Omega^{n-i} S^n f \mid 0 \leq i \leq n\}$. Clearly, $Q_n X$ and $Q_n f$ are objects and maps in $\mathcal{L}_n$. For the case $n = \infty$, we first define a functor $Q: \mathcal{T} \longrightarrow \mathcal{T}$ by letting $QX = \varinjlim \Omega^n S^n X$, where the limit is taken with respect to the inclusions

$$\Omega^n \mu^{-1}(1_{S^{n+1}X}): \Omega^n S^n X \longrightarrow \Omega^{n+1} S^{n+1} X$$

For $f: X \longrightarrow Y$, we define $Qf = \varinjlim \Omega^n S^n f: QX \longrightarrow QY$. It is clear that $QX = \Omega QSX$ and $Qf = \Omega QSf$. We can therefore define a functor $Q_\infty: \mathcal{T} \longrightarrow \mathcal{L}_\infty$ by $Q_\infty X = \{QS^i X \mid i \geq 0\}$ and $Q_\infty f = \{QS^i f \mid i \geq 0\}$.

<u>Proposition 1</u>

For each $n$, $1 \leq n \leq \infty$, there is an adjunction

$$\phi_n: \mathrm{Hom}_{\mathcal{T}}(X, U_n B) \longrightarrow \mathrm{Hom}_{\mathcal{L}_n}(Q_n X, B).$$

<u>Proof</u>. Observe first that the following two composites are the identity.

$$S^n X \xrightarrow{S^n \mu^{-n}(1_{S^n X})} S^n \Omega^n S^n X \xrightarrow{\mu^n(1_{\Omega^n S^n X})} S^n X, \; X \in \mathcal{T} \qquad (1.2)$$

$$\Omega^n X \xrightarrow{\mu^{-n}(1_{S^n \Omega^n X})} \Omega^n S^n \Omega^n X \xrightarrow{\Omega^n \mu^n(1_{\Omega^n X})} \Omega^n X, \; X \in \mathcal{T} \qquad (1.3)$$

In fact, since $\mu(f) = \mu(1_{\Omega Z}) \cdot Sf$ for any map $f: Y \longrightarrow \Omega Z$ in $T$, $\mu^n(1_{\Omega^n S^n X}) \cdot S^n \mu^{-n}(1_{S^n X}) = \mu^n \mu^{-n}(1_{S^n X}) = 1_{S^n X}$ ; this proves (1.2) and the proof of (1.3) is similar. Now define natural transformations $\Phi_n: Q_n U_n \longrightarrow 1_{L_n}$ and $\Psi_n: 1_T \longrightarrow U_n Q_n$ by

$$\Phi_n(B) = \{\Omega^{n-i}\mu^n(1_{B_0}) \mid 0 \le i \le n\}: Q_n U_n B \longrightarrow B \quad \text{if} \quad n < \infty; \qquad (1.4)$$

$$\Phi_\infty(B) = \{\varinjlim \Omega^j \mu^{i+j}(1_{B_0}) \mid i \ge 0\}: Q_\infty U_\infty B \longrightarrow B \quad \text{if} \quad n = \infty;$$

$$\Psi_n(X) = \mu^{-n}(1_{S^n X}): X \longrightarrow U_n Q_n X = \Omega^n S^n X \quad \text{if} \quad n < \infty; \qquad (1.5)$$

$$\Psi_\infty(X) = \varinjlim \mu^{-j}(1_{S^j X}): X \longrightarrow U_\infty Q_\infty X = QX \quad \text{if} \quad n = \infty.$$

We claim that (1.2) and (1.3) imply that the following two composites are the identity for all $n$.

$$Q_n X \xrightarrow{Q_n \Psi_n(X)} Q_n U_n Q_n X \xrightarrow{\Phi_n(Q_n X)} Q_n X, \; X \in T \qquad (1.6)$$

$$U_n B \xrightarrow{\Psi_n(U_n B)} U_n Q_n U_n B \xrightarrow{U_n \Phi_n(B)} U_n B, \; B \in L_n \qquad (1.7)$$

For $n < \infty$, (1.6) follows from (1.2) by application of $\Omega^{n-i}$ for $0 \le i \le n$ and (1.7) is just (1.3) applied to $X = B_n$, since $B_0 = U_n B = \Omega^n B_n$. For $n = \infty$, observe that $\Psi_\infty(X)$ factors as the composite

$$X \xrightarrow{\mu^{-1}(1_{SX})} \Omega SX \xrightarrow{\Omega \Psi_\infty(SX)} \Omega QSX = QX \quad .$$

It follows that $\Psi_\infty(X) = \mu^{-i}\Psi_\infty(S^i X)$ for all $i \ge 0$ since $\mu^{-i}\Psi_\infty(S^i X) = \mu^{-i}(\Omega\Psi_\infty(S^{i+1}X) \cdot \mu^{-1}(1_{S^{i+1}X})) = \mu^{-(i+1)}\Psi_\infty(S^{i+1}X)$. Observe also that

$$\Omega^j \Psi_\infty(S^{i+j}X): \Omega^j S^{i+j} X \longrightarrow \Omega^j Q S^{i+j} X = Q S^i X$$

is just the natural inclusion obtained from the definition of $QS^iX$ as $\varinjlim \Omega^j S^{i+j}X$. We therefore have that:

$$\Phi_\infty(Q_\infty X)_i \cdot Q_\infty\Psi_\infty(X)_i$$

$$= \varinjlim \Omega^j\mu^{i+j}(1_{QX}) \cdot \varinjlim \Omega^k S^{i+k}\mu^{-(i+k)}\Psi_\infty(S^{i+k}X)$$

$$= \varinjlim \Omega^j\mu^{i+j}(1_{QX}) \cdot \Omega^j S^{i+j}\mu^{-(i+j)}\Psi_\infty(S^{i+j}X)$$

$$= \varinjlim \Omega^j\Psi_\infty(S^{i+j}X) = 1_{QS^iX};$$

$$U_\infty\Phi_\infty(B) \cdot \Psi_\infty(U_\infty B) = \varinjlim \Omega^j\mu^j(1_{B_0}) \cdot \varinjlim \mu^{-k}(1_{S^kB_0})$$

$$= \varinjlim \Omega^j\mu^j(1_{B_0}) \cdot \mu^{-j}(1_{S^jB_0}) = \varinjlim 1_{B_0} = 1_{B_0}.$$

In both calculations, the second equality is an observation about the limit topology. The third equalities follow from formulas (1.2) and (1.3) respectively. Finally, define

$$\phi_n(f) = \Phi_n(B) \cdot Q_n f \quad \text{if} \quad f: X \longrightarrow U_nB \quad \text{is a map in} \quad \mathcal{T} \qquad (1.8)$$

$$\psi_n(g) = U_n g \cdot \Psi_n(X) \quad \text{if} \quad g: Q_nX \longrightarrow B \quad \text{is a map in} \quad \mathcal{L}_n \qquad (1.9)$$

It is a standard fact that $\phi_n$ is an adjunction with inverse $\psi_n$ since the composites (1.6) and (1.7) are each the identity.

If $B \in \mathcal{L}_n$, we define $H_*(B) = H_*(U_nB)$, where homology is taken with coefficients in any Abelian group $\Pi$. We regard $H_*$ as a functor defined on $\mathcal{L}_n$, but we deliberately do not specify a range category. Indeed, the problem of determining the homology operations on n-fold and (perfect) infinite loop spaces may be stated as that of obtaining an appropriate algebraic description of the range category. It follows easily from (1.2) and (1.5) of the proof above that $\Psi_n(X)_*: H_*(X) \longrightarrow H_*(U_nQ_nX)$ is a monomorphism. Since $Q_n$ is adjoint to $U_n$, the objects $Q_nX$ are, in a well-defined sense, free objects in the category $L_n$. It is therefore natural to expect $H_*(Q_nX)$ to be a functor of $H_*(X)$, with values in the appropriate range category. I have proven that this is the case if $\pi = Z_p$ and have computed the functor. By the previous proposition, if $B \in L_n$ then any map $f: X \longrightarrow U_nB$ in $T$ induces a map $\phi_n(f): Q_nX \longrightarrow B$ in $L_n$, and the functor describing $H_*(Q_nX)$ is geometrically free in the sense that $\phi_n(f)_*: H_*(Q_nX) \longrightarrow H_*(B)$ is determined by $f_* = U_n\phi_n(f)_*\Psi_n(X)_*: H_*(X) \longrightarrow H_*(U_nB)$ in terms of the homology operations that go into the definition of the functor. In this sense, we can geometrically realize enough free objects since $\phi_n(B)_*: H_*(Q_nU_nB) \longrightarrow H_*(B)$ is an epimorphism. All of these statements are analogs of well-known facts about the cohomology of spaces. The category of unstable algebras over the Steenrod algebra is the appropriate range category for cohomology with $Z_p$-coefficients. Products of $K(Z_p,n)$'s play the role analogous to that of the $Q_nX$ and their fundamental classes play the role analogous to that of $H_*(X) \subset H_*(Q_nX)$.

By use of Proposition 1, we can show the applicability of the method of acyclic models to the homology of iterated loop spaces. The applications envisaged are to natural transformations defined for iterated loop spaces but not for arbitrary spaces. The argument needed is purely categorical. Let $\mathcal{T}$ temporarily denote any category, let $\mathcal{A}$ denote the category of modules over a commutative ring $\Lambda$, and let $\mathcal{M}$ be a set of model objects in $\mathcal{T}$. Let $F: \mathcal{S} \longrightarrow \mathcal{A}$ be the free $\Lambda$-module functor, where $\mathcal{S}$ is the category of sets. If $R: \mathcal{T} \longrightarrow \mathcal{A}$ is any functor, define a functor $\tilde{R}: \mathcal{T} \longrightarrow \mathcal{A}$ by $\tilde{R}(X) = F[\bigcup_{M \in \mathcal{M}} \mathrm{Hom}_{\mathcal{T}}(M,X) \times R(M)]$ on objects and $\tilde{R}(f)(\nu,r) = (f \cdot \nu, r)$ on morphisms, where if $f: X \longrightarrow Y$, then $\nu \in \mathrm{Hom}_{\mathcal{T}}(M,X)$ and $r \in R(M)$. Define a natural transformation $\lambda: \tilde{R} \longrightarrow R$ by $\lambda(X)(\nu,r) = R(\nu)(r)$. Recall that $R$ is said to be representable by $\mathcal{M}$ if there exists a natural transformation $\xi: R \longrightarrow \tilde{R}$ such that $\lambda \cdot \xi: R \longrightarrow R$ is the identity natural transformation. With these notations, we have the following lemma.

<u>Lemma 2</u>

Let $\phi: \mathrm{Hom}_{\mathcal{T}}(X,UB) \longrightarrow \mathrm{Hom}_{\mathcal{L}}(QX,B)$ be an adjunction and let $R: \mathcal{T} \longrightarrow \mathcal{A}$ be a functor representable by $\mathcal{M}$. Define $Q\mathcal{M} = \{QM | M \in \mathcal{M}\}$ and let $S = R \cdot U: \mathcal{L} \longrightarrow \mathcal{A}$. Then $S$ is representable by $Q\mathcal{M}$.

<u>Proof.</u> Define a natural transformation $\eta: \tilde{R} \cdot U \longrightarrow \tilde{S}$ by $\eta(B)(\nu,r) = (\phi(\nu), R\phi^{-1}(1_{QM})(r))$ for $\nu: M \longrightarrow UB$, $r \in R(M)$. Write $\lambda'$ for the natural transformation $\tilde{S} \longrightarrow S$ defined as above for $\tilde{R}$. We have $\lambda'\eta = \lambda U: \tilde{R}U \longrightarrow RU = S$ since $\lambda'\eta(B)(\nu,r)$
$= S\phi(\nu)[R\phi^{-1}(1_{QM})(r)] = R[U\phi(\nu) \cdot \phi^{-1}(1_{QM})](r) = R(\nu)(r)$.

Therefore, if $\xi: R \longrightarrow \tilde{R}$ satisfies $\lambda\xi = 1: R \longrightarrow R$, then $\lambda'(\eta\xi U) = \lambda U \cdot \xi U = 1: S \longrightarrow S$, and this proves the result.

Of course, if $\phi$ is an adjunction as in the lemma and if $T^j$ denotes the product of $j$ factors $T$, then $\phi^j: \mathrm{Hom}_{T^j}(X, U^jB) \longrightarrow \mathrm{Hom}_{L^j}(Q^jX, B)$ is also an adjunction $(X \in T^j,\ B \in L^j)$. Thus the lemma applies to functors $R: T^j \longrightarrow A$ and $RU^j: L^j \longrightarrow A$.

Returning to topology, let $C_*: T \longrightarrow A$ be the singular chain complex functor, with coefficients in $\Lambda$. The lemma applies to $C_*U_n: L_n \longrightarrow A$ for $1 \le n \le \infty$ and, by the remark above, to the usual related functors on $L_n^j$ (tensor and Cartesian products of singular chain complexes). With $M = \{\Delta_m\}$, the standard set of models in $T$, we have $U_nQ_n\Delta_m = \Omega^nS^n\Delta_m$ if $n < \infty$ and $U_\infty Q_\infty \Delta_m = Q\Delta_m$; these spaces are contractible and the model objects $\{Q_n\Delta_m\} \subset L_n$ are therefore acyclic. We conclude that the method of acyclic models [4] is applicable to the study of the homology of n-fold and perfect infinite loop spaces.

## 2 COMPARISONS OF CATEGORIES OF SPECTRA

The work of the previous section shows that the category $L$ is a reasonable object of study conceptually, but it is not obvious that $L$ is large enough to be of topological interest. For example, it is not clear that the infinite classical groups are homotopy equivalent to perfect infinite loop spaces. We shall show that, from the point of view of homotopy theory, $\mathcal{L}$ is in fact equivalent to the usual category of (bounded) $\Omega$-spectra. To do this, we shall have to proceed by stages through a sequence of successively more restrictive categories of spectra.

By a spectrum, we shall mean a sequence $B = \{B_i, f_i \mid i \geq 0\}$, where $B_i$ is a space and $f_i \colon B_i \longrightarrow \Omega B_{i+1}$ is a map. By a map $g \colon B \longrightarrow B'$ of spectra we shall mean a sequence of maps $g_i \colon B_i \longrightarrow B_i'$ such that the following diagrams are homotopy commutative, $i \geq 0$.

$$\begin{array}{ccc} B_i & \xrightarrow{g_i} & B_i' \\ {\scriptstyle f_i}\downarrow & & \downarrow{\scriptstyle f_i'} \\ \Omega B_{i+1} & \xrightarrow{\Omega g_{i+1}} & \Omega B_{i+1}' \end{array} \quad . \qquad (2.1)$$

We call the resulting category $\mathcal{S}$. We say that $B \in \mathcal{S}$ is an inclusion spectrum if each $f_i$ is an inclusion. We obtain the category $\mathcal{I}$ of inclusion spectra by letting a map in $\mathcal{I}$ be a map in $\mathcal{S}$ such that the diagrams (2.1) actually commute on the nose for each $i \geq 0$. (Thus, $\mathcal{I}$ is not a full subcategory of $\mathcal{S}$.) We say that $B \in \mathcal{S}$ is an $\Omega$-spectrum if each $f_i$ is a homotopy equivalence. We let $\Omega\mathcal{S}$ be the full subcategory of $\mathcal{S}$ whose objects are the $\Omega$-spectra, and we let $\Omega\mathcal{I} = \mathcal{I} \cap \Omega\mathcal{S}$ be the full subcategory of $\mathcal{I}$ whose objects are the inclusion $\Omega$-spectra. A spectrum $B \in \Omega\mathcal{I}$ will be said to be a retraction spectrum if $B_i$ is a deformation retract of $\Omega B_{i+1}$ for all $i$. We let $\mathcal{R}$ denote the full subcategory of $\Omega\mathcal{I}$ whose objects are the retraction spectra. Clearly, $\mathcal{L}$ is a full subcategory of $\mathcal{R}$, since if $B \in \mathcal{L}$ we may take $f_i = 1$ and then any map in $\mathcal{R}$ between objects of $\mathcal{L}$ will be a map in $\mathcal{L}$ by the commutativity of the diagrams (2.1). Thus we have the following categories and inclusions

$$\mathcal{L} \subset \mathcal{R} \subset \Omega\mathcal{I} \subset \Omega\mathcal{S} \quad \text{and} \quad \mathcal{I} \subset \mathcal{S}\,. \tag{2.2}$$

For each of these categories $\mathcal{C}$, if $g,g'\colon B \longrightarrow B'$ are maps in $\mathcal{C}$, then we say that $g$ is homotopic to $g'$ if $g_i$ is homotopic to $g_i'$ in $\mathcal{T}$ for each $i$. We say that $g$ is a (weak) homotopy equivalence if each $g_i$ is a (weak) homotopy equivalence. Now each $\mathcal{C}$ has a homotopy category $H\mathcal{C}$ and a quotient functor $H\colon \mathcal{C} \longrightarrow H\mathcal{C}$. The objects of $H\mathcal{C}$ are the same as those of $\mathcal{C}$ and the maps of $H\mathcal{C}$ are homotopy equivalence classes of maps in $\mathcal{C}$. Note that each of the inclusions of (2.2) is homotopy preserving in the sense that if $\mathcal{C} \subset \mathcal{D}$ and $g \simeq g'$ in $\mathcal{C}$, then $g \simeq g'$ in $\mathcal{D}$. We therefore have induced functors $H\mathcal{C} \longrightarrow H\mathcal{D}$ and these are still inclusions since if $g,g' \in \mathcal{C}$ and $g \simeq g'$ in $\mathcal{D}$, then $g \simeq g'$ in $\mathcal{C}$.

The following definitions, due to Swan [11], will be needed in order to obtain precise comparisons of our various categories of spectra.

<u>Definitions 3</u>

(i) A category $\mathcal{C}$ is an H-category if there is an equivalence relation $\simeq$ , called homotopy, on its hom sets such that $f \simeq f'$ and $g \simeq g'$ implies $fg \simeq f'g'$ whenever $fg$ is defined. We then have a quotient category $HC$ and a quotient functor $H: C \longrightarrow HC$.

(ii) Let $C$ be any category and $D$ an H-category. A prefunctor $T: C \longrightarrow D$ is a function, on objects and maps, such that $HT: C \longrightarrow HD$ is a functor. This amounts to requiring $T(1_C) \simeq 1_{T(C)}$ for each $C \in C$ and $T(fg) \simeq T(f)T(g)$ whenever $fg$ is defined in $C$. If $C$ is also an H-category, we say that a prefunctor $T: C \longrightarrow D$ is homotopy preserving if $f \simeq g$ in $C$ implies $T(f) \simeq T(g)$ in $D$. Clearly, $T$ is homotopy preserving if and only if $T$ determines a functor $T_*: HC \longrightarrow HD$ such that $HT = T_*H$.

(iii) Let $S,T: C \longrightarrow D$ be prefunctors. A natural transformation of prefunctors $\eta: S \longrightarrow T$ is a collection of maps $\eta(C): S(C) \longrightarrow T(C)$, $C \in C$, such that $T(f)\eta(C) \simeq \eta(C')S(f)$ in $D$ for each map $f: C \longrightarrow C'$ in $C$. $\eta$ is said to be a natural equivalence of prefunctors if there exists a natural transformation of prefunctors $\xi: T \longrightarrow S$ such that $\eta(C)\xi(C) \simeq 1_{T(C)}$ and $\xi(C)\eta(C) \simeq 1_{S(C)}$ for each $C \in C$. A natural transformation of prefunctors $\eta: S \longrightarrow T$ determines a natural transformation of functors $H\eta: HS \longrightarrow HT$ and, if $S$ and $T$ are homotopy preserving, a natural transformation of functors $\eta_*: S_* \longrightarrow T_*$ such that $\eta_*H = H\eta$; if $\eta$ is a natural equivalence of prefunctors, then $H\eta$ and, if defined, $\eta_*$ are natural equivalences of functors.

(iv) If $S: \mathcal{D} \longrightarrow \mathcal{C}$ and $T: \mathcal{C} \longrightarrow \mathcal{D}$ are homotopy preserving prefunctors between H-categories, we say that T is adjoint to S if there exist natural transformations of prefunctors $\Phi: TS \longrightarrow 1_{\mathcal{D}}$ and $\Psi: 1_{\mathcal{C}} \longrightarrow ST$ such that for each $D \in \mathcal{D}$ the composite $S\Phi(D)\Psi(SD): SD \longrightarrow SD$ is homotopic in $\mathcal{C}$ to the identity map of SD and for each $C \in \mathcal{C}$ the composite $\Phi(TC) \cdot T\Psi(C): TC \longrightarrow TC$ is homotopic in $\mathcal{D}$ to the identity map of TC. If S and T are adjoint prefunctors, then $S_*: H\mathcal{D} \longrightarrow H\mathcal{C}$ and $T_*: H\mathcal{D} \longrightarrow H\mathcal{C}$ are adjoint functors, with adjunction $\phi_* = \Phi_* T_*: \mathrm{Hom}_{H\mathcal{C}}(A, S_*B) \longrightarrow \mathrm{Hom}_{H\mathcal{D}}(T_*A, B)$.

We can now compare our various categories of spectra. The following theorem implies that $\mathcal{I}$ is equivalent to $\mathcal{S}$ for the purposes of homotopy theory in the sense that no homotopy invariant information is lost by restricting attention to spectra and maps of spectra in $\mathcal{I}$, and that $\Omega\mathcal{I}$ is equivalent to $\Omega\mathcal{S}$ in this sense. Under restrictions on the types of spaces considered, it similarly compares $\mathcal{R}$ to $\Omega\mathcal{S}$. To state the restrictions, let $\mathcal{C}$ denote the full subcategory of $\mathcal{S}$ whose objects are those spectra $\{B_i, f_i\}$ such that each $B_i$ is a locally finite countable simplicial complex and each $\mu(f_i): SB_i \longrightarrow B_{i+1}$ is simplicial. Observe that if $\mathcal{W}$ is the full subcategory of $\mathcal{S}$ whose objects are those spectra B such that each $B_i$ has the homotopy type of a countable CW-complex, then every object of $\mathcal{W}$ is homotopy equivalent (in $\mathcal{S}$) to an object of $\mathcal{C}$. In fact, if $\{B_i, f_i\} \in \mathcal{W}$, then each $B_i$ is homotopy equivalent to a locally finite simplicial complex $B_i'$ by [9, Theorem 1]; if $f_i'$ is the composite $B_i' \longrightarrow B_i \xrightarrow{f_i} \Omega B_{i+1} \longrightarrow \Omega B_{i+1}'$ determined by chosen homotopy equivalences $B_i \rightleftarrows B_i'$ and if $\mu(f_i'')$ is a simplicial approximation to $\mu(f_i')$, then $\{B_i, f_i\}$ is homotopy equivalent to $\{B_i', f_i'\}$ and therefore to $\{B_i', f_i''\} \in C$.

Theorem 4

There is a homotopy preserving prefunctor $M\colon S \longrightarrow I$ such that

(i) There exists a natural equivalence of prefunctors $\eta\colon 1_S \longrightarrow JM$, with inverse $\xi\colon JM \longrightarrow 1_S$, where $J\colon I \longrightarrow S$ is the inclusion. Therefore $J_*M_*$ is naturally equivalent to the identity functor of $HS$.

(ii) $MJ\colon I \longrightarrow I$ is a functor, $\xi(JB)\colon JMJB \longrightarrow JB$ is a map in $I$ if $B \in I$, and if $\zeta\colon MJ \longrightarrow 1_I$ is defined by $\zeta(B) = \xi(JB)$, then $\zeta$ is a natural transformation of functors.

(iii) $\eta$ and $\zeta$ establish an adjoint prefunctor relationship between $J$ and $M$. Therefore $\phi_*\colon \mathrm{Hom}_{HS}(A, J_*B) \longrightarrow \mathrm{Hom}_{HI}(M_*A, B)$ is an adjunction, where $\phi_*(f) = \zeta_*(B)M_*f$, $f\colon A \longrightarrow J_*B$, and $\phi_*^{-1}(g) = J_*g \cdot \eta_*(A)$, $g\colon M_*A \longrightarrow B$.

(iv) By restriction, $M$ induces a homotopy preserving prefunctor $\Omega S \longrightarrow \Omega I$ which satisfies (i) through (iii) with respect to the inclusion $\Omega I \longrightarrow \Omega S$.

(v) By restriction, $M$ induces a homotopy

preserving prefunctor $\Omega\mathcal{S} \cap \mathcal{C} \longrightarrow \mathcal{R} \cap \mathcal{C}$ which satisfies (i) through (iii) with respect to the inclusion $\mathcal{R} \cap \mathcal{C} \longrightarrow \Omega\mathcal{S} \cap \mathcal{C}$.

Proof. We first construct $M$ and prove (i) and (ii) simultaneously. Let $B = \{B_i, f_i\} \in \mathcal{S}$. Define $MB = \{M_iB, M_if\} \in \mathcal{T}$ by induction on $i$ as follows. Let $M_0B = B_0$. Assume that $M_jB$, $j \leq i$, and $M_jf$, $j < i$, have been constructed. Let $\eta_0 = 1 = \xi_0$ and assume further that $\eta_j \colon B_j \longrightarrow M_jB$ and $\xi_j \colon M_jB \longrightarrow B_j$ have been constructed such that

(a) $\xi_j\eta_j = 1 \colon B_j \longrightarrow B_j$ and $\eta_j\xi_j \simeq 1 \colon M_jB \longrightarrow M_jB$ ;

(b) $\Omega\xi_j \cdot M_{j-1}f = f_{j-1} \cdot \xi_{j-1}$ and $\Omega\eta_j \cdot f_{j-1} \simeq M_{j-1}f \cdot \eta_{j-1}$.

Define $M_{i+1}B$ to be the mapping cylinder of the map $\mu(f_i) \cdot S\xi_i \colon SM_iB \longrightarrow B_{i+1}$, let $k_i \colon SM_iB \longrightarrow M_{i+1}B$ denote the standard inclusion, and define $M_if = \mu^{-1}(k_i) \colon M_iB \longrightarrow \Omega M_{i+1}B$. Clearly $M_if$ is then an inclusion. Consider the diagram

$$\begin{array}{ccc} SM_iB & \overset{S\xi_i}{\underset{S\eta_i}{\rightleftarrows}} & SB_i \\ {\scriptstyle k_i=\mu(M_if)}\big\downarrow & & \big\downarrow{\scriptstyle \mu(f_i)} \\ M_{i+1}B & \overset{\xi_{i+1}}{\underset{\eta_{i+1}}{\rightleftarrows}} & B_{i+1} \end{array}$$

Here $\eta_{i+1}$ and $\xi_{i+1}$ are the inclusion and retraction obtained by the standard properties of mapping cylinders, hence (a) is satisfied for $j = i + 1$. It is standard that $\xi_{i+1} \cdot \mu(M_if) = \mu(f_i)S\xi_i$, and $\Omega\xi_{i+1} \cdot M_if = f_i\xi_i$ follows by application of $\mu^{-1}$. Now $\Omega\eta_{i+1} \cdot f_i \simeq M_i f \cdot \eta_i$ is obtained by a simple chase of the diagram. This proves (b) for $j = i + 1$ and thus constructs $M$ on objects and constructs maps $\eta(B): B \longrightarrow JMB$ and $\xi(B): JMB \longrightarrow B$ in $S$. If $B \in I$, then $\xi(JB)$ is a map in $I$ by (b) and we can define $\zeta(B) = \xi(JB): MJB \longrightarrow B$. We next construct $M$ on maps. Let $g: B \longrightarrow B'$ be a map in $S$. Define $M_0 g = g_0$ and assume that $M_j g$ have been found for $j \leq i$ such that (with $\eta' = \eta(B')$, etc.)

(c) $\eta_j' g_j = M_j g \cdot \eta_j$; $\xi_j' \cdot M_j g \simeq g_j \xi_j$ with equality if $g \in I$;

(d) $\Omega M_j g \cdot M_{j-1} f = M_{j-1} f' \cdot M_{j-1} g$.

Then, by (c) and the definition of maps in the categories $S$ and $I$, $f_i^! \xi_i^! M_i g \simeq f_i^! g_i \xi_i \simeq \Omega g_{i+1} f_i \xi_i: M_i B \longrightarrow \Omega B_{i+1}^!$, with equalities if $g \in I$. Applying $\mu$, we see that there exists a homotopy $h_i: SM_i B \times I \longrightarrow B_{i+1}^!$ from $\mu(f_i^!)S\xi_i^! SM_i g$ to $g_{i+1}\mu(f_i)S\xi_i$, and we agree to choose $h_i$ to be the constant homotopy if $g \in I$. Write $[x,t]$ and $[y]$ for the images of $(x,t) \in SM_i B \times I$ and $y \in B_{i+1}$ in the mapping cylinder $M_{i+1}B$ of $\mu(f_i)S\xi_i$, and similarly for $M_{i+1}B'$. Define $M_{i+1}g: M_{i+1}B \longrightarrow M_{i+1}B'$ by

(e) $$M_{i+1}g[x,t] = \begin{cases} [SM_i g(x), 2t], & 0 \leq t \leq 1/2 \\ [h_i(x, 2t-1)], & 1/2 \leq t \leq 1. \end{cases}$$

$$M_{i+1}g[y] = [g_{i+1}(y)] .$$

It is trivial to verify that $M_{i+1}g$ is well-defined and continuous. Now consider the following diagram:

$$\begin{array}{ccccc}
SM_iB & \xrightarrow{\mu(M_if)} & M_{i+1}B & \underset{\eta_{i+1}}{\overset{\xi_{i+1}}{\rightleftarrows}} & B_{i+1} \\
\downarrow{\scriptstyle SM_ig} & & \downarrow{\scriptstyle M_{i+1}g} & & \downarrow{\scriptstyle g_{i+1}} \\
SM_iB' & \xrightarrow{\mu(M_if')} & M_{i+1}B' & \underset{\eta'_{i+1}}{\overset{\xi'_{i+1}}{\rightleftarrows}} & B'_{i+1}
\end{array}$$

Since $\eta_{i+1}(y) = [y]$, $\eta'_{i+1} \cdot g_{i+1} = M_{i+1}g \cdot \eta_{i+1}$ is obvious, and $\xi'_{i+1} \cdot M_{i+1}g \simeq g_{i+1}\xi_{i+1}$ then follows from (a) and a simple chase of the right-hand square. If the map $g$ is in $I$, then $\xi'_{i+1}M_{i+1}g = g_{i+1}\xi_{i+1}$ is easily verified by explicit computation since $h_i(x,t) = g_{i+1}\mu(f_i)S\xi_i(x)$ for all $t$. This proves (c) for $j = i + 1$. To prove (d) for $j = i + 1$, merely observe that the left-hand square clearly commutes, since $\mu(M_if)(x) = [x,0]$, and apply $\mu^{-1}$ to this square. Of course, (d) proves that $M_g$ is a map in $I$, and (c) completes the proof of (ii) of the theorem since $MJ\colon I \longrightarrow I$ is clearly a functor. If $\ell\colon M_{i+1}B \longrightarrow M_{i+1}B'$ is any map whatever such that $\ell\eta_{i+1} \simeq \eta'_{i+1}g_{i+1}$, then

$$M_{i+1}g \simeq \eta'_{i+1}\xi'_{i+1}M_{i+1}g \simeq \eta'_{i+1}g_{i+1}\xi_{i+1} \simeq \ell\eta_{i+1}\xi_{i+1} \simeq \ell .$$

It follows that the homotopy class of $M_{i+1}g$ is independent of the choice of $h_i$, and from this it follows easily that $M\colon S \longrightarrow I$ is a prefunctor. $M$ is homotopy preserving since if $g \simeq g'\colon B \longrightarrow B'$ in $S$, then

$$M_ig \simeq M_ig \cdot \eta_i\xi_i = \eta'_ig_i\xi_i \simeq \eta'_ig'_i\xi_i = M_ig' \cdot \eta_i\xi_i \simeq M_ig' ,$$

$i \geq 0$. Now (i) of the theorem follows immediately from (a), (b), and (c).

(iii) To prove (iii), we must show that the following

two composites are homotopic to the identity map.

(f) $JB \xrightarrow{\eta(JB)} JMJB \xrightarrow{J\zeta(B)} JB$, $B \in \mathcal{I}$

(g) $MB \xrightarrow{M\eta(B)} MJMB \xrightarrow{\zeta(MB)} MB$, $B \in \mathcal{S}$.

By (a) and $\zeta(B) = \xi(JB)$, the composite (f) is the identity map. For (g), note that $\xi(JMB)\eta(JMB) = 1 \simeq \eta(B)\xi(B): JMB \longrightarrow JMB$. By the uniqueness proof above for the homotopy class of $M_{i+1}g$ applied to the case $g = \xi(B)$, we have $M\xi(B) \simeq \xi(JMB) = \zeta(MB)$. Since $M\xi(B)M\eta(B) \simeq 1$ by the fact that $M$ is a prefunctor, this proves that the composite (g) is homotopic to the identity.

(iv) Since $\Omega\mathcal{S}$ and $\Omega\mathcal{I}$ are full subcategories of $\mathcal{S}$ and $\mathcal{I}$, it suffices for (iv) to prove that $MB \in \Omega\mathcal{I}$ if $B \in \Omega\mathcal{S}$, and this follows from (a) and (b) which show that if $g_j: \Omega B_{j+1} \longrightarrow B_j$ is a homotopy inverse to $f_j$, then $\eta_j g_j \Omega\xi_{j+1}: \Omega M_{j+1}B \longrightarrow M_jB$ is a homotopy inverse to $M_jf$.

(v) Again, it suffices to show that $MB \in \mathcal{R} \cap \mathcal{C}$ if $B \in \Omega\mathcal{S} \cap \mathcal{C}$. By induction on $i$, starting with $M_0B = B$ and $\eta_0 = 1 = \xi_0$, we see that each $M_iB$ is a locally finite countable simplicial complex and that each map $\mu(M_{i-1}f)$, $\eta_i$, and $\xi_i$ is simplicial, since $M_{i+1}B$ is the mapping cylinder of the simplicial map $\mu(f_i)S\xi_i: SM_iB \longrightarrow B_{i+1}$ [10, p. 151]. By Hanner [5, Corollary 3.5], every countable locally finite simplicial complex is an absolute neighborhood retract (ANR) and, by Kuratowski [7, p. 284], the loop space of an ANR is an ANR. Since the image of $M_if$ is a closed subspace of the

ANR $\Omega M_{i+1}B$, $M_if$ has the homotopy extension property with respect to the ANR $M_iB$ [6, p. 86], and therefore $M_iB$ is a deformation retract of $\Omega M_{i+1}B$ [10, p. 31]. This proves that $MB \in R \cap C$, as was to be shown.

The category $I$ is not only large and convenient. It is also conceptually satisfactory in view of the following observation relating maps in $T$ to maps in $I$. We can define a functor $\Sigma\colon T \longrightarrow I$ by letting $\Sigma X$ be the suspension spectrum of $X$, $\Sigma_i X = S^i X$ and $f_i = \mu^{-1}(1_{S^{i+1}X})$. If $g\colon X \longrightarrow Y$ is a map in $T$, define $\Sigma_i g = S^i g$; it is clear that $\Sigma g$ is in fact a map in $I$. Let $U = U_I\colon I \longrightarrow T$ be the forgetful functor, $UB = B_0$ and $Ug = g_0$. Observe that $U\Sigma\colon T \longrightarrow T$ is the identity functor. With these notations, we have the following proposition.

<u>Proposition 5</u>

$$U\colon \mathrm{Hom}_I(\Sigma X, B) \longrightarrow \mathrm{Hom}_T(X, UB) \text{ is an adjunction.}$$

<u>Proof</u>. If $B = \{B_i, f_i\} \in I$, define $f^i\colon B_0 \longrightarrow \Omega^i B_i$ inductively by $f^0 = 1$, $f^1 = f_0$, and $f^{i+1} = \Omega^i f_i \cdot f^i$ if $i > 0$. Define a natural transformation $\Phi\colon \Sigma U \longrightarrow 1_I$ by $\Phi(B) = \{\mu^i(f^i)\}\colon \Sigma UB \longrightarrow B$. Since $\Omega\mu^{i+1}(f^{i+1}) \cdot \mu^{-1}(1_{S^{i+1}B_0})$ $= \mu^i(f^{i+1}) = \mu^i(\Omega^i f_i \cdot f^i) = f_i\mu^i(f^i)$, $\Phi(B)$ is a map in $I$. For $g\colon X \longrightarrow UB$, define $\phi(g) = \Phi(B)\Sigma g$. Clearly $U\phi(g) = \mu^0(f^0)\Sigma_0 g$ $= g$. Now $f^i$ for $\Sigma X$ is easily verified to be $\mu^{-i}(1_{S^iX})\colon X \longrightarrow \Omega^i S^i X$. Therefore $\Phi(\Sigma X) = 1\colon \Sigma X \longrightarrow \Sigma X$; since

we obviously have $\Sigma U(1_{\Sigma X}) = 1: \Sigma X \longrightarrow \Sigma U \Sigma X = \Sigma X$, this implies that $\phi U = 1$.

Finally, we compare $\mathcal{L}$ to the categories $\mathcal{I}$, $\Omega\mathcal{I}$, and $\mathcal{R}$. The following theorem shows that $\mathcal{L}$ is nicely related conceptually to $\mathcal{I}$ and is equivalent for the purposes of weak homotopy theory to $\Omega\mathcal{I}$ in the sense that no weak homotopy invariant information is lost by restricting attention to spectra and maps of spectra in $\mathcal{L}$; coupled with the remarks preceding Theorem 4, it also shows that $\mathcal{L} \cap \mathcal{W}$ is equivalent to $\mathcal{R} \cap \mathcal{W}$ for the purposes of homotopy theory.

Theorem 6

There is a functor $L: \mathcal{I} \longrightarrow \mathcal{L}$ and a natural transformation of functors $\eta: 1_{\mathcal{I}} \longrightarrow KL$, where $K: L \longrightarrow \mathcal{I}$ is the inclusion, such that

(i) $LK: \mathcal{L} \longrightarrow \mathcal{L}$ is the identity functor and

$$L: \mathrm{Hom}_{\mathcal{I}}(A,KB) \longrightarrow \mathrm{Hom}_{\mathcal{L}}(LA,B)$$

is an adjunction with $L^{-1}(g) = Kg \cdot \eta(A)$ for $g: LA \longrightarrow B$.

(ii) If $g \simeq g'$ in $\mathcal{I}$, then $Lg$ is weakly homotopic to $Lg'$ in $\mathcal{L}$, and if $B \in \Omega\mathcal{I}$, then $\eta(B): B \longrightarrow KLB$ is a weak homotopy equivalence.

(iii) Let $B \in \mathcal{R} \cap \mathcal{C}$; then $\eta(B): B \longrightarrow KLB$ is a homotopy equivalence and if $g \simeq g': B \longrightarrow B'$ in $\mathcal{I}$, then $Lg \simeq Lg': LB \longrightarrow LB'$ in $\mathcal{L}$.

Proof. Let $B = \{B_i, f_i\} \in \mathcal{I}$. Since each $f_i$ is an inclusion, we can define $L_iB = \varinjlim \Omega^i B_{i+j}$, where the limit is taken with respect to the inclusions $\Omega^j f_{i+j}\colon \Omega^j B_{i+j} \longrightarrow \Omega^{j+1} B_{i+j+1}$. Clearly $\Omega L_{i+1} B = L_i B$, hence $LB \in \mathcal{L}$. If $g\colon B \longrightarrow B'$ is a map in $\mathcal{I}$, define $L_i g = \varinjlim \Omega^j g_{i+j}\colon L_i B \longrightarrow L_i B'$ ; the limit makes sense since $\Omega^j f'_{i+j} \Omega^j g_{i+j} = \Omega^{j+1} g_{i+j+1} \Omega^j f_{i+j}$ by the definition of maps in $\mathcal{I}$. Clearly $\Omega L_{i+1} g = L_i g$, hence $Lg \in \mathcal{L}$. Define $\eta\colon 1_{\mathcal{I}} \longrightarrow KL$ by letting $\eta_i(B)\colon B_i \longrightarrow L_i B$ be the natural inclusion; $\eta(B)$ is obviously a map in $\mathcal{I}$ since $\Omega\eta_{i+1}(B) \cdot f_i = \eta_i(B)$. Now (ii) of the theorem is a standard consequence of the definition of the limit topology. The fact that $LK$ is the identity functor of $\mathcal{L}$ is evident, and $\eta K\colon K \longrightarrow KLK$ and $L\eta\colon L \longrightarrow LKL$ are easily verified to be the identity natural transformations. This implies (i) and it remains to prove (iii). If $B \in \mathcal{R}$, with retractions $r_i\colon \Omega B_{i+1} \longrightarrow B_i$, define maps $r^{ij}\colon \Omega^j B_{i+j} \longrightarrow B_i$ inductively by $r^{i0} = 1$, $r^{i1} = r_i$, and $r^{i,j+1} = r^{ij}\Omega^j r_{i+j}$ if $j > 0$. Since $r_{i+j} f_{i+j} = 1$, we have $r^{i,j+1}\Omega^j f_{i+j} = r^{ij}$. We can therefore define maps

$\xi_i = \varinjlim r^{ij}\colon L_i B \longrightarrow B_i$. Obviously $\xi_i \eta_i\colon B_i \longrightarrow B_i$ is the identity map. Suppose further that $B \in \mathcal{C}$. Then we claim that $\eta_i \xi_i \simeq 1\colon L_i B \longrightarrow L_i B$. As in the proof of (v) of Theorem 4, each $\Omega^j B_{i+j}$ is now an ANR. Let us identify $\Omega^j B_{i+j}$ with its image under $\Omega^j f_{i+j}$ in $\Omega^{j+1} B_{i+j+1}$ for all $i$ and $j$ and omit the inclusion maps $\Omega^j f_{i+j}$ from the notation. Then the inclusion

$$\Omega^j B_{i+j} \times I \cup \Omega^{j+1} B_{i+j+1} \times \dot{I} \subset \Omega^{j+1} B_{i+j+1} \times I$$

is that of a closed subset in an ANR, and it therefore has the homotopy extension property with respect to the ANR $\Omega^{j+1} B_{i+j+1}$.

In particular, by [10, p. 31], each $B_i$ is a strong deformation retract of $\Omega B_{i+1}$, and we assume given homotopies $k_i\colon \Omega B_{i+1} \times I \longrightarrow \Omega B_{i+1}$, $k_i\colon 1 \simeq r_i$ rel $B_i$. The $k_i$ induce homotopies: $k_{ij}\colon \Omega^{j+1} B_{i+j+1} \times I \longrightarrow \Omega^{j+1} B_{i+j+1}$, $k_{ij}\colon 1 \simeq \Omega^j r_{i+j}$ rel $\Omega^j B_{i+j}$, in the obvious fashion ($k_{ij,t} = \Omega^j k_{i+j,t}$). We claim that, by induction on $j$, we can choose homotopies $h_{ij}\colon \Omega^j B_{i+j} \times I \longrightarrow \Omega^j B_{i+j}$, $h_{ij}\colon 1 \simeq r^{ij}$ rel $B_i$, such that $h_{i,j+1} = h_{ij}$ on $\Omega^j B_{i+j} \times I$. To see this, let $h_{i0}$ be the constant homotopy, let $h_{i1} = k_i = k_{i0}$, and suppose given $h_{ij}$ for some $j > 0$. Consider the following diagram:

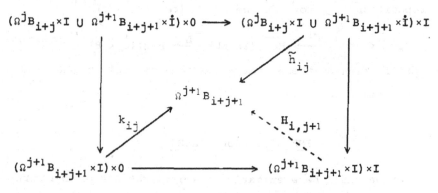

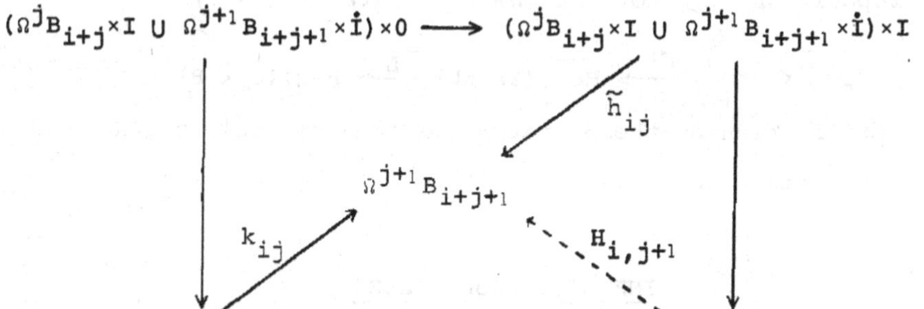


The unlabeled arrows are inclusions, and $\tilde{h}_{ij}$ is defined by $\tilde{h}_{ij}(x,s,t) = h_{ij}(x,st)$ if $x \in \Omega^j B_{i+j}$; and $\tilde{h}_{ij}(y,0,t) = y$, $\tilde{h}(y,1,t) = h_{ij}(\Omega^j r_{i+j}(y),t)$ if $y \in \Omega^{j+1} B_{i+j+1}$. It is easily verified that $\tilde{h}_{ij}$ is well-defined and continuous and that $\tilde{h}_{ij} = k_{ij}$ on the common parts of their domains. We can therefore obtain $H_{i,j+1}$ such that the diagram commutes. Define $h_{i,j+1}(x,s) = H_{i,j+1}(x,s,1)$. It is trivial to verify that

$h_{i,j+1}$ has the desired properties. Now $\underrightarrow{\lim}\, h_{ij}: L_iB \times I \longrightarrow L_iB$ is defined and is clearly a homotopy from $1$ to $\eta_i\xi_i$. Finally, if $g \simeq g': B \longrightarrow B'$ in $\mathcal{I}$ and $B \in \mathcal{R} \cap \mathcal{C}$, then

$$L_ig \simeq L_ig\eta_i\xi_i = \eta_i^! g_i\xi_i \simeq \eta' g_i^! \xi_i = L_ig'\eta_i\xi_i \simeq L_ig' \, , \; i \geq 0 \, .$$

This completes the proof of (iii) and of the theorem.

We remark that the categorical relationships of Propositions 1 and 5 and of the theorem are closely related. In fact, the composite functor $L\Sigma: \mathcal{T} \longrightarrow \mathcal{L}$ is precisely $Q_\infty$, and the adjunction

$$\phi_\infty: \mathrm{Hom}_{\mathcal{T}}(X, U_\infty B) \longrightarrow \mathrm{Hom}_{\mathcal{L}}(Q_\infty X, B)$$

of Proposition 1 factors as the composite $(U_\infty = U)$

$$\mathrm{Hom}_{\mathcal{T}}(X, UKB) \xrightarrow{U^{-1}} \mathrm{Hom}_{\mathcal{I}}(\Sigma X, KB) \xrightarrow{L} \mathrm{Hom}_{\mathcal{L}}(Q_\infty X, B) .$$

The verification of these statements requires only a glance at the definitions.

## 3 INFINITE LOOP SPACES

We shall here summarize the implications of the work of the previous section for infinite loop spaces and give the promised applications. We then make a few remarks about the extension of our results to unbounded spectra and point out an interesting collection of connective cohomology theories.

It is customary to say that $X \in \mathcal{T}$ is an infinite loop space if $X$ is the initial space $B_0$ of an $\Omega$-spectrum $B$. If $X$ is given as an H-space, it is required that its product be homotopic to the product induced from the homotopy equivalence $X \longrightarrow \Omega B_1$. Similarly, a map $f \in \mathcal{T}$ is said to be an infinite loop map if $f$ is the initial map $g_0$ of a map of $\Omega$-spectra $g$. The functor $M: \Omega\mathcal{S} \longrightarrow \Omega\mathcal{I}$ of Theorem 4 satisfies $M_0 B = B_0$ and $M_0 g = g_0$. We therefore see that the identical infinite loop spaces and maps are obtained if we restrict attention to inclusion $\Omega$-spectra and maps in $\mathcal{I}$. If $f: X \longrightarrow X'$ is any infinite loop map, then Theorem 6 implies the existence of a commutative diagram of infinite loop maps

$$\begin{array}{ccc} X & \xrightarrow{g} & Y \\ {\scriptstyle f}\downarrow & & \downarrow{\scriptstyle f'} \\ X' & \xrightarrow{g'} & Y' \end{array} \qquad (3.1)$$

such that $f'$ is a perfect infinite loop map between perfect infinite loop spaces and $g$ and $g'$ are weak homotopy equivalences.

If $X$ is an infinite loop space of the homotopy type of a countable CW-complex, then it follows from arguments of Boardman and Vogt [1, p. 15] that there is an infinite loop map $g: X \longrightarrow Y$ such that $g$ is a homotopy equivalence and $Y$ is the initial space of a spectrum in $\Omega\mathcal{S} \cap \mathcal{W}$. Combining this fact with (v) of Theorem 4, the remarks preceding that theorem, and (iii) of Theorem 6, we see that if $f: X \longrightarrow X'$ is any infinite loop map between spaces of the homotopy type of countable CW-complexes, then there is a homotopy commutative diagram of infinite loop maps, of the form given in (1), such that $f'$ is a perfect infinite loop map and $g$ and $g'$ are homotopy equivalences.

Therefore nothing is lost for the purposes of weak homotopy theory if the notions of infinite loop spaces and maps are replaced by those of perfect infinite loop spaces and maps, and similarly for homotopy theory provided that we restrict attention to spaces of the homotopy type of countable CW-complexes.

The promised comparison of stable and unstable homotopy groups of infinite loop spaces is now an easy consequence of Proposition 1. In fact, if $Y$ is an infinite loop space, say $Y = B_0$ where $B \in \Omega\mathcal{S}$, then that proposition gives a map $\Phi_\infty(LMB): Q_\infty L_0 MB \longrightarrow LMB$ in $\mathcal{L}$, and Theorem 6 gives a map $\eta(MB): MB \longrightarrow LB$ in $\mathcal{I}$. Define maps

$$QY \xrightarrow{\alpha} QL_0MB \xrightarrow{\beta} L_0MB \xleftarrow{\gamma} Y$$

by $\alpha = Q\eta_0(MB)$, $\beta = \Phi_{\infty,0}(LMB)$, and $\gamma = \eta_0(MB)$. $\gamma$ is clearly a weak homotopy equivalence, and therefore so is $\alpha$ since $Q: \mathcal{T} \longrightarrow \mathcal{T}$ is easily verified to preserve weak homotopy equivalences. Since $\Phi_{\infty,0}(LMB) \cdot \Psi_\infty(L_0MB)$ is the identity map of $L_0MB$, $\beta_*$ is an epimorphism on homotopy. If $X \in \mathcal{T}$, then $\pi_n(QX) = \pi_n^S(X)$, the $n^{\underline{th}}$ stable homotopy group of $X$. Therefore $\rho(Y) = \gamma_*^{-1}\beta_*\alpha_*: \pi_*(QY) \longrightarrow \pi_*(Y)$ gives an epimorphism $\pi_*^S(Y) \longrightarrow \pi_*(Y)$. It is clear that if $f: Y \longrightarrow Y'$ is any infinite loop map, then

$$\rho(Y')(Qf)_* = f_*\rho(Y): \pi_n^S(Y) \longrightarrow \pi_n(Y') .$$

It should be observed that the notions of infinite loop spaces and maps are not very useful from a categorical point of view since the composite of infinite loop maps need not be an infinite loop map. In fact, given infinite loop maps $f: X \longrightarrow Y$ and $g: Y \longrightarrow Z$, there need be no spectrum $B$ with $B_0 = Y$ which is simultaneously the range of a map of spectra giving $f$ and the domain of a map of spectra giving $g$. One can get around this by requiring infinite loop spaces to be topological monoids and using a classifying space argument to allow composition of maps, but this is awkward. These conditions motivate the use of $L$ in the definition of homology in section 1.

The following application of our results, which will be used in the computation of $H^*(BF)$, illustrates the technical convenience of the category $L$. Let $\tilde{F}(n) = \mathrm{Hom}_T(S^n, S^n)$ and let $\tilde{F} = \varinjlim \tilde{F}(n)$, where the limit is taken with respect to suspension of maps $S: \tilde{F}(n) \longrightarrow \tilde{F}(n+1)$. $\tilde{F}(n)$ and $\tilde{F}$ are topological monoids under composition of maps. If $X \in T$, define $\gamma: \Omega^n X \times \tilde{F}(n) \longrightarrow \Omega^n X$ by $\gamma(x,f) = \mu^{-n}(\mu^n(x) \cdot f)$, that is, with $\Omega^n X$ identified with $\mathrm{Hom}_T(S^0, \Omega^n X)$, by the composite

$$\Omega^n X \times \tilde{F}(n) \xrightarrow{\mu^n \times 1} \mathrm{Hom}_T(S^n, X) \times \tilde{F}(n) \xrightarrow{\text{composition}} \mathrm{Hom}_T(S^n, X) \xrightarrow{\mu^{-n}} \Omega^n X.$$

This defines an operation of $\tilde{F}(n)$ on $\Omega^n X$. Now let $B = \{B_i, f_i\} \in \Omega S$, and let $g_i: \Omega B_{i+1} \longrightarrow B_i$ be a homotopy inverse to $f_i$. Define homotopy equivalences $f^n: B_0 \longrightarrow \Omega^n B_n$ and $g^n: \Omega^n B_n \longrightarrow B_0$ in the obvious inductive manner and define

$$\gamma_n = g^n\gamma(f^n \times 1): B_0 \times \widetilde{F}(n) \longrightarrow B_0 .$$

Observe that $\gamma_n$ fails to define an operation of $\widetilde{F}(n)$ on $B_0$ since the associativity condition $(xf)g = x(fg)$ is lost. Of course, $\gamma_n$ coincides with $\gamma$ on $\Omega^n B_n$ if $B \in L$, and associativity is then retained. Now consider the following diagram:

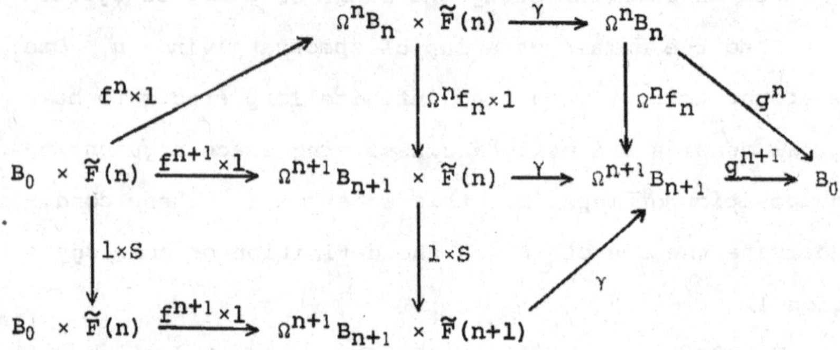


The left-hand triangle and square commute trivially. Clearly $\gamma$ is natural on n-fold loop maps, hence $\Omega^n f_n \gamma = \gamma(\Omega^n f \times 1)$. $\gamma(1 \times S) = \gamma$ since

$$\mu^{-n}(\mu^n(x)f) = \mu^{-(n+1)}\mu(\mu^n(x)f) = \mu^{-(n+1)}(\mu^{n+1}(x) \cdot Sf) .$$

$g^n$ is homotopic to $g^{n+1}\Omega^n f_n$, and if $B \in R$ and the $g_i$ are chosen retractions, then $g^n = g^{n+1}\Omega^n f_n$ . Thus if $B \in R$ we have $\gamma_n = \gamma_{n+1}(1 \times S)$ and we can define $\gamma = \varinjlim \gamma_n : B_0 \times \widetilde{F} \longrightarrow B_0$ . Since the right-hand triangle is not transformed naturally by maps in $R$, the map $\gamma$ is not natural on $R$. For $B \in L$, the f's and g's are the identity maps, and the diagram trivializes. Therefore, for each $B \in L$, we have an operation $\gamma: B_0 \times \widetilde{F} \longrightarrow B_0$ and if $h: B \longrightarrow B'$ is a map in $L$, then $h_0(xf) = h_0(x)f$ for $x \in B_0$ and $f \in \widetilde{F}$.

Stasheff [unpublished] has generalized work of Dold and Lashof [2] to show that if a topological monoid $M$ operates on a space $X$, then there is a natural way to form an associated quasifibration $X \longrightarrow X \times_M EM \longrightarrow BM$ to the classifying principal quasifibration $M \longrightarrow EM \longrightarrow BM$. As usual, let $F \subset \tilde{F}$ consist of the homotopy equivalences of spheres. By restriction, if $B \in \mathcal{L}$ and $Y = B_0$, we have an operation of $F$ on $Y$ and we can therefore form $Y \times_F EF$. Of course, this construction is natural on $\mathcal{L}$.

Boardman and Vogt [1] have proven that the standard inclusions $U \subset O \subset PL \subset Top \subset F$ are all infinite loop maps between infinite loop spaces with respect to the H-space structures given by Whitney sum (on $F$, this structure is weakly homotopic to the composition product used above). We now know that we can pass to $\mathcal{L}$ and obtain natural operations of $F$ on (spaces homotopy equivalent to) each of these sub H-spaces $G$ of $F$. The same is true for their classifying spaces $BG$. Observe that the resulting operation of $F$ on $F$ is not equivalent to its product. (In fact, if $\phi \in B_0$ is the identity under the loop product of $\Omega B_1$, where $B \in \mathcal{L}$, then $\phi f = \phi$ for all $f \in \tilde{F}$ since composing any map with the trivial map gives the trivial map.) It would be of interest to understand the geometric significance of these operations by $F$ on its various sub H-spaces and of the spaces $G \times_F EF$ and $BG \times_F EF$.

I shall show elsewhere that, with mod p coefficients, $\gamma_*: H_*(B) \otimes H_*(\tilde{F}) \longrightarrow H_*(B)$ gives $H_*(B)$ a structure of Hopf algebra over $H_*(\widetilde{F})$ for $B \in \mathcal{L}$ (and, a fortiori, for $B \in \Omega\mathcal{S}$), where $H_*(B) = H_*(B_0)$ as in section 1. $H_*(B)$ is also a Hopf algebra over the opposite algebra of the Steenrod algebra and over the Dyer-Lashof algebra, which is defined in terms of the homology operations introduced by Dyer and Lashof in [3]. These operations are all natural on $\mathcal{L}$. The appropriate range category for $H_*\colon \mathcal{L} \longrightarrow ?$ is determined by specifying how these three types of homology operations commute, and, coupled with known information, these commutation formulas are all that is required to compute $H^*(BF)$.

Finally, we observe that there is a natural way to extend our results of section 2 to unbounded spectra. Let $\overline{\mathcal{S}}$ denote the category whose objects are sequences $\{B_i, f_i \mid i \in Z\}$ such that $\{B_i, f_i \mid i \geq 0\} \in \mathcal{S}$ and $B_i = \Omega^{-i}B_0$ and $f_i\colon B_i \longrightarrow \Omega B_{i+1}$ is the identity map for $i < 0$. The maps in $\overline{\mathcal{S}}$ are sequences $g = \{g_i \mid i \in Z\}$ such that $\{g_i \mid i \geq 0\} \in \mathcal{S}$ and $g_i = \Omega g_{i+1}$ if $i < 0$. We have an obvious completion functor $C\colon \mathcal{S} \longrightarrow \overline{\mathcal{S}}$ defined on objects by $C_iB = B_i$ if $i \geq 0$ and $C_iB = \Omega^{-i}B_0$ if $i < o$, with $C_if = f_i$ for $i \geq 0$ and $C_if = 1$ for $i < 0$, and defined similarly on maps. $C$ is an isomorphism of categories with inverse the evident forgetful functor $\overline{\mathcal{S}} \longrightarrow \mathcal{S}$. For each of our previously defined subcategories $\mathcal{D}$ of $\mathcal{S}$ define $\overline{\mathcal{D}}$ to be the image of $\mathcal{D}$ under $C$ in $\overline{\mathcal{S}}$. $\overline{\mathcal{L}}$ is of particular interest. Its objects and maps are sequences $\{B_i \mid i \in Z\}$ and $\{g_i \mid i \in Z\}$ such that $B_i = \Omega B_{i+1}$ for all $i$ and $g_i = \Omega g_{i+1}$ for all i. Clearly all of the results of section 2 remain valid for the completed categories.

Our results show that any reasonable cohomology theory, by which we mean any cohomology theory determined by a spectrum $B \in \Omega\mathcal{S} \cap \mathcal{W}$, is isomorphic to a cohomology theory determined by a spectrum in $\mathcal{T} \cap \mathcal{W}$ and that any transformation of such theories determined by a map $g: B \longrightarrow B'$ in $\Omega\mathcal{S} \cap \mathcal{W}$ is naturally equivalent to a transformation determined by a map in $\mathcal{T} \cap \mathcal{W}$. Recall that

$$H^n(X,A;B) = \mathrm{Hom}_{H\mathcal{T}}(X/A, B_n)$$

defines the cohomology theory determined by $B \in \Omega\mathcal{S}$ on CW pairs $(X,A)$. Call such a theory connective if $H^n(P;B) = 0$ for $n > 0$, where $P$ is a point. Of course, $H^{-n}(P;B) = \Pi_0(\Omega^n B_0) = \Pi_n(B_0)$. Any infinite loop space $Y$ determines a connective (additive) cohomology theory since, by a classifying space argument, we can obtain $CB \in \Omega\mathcal{S}$ such that $B_0$ is homotopy equivalent to $Y$ and $\Pi_0(B_n) = 0$ for $n > 0$; according to Boardman and Vogt [1], any such cohomology theory is so obtainable and determines $Y$ up to homotopy equivalence of infinite loop spaces. If $X \in \mathcal{T}$, then $CQ_\infty X$ determines a connective cohomology theory, since $C_n Q_\infty X = QS^n X$ for $n > 0$, and $H^{-n}(P;CQ_\infty X) = \Pi_n(QX) = \Pi_n^S(X)$ if $n \geq 0$. In view of Proposition 1, these theories play a privileged role among all connective cohomology theories, and an analysis of their properties might prove to be of interest. Observe that if $B \in \mathcal{L}$, then $C\Phi_\infty B: CQ_\infty B_0 \longrightarrow CB$ determines a natural transformation of cohomology theories $H^*(X,A;CQ_\infty B_0) \longrightarrow H^*(X,A;CB)$ and, if the theory determined by CB is connective, this transformation is epimorphic on the cohomology of a point.

# BIBLIOGRAPHY

[1] Boardman, J. M. and Vogt, R. M. "Homotopy Everything H-spaces", Mimeographed Notes, Math. Inst., Univ. of Warwick, (1968).

[2] Dold, A. and Lashof, R., "Principal Quasifibrations and Fibre Homotopy Equivalence of Bundles", *Ill. J. Math.*, 3; 285-305, (1959).

[3] Dyer, E. and Lashof, R., "Homology of Iterated Loop Spaces", *Amer. J. Math.*, 84; 35-88, (1962).

[4] Eilenberg, S. and MacLane, S., "Acyclic Models", *Amer. J. Math.*, 75; 189-199, (1953).

[5] Hanner, O., "Some Theorems on Absolute Neighborhood Retracts", *Ark. Mat.*, 1; 389-408, (1950).

[6] Hurewicz, W. and Wallman, H., *Dimension Theory*, Princeton Univ. Press, (1961).

[7] Kuratowski, C., "Sur Les Espaces Localement Connexes et Péaniens en Dimension n", *Fund. Math.*, 24; 269-287, (1935).

[8] Milgram, R. J., "Iterated Loop Spaces", *Annals of Math.*, 84; 386-403, (1966).

[9] Milnor, J., "On Spaces Having the Homotopy Type of a CW-complex", *Trans. A.M.S.*, 90; 272-280, (1959).

[10] Spanier, E. H., *Algebraic Topology*, McGraw-Hill, Inc., (1966).

[11] Swan, R. G., "The Homology of Cyclic Products", *Trans. A.M.S.*, 95; 27-68, (1960).

# HOMOLOGY OF SQUARES AND FACTORING OF DIAGRAMS*

Paul Olum

## 1. Introduction

Our purpose here is to present some technique in the study of mappings of spaces which seems to have a number of useful applications and which should generalize to other categories than the topological. We shall illustrate the use of the method by applying it to derive some well-known results in §4 below; more extensive applications will appear in a later work.

Let us look at the diagram (of solid arrows):

(1.1) 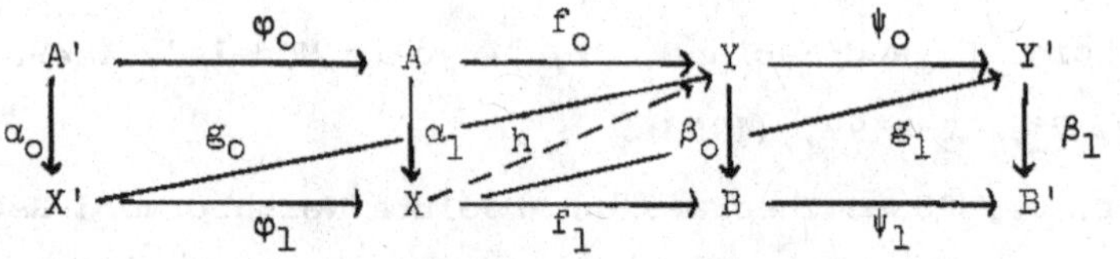


where commutativity holds everywhere. We want to know under what circumstances there will exist a map $h: X \to Y$, shown by the dotted arrow, such that the diagram with h present will continue to be commutative or, at least, as nearly so as possible. Such a map h will be said to "factor" diagram (1.1).

In a systematic treatment of this problem we would first give a precise definition of what we require of this h in order that it be a factorization. We would then develop an obstruction theory for the existence of h and study the properties of these obstructions. We will not do this here, however, but will defer the systematic account and all proofs to a later work.

What we shall do in the present discussion is give the principal consequences of the definition of a factorization and indicate

*A portion of this work was done under NSF grant GP-7905.

the groups in which the obstructions lie as well as their (customary) main properties. For the applications we have presently in mind this is all that will be needed.

## 2. Some properties of a factorization

All of the spaces in diagram (1.1) are taken to be path-connected and to have a base point $*$; all mappings and homotopies are to preserve base points. Apart from this we make only the following assumptions:

(2.1a) Each of A', A, X', X has the homotopy type of a CW-complex and the base points are non-degenerate.

(2.1b) Either $\beta_{1\#} : \pi_1(Y') \to \pi_1(B')$ or $\psi_{1\#} : \pi_1(B) \to \pi_1(B')$ is onto.

By way of notation, a homotopy $\Gamma : X \times I \to Y$ will be called "rel $\alpha_1$" or "rel $\beta_o$" if $\Gamma(\alpha_1 \times 1)$ or $\beta_o\Gamma$ is stationary. Given two homotopies $\Gamma_1, \Gamma_2 : X \times I \to Y$ with $\Gamma_1(x,1) = \Gamma_2(x,0)$, we write $\Gamma_1 \cdot \Gamma_2$ for the homotopy which results from $\Gamma_1$ followed by $\Gamma_2$. We adopt also the following convention:

(2.2) A homotopy of homotopies, e.g., $\Gamma_1 \cong \Gamma_2$ will always be assumed to be <u>proper</u>, that is, to be stationary on $X \times 0 \cup X \times 1$.

As indicated in §1, we will not give the definition of a factorization here, but the following two theorems contain its main properties:

Theorem 2.3 *A factorization of (1.1) gives rise in a canonical way to a mapping* $h : X \to Y$ *and four homotopies*

$$(2.4)\qquad \begin{array}{ll} \Lambda_1\colon f_o \cong h\alpha_1 & \Lambda_2\colon g_o \cong h\varphi_1 \\ \Gamma_1\colon f_1 \cong \beta_o h & \Gamma_2\colon g_1 \cong \psi_o h \end{array}$$

*such that in the diagram*

(2.5)

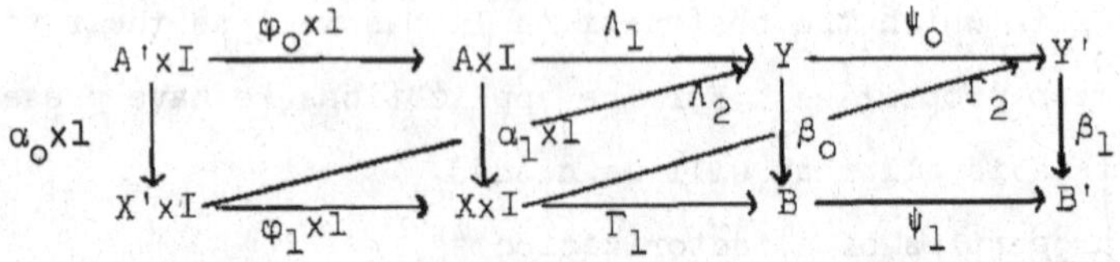


*all mapping squares are properly homotopy commutative (in the sense of (2.2)).* (*We shall say the factorization induces* $\theta: \pi_1(X) \to \pi_1(Y)$ *if* h *induces this* $\theta$.)

For the next theorem, "fibration" means regular Hurewicz fibration, that is, the homotopy lifting property for any space, with the lifted homotopy stationary wherever the original one is. If the spaces A', A, X', X are CW-complexes and the maps, $\alpha_o$, $\alpha_1$, $\varphi_o$, $\varphi_1$ are cellular, then fibration may be taken to mean weak (or "Serre") fibration.

*Theorem 2.6* *For each of the conditions listed below, if (1.1) can be factored then the factorization can be so chosen as to make the accompanying properties hold in addition to those given in Theorem 2.3. For any combination of the conditions this can be done so that the corresponding properties hold simultaneously:*

(a) $\alpha_1$ *cofibration*: $f_o = h\alpha_1$ *and* $\Lambda_1$ *is stationary*; $\Gamma_1$ *and* $\Gamma_2$ *are rel* $\alpha_1$ *and the homotopy* $\psi_1\Gamma_1 \cong \beta_1\Gamma_2$ *is rel* $(\alpha_1 \times 1)$

(b) $\alpha_o$ *and* $\alpha_1$ *cofibrations*: $\Lambda_2$ *is rel* $\alpha_o$

(c) $\beta_o$ *fibration*: $f_1 = \beta_o h$ *and* $\Gamma_1$ *is stationary*; $\Lambda_1$ *and* $\Lambda_2$ *are rel* $\beta_o$ *and the homotopy* $\Lambda_1(\varphi_o \times 1) \cong \Lambda_2(\alpha_o \times 1)$ *is rel* $\beta_o$

(d) $\beta_o$ *and* $\beta_1$ *fibrations*: $\Gamma_2$ *is rel* $\beta_1$.

*Remark 2.7* We can replace $\alpha_o$, $\alpha_1$, $f_o$ in (a) and (b) above by $\varphi_o$, $\varphi_1$, $g_o$; this is clear from the symmetry of the diagram; similarly for $\psi_o$, $\psi_1$, $g_1$ instead of $\beta_o$, $\beta_1$, $f_1$ in (c) and (d).

But we can not add these to the theorem since, for example, if $\alpha_1$ and $\varphi_1$ are both cofibrations it need not be true that $f_o = h\alpha_1$ and $g_o = h\varphi_1$ for the same h.

Remark 2.8 There is, as one would expect, an appropriate notion of the homotopy of two factorizations of (1.1), and analogues of Theorems 2.3 and 2.6 for this notion. We shall omit this here; it will be found in the later account promised in the introduction.

## 3. Cohomology and homotopy of squares; obstructions

Let $S_1$ denote the mapping square $\alpha_o$, $\alpha_1$, $\varphi_o$, $\varphi_1$ and $S_2$ the mapping square $\beta_o$, $\beta_1$, $\psi_o$, $\psi_1$. As a setting for our obstructions we shall need the cohomology groups $H^k(S_1; G)$(where G is a coefficient group) and the homotopy groups $\pi_k(S_2)$. For our purposes the most important property of these groups is that there are exact sequences (we omit coefficients):

$$(3.1) \quad \to H^k(\alpha_1) \xrightarrow{\varphi^*} H^k(\alpha_o) \xrightarrow{\delta} H^{k+1}(S_1) \xrightarrow{j} H^{k+1}(\alpha_1) \to$$

$$(3.2) \quad \to \pi_k(\beta_o) \xrightarrow{\psi_\#} \pi_k(\beta_1) \xrightarrow{\iota} \pi_k(S_2) \xrightarrow{\partial} \pi_{k-1}(\beta_o) \to$$

and the same with $\alpha_o$, $\alpha_1$ replaced by $\varphi_o$, $\varphi_1$ and $\beta_o$, $\beta_1$ replaced by $\psi_o$, $\psi_1$.

Definitions of these groups and proofs of exactness for (3.1) and (3.2) are due to Eckmann-Hilton; see [2,Chap.9].

It is easy to see that the homotopy groups $\pi_k(S_2)$ are local groups at the base point in Y (i.e., $\pi_1(Y)$ operates on $\pi_k(S_2)$) and any homomorphism $\theta$: $\pi_1(X) \to \pi_1(Y)$ induces $\pi_k(S_2)$ as a local group in X and hence in the square $S_1$; we denote this induced local group in $S_1$ by $\theta^*\pi_k(S_2)$. As indicated in §1, we shall omit the definition of the obstructions, but the following theorem gives all the information we need about them here:

Theorem 3.3 For $n \geq 3$, the n-th obstruction $\mathcal{O}_\theta^n$ to a factorization of (1.1) inducing a given $\theta\colon \pi_1(X) \to \pi_1(Y)$ is a subset (possibly void) of $H^n(S_1; \theta^*\pi_{n+1}(S_2))$. It has the following properties:

(i) $0 \in \mathcal{O}_\theta^n$ if and only if $\mathcal{O}_\theta^{n+1}$ is non-void

(ii) Suppose $H^n(S_1; \theta^*\pi_{n+1}(S_2)) = 0$ for all sufficiently large n. Then there is a factorization of (1.1) inducing $\theta$ if and only if $0 \in \mathcal{O}_\theta^n$ for all $n \geq 3$.

To complement this theorem we need conditions which will imply that $\mathcal{O}_\theta^3$ is not void. Proposition 3.5 below gives some sufficient conditions which are adequate for our present needs.

We require some notation for this. Let $\mathfrak{J}(\alpha_o, \varphi_o)$ be the free product $\pi_1(X')*\pi_1(A)$ modulo the normal subgroup N generated by the set $\{\alpha_o(a)\varphi_o(a^{-1}) \mid a \in \pi_1(A')\}$, i.e., the "reduced" free product. The maps $\varphi_1$, $\alpha_1$ together clearly define a homomorphism

(3.4) $$(\varphi_1, \alpha_1)_\# : \mathfrak{J}(\alpha_o, \varphi_o) \to \pi_1(X)$$

Then we have

Proposition 3.5 Suppose that either (a) or (b) holds for diagram (1.1):

(a) $(\varphi_1, \alpha_1)_\#$ in (3.4) is an isomorphism; $\varphi_{1\#}(\pi_2(X'))$ and $\alpha_{1\#}(\pi_2(A'))$ together generate $\pi_2(X)$.

(b) $\beta_{o\#}\colon \pi_1(Y) \to \pi_1(B)$ is an epimorphism; $\psi_\#\colon \pi_i(\beta_o) \to \pi_i(\beta_1)$ is an isomorphism for $i = 2$ and an epimorphism for $i = 3$. (Recall also 2.1b.)

Then there is a unique $\theta\colon \pi_1(X) \to \pi_1(Y)$ for which $\mathcal{O}_\theta^3$ is non-void.

Remark 3.6 For the homotopy problem (see Remark 2.8) the obstructions lie in the groups $H^n(S_1; \theta^* \pi_{n+2}(S_2))$ and there are obvious analogues of Theorem 3.3 and Proposition 3.5.

## 4. Examples

We give three examples to illustrate the application of the material above.

1) Our first example is a theorem of James. For this we recall that a *loop* is a "non-associative group", that is, a set M with multiplication and a two-sided identity, and such that the equations

$$xa = b, \quad ay = b \qquad\qquad a, b \text{ in } M$$

admit one and only one pair of solutions x, y in M. The following is Theorem 1.1 of [4].

Theorem 4.1 *Let* X *have the homotopy type of a CW-complex and let* Y *be a connected H-space. Then the homotopy classes of maps of* X *into* Y *form a loop with multiplication inherited from* Y.

*Proof.* The diagram is the following special case of (1.1):

(4.2)

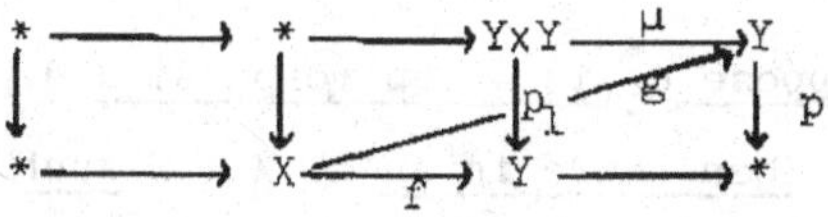


where f and g are given maps, μ is the multiplication in Y and $p_1$ is projection on the first factor. It is clear that what has to be proved is the existence of a factorization h of this diagram, unique up to homotopy, and the same for (4.2) with $p_1$ replaced by the projection $p_2$ on the second factor.

Obviously $\mu$ induces an isomorphism $\pi_q(p_1) \approx \pi_q(p)$ for all q and therefore (by (3.2)) $\pi_q(S_2) = 0$ for all q; similarly for $p_2$. The existence of the factorization now follows from Proposition 3.5 and Theorem 3.3. The uniqueness follows similarly from the analogous results for the homotopy of two factorizations; see Remarks 2.8 and 3.6 above.

The other theorems of [4, §4] follow in the same way from similar diagrams.

2) Our second example is a result of Hilton [3]. We consider maps of path-connected spaces:

(4.3) $$F \xrightarrow{i} E \xrightarrow{p} X$$

where each of F,E, X has the homotopy type of a CW-complex, i is a cofibration and pi = *; here * is a non-degenerate base point in X. Let W be a 1-connected space and suppose there is given a map

$$f: (E, F) \to (W, *).$$

Denote by

$$p_i^* : H^{n+i}(X; \pi_n(W)) \to H^{n+i}(E,F; \pi_n(W))$$

the cohomology homomorphisms induced by p. The following then contains the main theorem in Hilton [3,p.77]:

Theorem 4.4 (a) *Suppose $p_o^*$ is an epimorphism and $p_1^*$ a monomorphism for all $n \geq 2$. Then there is an h: $X \to W$ such that* $hp \simeq f$ rel F.

(b) *Suppose $p_1^*$ is an epimorphism and $p_o^*$ a monomorphism for all* $n \geq 2$. *Suppose* h, h' : $X \to W$ *satisfy* $hp \simeq h'p$ rel F. *Then* $h \simeq h'$.

(An immediate consequence of (a) here is a well-known result proved by several authors (Ganea, Hilton, Mayer, Nomura) to the effect that if (4.3) is a fibration with X (m - 1)-connected ($m \geq 2$) and with the homotopy groups of $F = \Omega W$ zero outside a band of width m - 1, then the fibration is equivalent to one induced by a map $h: X \to W$; see [3, p. 81].)

<u>Proof</u>. The diagram for 4.4 (a) is the following special case of (1.1).

(4.5)

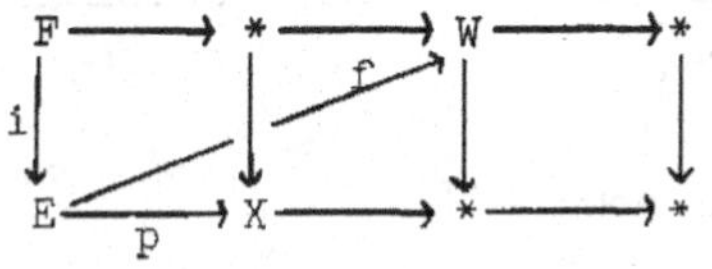


By Prop. 3.5(b), since W is 1-connected, $\mathcal{O}^3_\theta$ is non-void, where $\theta : \pi_1(X) \to \pi_1(W)$ is necessarily trivial. The vanishing of all obstructions follows at once then from Theorem 3.3, the hypotheses of (a) above and the exact sequences (3.1) and (3.2), so that diagram (4.5) has a factorization h. By Theorems 2.3 and 2.6 (since i is a cofibration), $hp \cong f$ rel F.

Part (b) follows in corresponding fashion from the analogous results for the homotopy of two factorizations; see Remarks 2.8 and 3.6 above.

The other theorems in [3] follow similarly from appropriate specializations of diagram (1.1).

3) Finally, we look at a theorem of Dold [1] on fiber homotopy equivalences:

<u>Theorem 4.6</u> <u>Let $\alpha$ be a map of one regular Hurewicz fibration into another over the same base</u>

$$(4.7)\qquad \begin{array}{ccc} F & \xrightarrow{\alpha|F} & F' \\ {\scriptstyle i}\downarrow & & \downarrow{\scriptstyle i'} \\ Y & \xrightarrow{\alpha} & Y' \\ {\scriptstyle \beta}\downarrow & & \downarrow{\scriptstyle \beta'} \\ B & \xrightarrow{1} & B \end{array}$$

<u>We suppose all spaces path-connected and also that</u> Y <u>and</u> Y' <u>have the homotopy-type of</u> CW-<u>complexes</u>. <u>If</u> $\alpha|F : F \to F'$ <u>induces isomorphisms</u> $\pi_j(F) \approx \pi_j(F')$ <u>for all</u> j, <u>then</u> $\alpha$ <u>is a fiber homotopy equivalence</u>.

<u>Proof</u>. The diagram is now

(4.8)

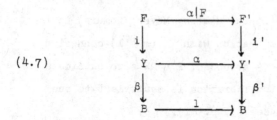

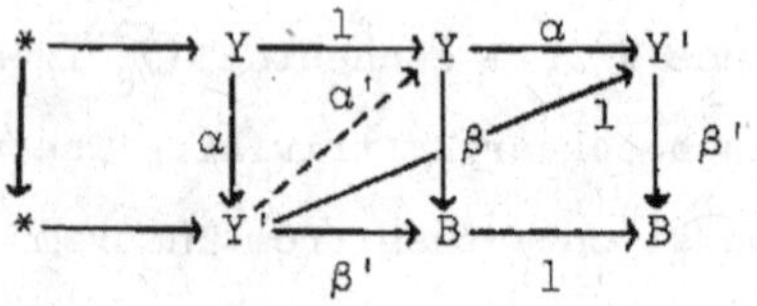


Since we may identify $\pi_{j+1}(\beta) = \pi_j(F)$, $\pi_{j+1}(\beta') = \pi_j(F')$, $\alpha$ induces isomorphisms $\pi_j(\beta) \approx \pi_j(\beta')$ for all j, and therefore (by (3.2)) all homotopy groups of the right hand square vanish. The existence of the factorization $\alpha'$ as shown by the dotted line exists then by Prop. 3.5(b) and Theorem 3.3.

By (c) and (d) of Theorem 2.6, $\alpha'$ may be so chosen that $\beta\alpha' = \beta'$, $\alpha'\alpha \cong 1$ rel $\beta$ and $\alpha\alpha' \cong 1$ rel $\beta'$. This is precisely the assertion of the theorem.

<u>Remark 4.9</u> If Y and Y' are CW-complexes and $\alpha$ is cellular then by the same argument (see the remarks preceding Theorem 2.6) it is enough to suppose that $\beta$ and $\beta'$ are weak fibrations here.

## References

[1] A. Dold, Über fasernweise Homotopieäquivalenz von Faserräumen, Math. Zeit. 62(1955), 111-136.

[2] P. J. Hilton, Homotopy Theory and Duality, Gordon and Breach, New York, 1965.

[3] P. J. Hilton, On excision and principal fibrations, Comment. Math. Helv. 35(1961), 77-84.

[4] I. M. James, On H-spaces and their homotopy groups, Quart. J. Math. Oxford (2), 11(1960), 161-179.

Offsetdruck: Julius Beltz, Weinheim/Bergstr.